POWER SYSTEM AUTOMATION

BUILD SECURE POWER SYSTEM SCADA & SMART GRIDS

K S MANOJ

INDIA • SINGAPORE • MALAYSIA

Notion Press

No. 8, 3rd Cross Street
CIT Colony, Mylapore
Chennai, Tamil Nadu – 600004

First Published by Notion Press 2021
Copyright © K S Manoj 2021
All Rights Reserved.

ISBN 978-1-63669-656-0

This book has been published with all efforts taken to make the material error-free after the consent of the author. However, the author and the publisher do not assume and hereby disclaim any liability to any party for any loss, damage, or disruption caused by errors or omissions, whether such errors or omissions result from negligence, accident, or any other cause.

While every effort has been made to avoid any mistake or omission, this publication is being sold on the condition and understanding that neither the author nor the publishers or printers would be liable in any manner to any person by reason of any mistake or omission in this publication or for any action taken or omitted to be taken or advice rendered or accepted on the basis of this work. For any defect in printing or binding the publishers will be liable only to replace the defective copy by another copy of this work then available.

Dedicated

**to
Sarada Devi of Sivagiri Mutt
and my wife Dr. Shreelekshmi**

PREFACE

Author's Note

As the term suggests, is the Power System Automation is the execution of various tasks in substations, generating stations and power distribution without periodic interference. It aims to minimize and gradually keep away the human intervention by simplifying the complicated tasks by reducing them to a single instance. Today, automation has the effect of making human resources available for more engaging tasks, leading to greater efficiency and improved job satisfaction. Furthermore, tasks that were beyond human capabilities can be automated. Automatic devices monitor steps on their own, resulting in cost optimization and timeline management. This also facilitates identification of problems and subsequent mitigation to ensure seamless performance.

While closely working with Power Distribution Automation, it has been comprehended that the electrical engineers and the ICT professionals move forward parallel and hard to converge in attaining the desired goal. Hence this book is mainly intended reconciles the aperture between the fields of Power System Engineering and Information Technology to fast-track the Power System Automation. Further to equip the Energy Engineers and Electrical Engineering students with the basic concepts required to design and implement a safe and secure Smart Grid. Power sector being a super critical infrastructure of a Nation, the security especially physical-cyber-security of the Critical Infrastructure must be given utmost importance. Hence emphasis has been given to describe the communication technology with necessary cyber-security features.

This book is an outcome of my long cherished dream since 2011 and written with the intention to assist the academia and the industry alike. Another objective of this book is to equip the readers with a working knowledge of fundamentals, design tools, Power System Automation research methodology, and various solutions to the critical issues in the development and deployment of Automation. The material presented in its ten chapters is an outgrowth of numerous talks, seminars, training, and industry debates with my friends and colleagues on how to automate the present grid to transform to an intelligent grid.

The major challenge of this endeavor was bringing out the diversified technologies like Data Acquisition System, Power System SCADA and its design considerations, Smart Power Transmission, Smart Power Distribution, secure M2M Communication and Computation, Smart Grid Concepts, SCADA Protocols, Physical-Cyber-Security etc. under a single umbrella in a succinct but palatable manner to both energy engineers and ICT professionals. Sincere attempts have been made to elaborate important topics appropriately while others are cited briefly. Each chapter is accompanied by a summary which will be useful for a quick revision. As the present domain of expertise of most of the practicing energy engineers may be quite different, most of them may not be equipped enough to conceive the concepts of automation in its true sense because of the diversified requirements of advanced ICT based systems. Hence this book is written with an intention to abridge this gap between ICT and energy system engineering to a considerable extent.

Intended Audience

This book is intended for a wide range of readers who will benefit from an understanding of Power System Automation, Data Acquisition Systems, Power System SCADA (PSS), Smart Power Transmission, Smart Power Distribution, secure Industrial Communication and Networking, Smart Grid Concepts, SCADA Protocols, Physical-Cyber-Security, etc. This includes Power System Engineers who are already engaged with design and implementation of Power System SCADA, Microgrid, Smart Grid, etc. and the Electrical, Electronics, Instrumentation, and Computer Engineering students who wish to get involved in the design and development of Substation Automation Systems (SAS), evolving Smart Grid and Microgrid. For the professionals who are already engaged in the design of SAS with interoperable protocols, this book serves as a ready reference and is suitable for self-study. As an academic text book, it is most useful for the senior under graduate and post graduate students. All the chapters of the book are very structured and modular to provide a considerable deal of flexibility for the design of courses in Power System Automation and Smart Grid.

Salient Features

All basic knowledge, is provided for practicing Power System Engineers and Electrical, Electronics, Computer and Instrumentation Engineering students who work or wish to work in the challenging and complex field of Power

System Automation. This book specifically aims to narrow the gap created by fast changing technologies impacting on a series of legacy principles related to how Power Systems are conceived and implemented. It is specifically written with the intention to help those who have to define, design and implement Power Automation Systems, while conforming to the current industry best practice standards.

Key features
- Strong practical oriented approach with strong theoretical backup to all practical aspects of project design, development and implementation of Power System Automation.
- Exclusively focuses on the rapidly changing control aspect of power system engineering, using swiftly advancing communication technologies with Intelligent Electronic Devices.
- Covers the complete chain of Power System Automation components and related equipment.
- Explains significantly to understand the commonly used and standard protocols such as IEC 61850, IEC 60870, DNP3, ICCP TASE 2, etc which are viewed as a black box for a significant number of energy engineers.
- Provides the reader with a essential understanding of both physical-cyber security and computer networking.
- Explores the SCADA communication from conceptualization to realization.
- Presents the complexity and operational requirements of the Power System Automation to the ICT professional and presents the same for ICT to the power system engineers.
- Is a suitable material for the undergraduate and post graduate students of electrical engineering to learn Power System Automation.

In general this book is a valuable resource for engineers operating within different Power System Automation project stages including the specification process, design and engineering process, integration process, testing process and the operation and maintenance process.

ACKNOWLEDGEMENTS

First and foremost, let me contemplate over the kindness of the Almighty for giving me the opportunity and ambience to write this complex technical book. A relaxed mind and inspiring friends are most essential to have a tireless body to write a technical book. Hence the availability of technically cognizant colleagues and friends who are willing to spend time both to discuss complex technical matters and suggest remarks on the structure and sections of this book, are of great advantage to the author. Else it would have been a fretting bustle. I also remember my parents with respect and sincerely thank to all my friends, colleagues and classmates who helped me with many technical debates and especially for reading and commenting on presentation of particular sections of this book.

At this moment of completion, I do recognize the contribution of my wife Dr. Shreelekshmi and my sons Harishankar and Harikrishnan. I express my profound gratitude for the spiritual and ethical support of Swami Rithmbarananda of Sivagiri Mutt without which this attempt would not be materialized. I remember my dear professors with great respect especially Prof. S. Sooryadas, Prof. P.Saraswathy, Dr. M.S.Valiyathan, Dr. G.S.Bhuvaneshwer, Dr. M.Harisankar, Dr. K.G.Nair, Dr. C.S.Sridhar, Dr. P.Mohanan, Dr. Tessamma Thomas, Dr. R.Gopikakumari and Dr. P.Sethumadhavan for their support and blessings. I also remember my cousins especially Sri. Saji Natarajan, Sri.R.Ranjith, Smt. D.Bindu Chitharanjan, Sri.K.S. Jayamohan and my brother-in-law Maj.G.Chitharanjan for their genuine love and support which have also been very crucial in non-academic aspects.

Above all, sincere gratitude to my classmate Smt. Ananda Parvathy T.G, for making my words read better. Finally I dedicate this work to the lotus feet of Goddess Sarada of Sivagiri Mutt and to my wife Dr. Shreelekshmi.

In spite of all my efforts, there may be quite a few errors remaining in the book, and there would have been many more. Without the help of the expert reviewers, they may not be corrected and any suggestions for improvement of this book are always welcome. The author will be definitely privileged, if the readers get the intended sense which is the sole aim of this effort. My sincere thanks to Notion team for their support and co-operation on this project.

– **K S Manoj**

ABOUT THE AUTHOR

K S Manoj is an electronics and communication engineer with research interest in physical-cyber security of ICS who continues to explore the adoption of new security strategies in order to promote safer and more reliable automation infrastructures. Initially he specialized in design and development of ICS, by focusing on monitoring and control of SCADA and DCS networks. However after realizing the importance of the industrial cyber security, he later focused towards the threat vectors of advanced attacks in these environments and framing mitigation strategies. He holds a Masters in Technology in Communication Electronics from Department of Electronics, Cochin University of Science and Technology, India and another Masters in Solid State Electronics from University of Kerala, India. He has over 30 years of experience in Automation and Energy Engineering specializing in power system automation and critical infrastructure security. He has authored three more titles viz. Industrial Automation with SCADA, Cyber Security in Industrial Automation and Smart Grid-Concepts to Design. His research interests include Machine Learning and its application in Clinical Diagnostic Support Systems (CDSS). He may be contacted *ksmanoj321@gmail.com*.

CONTENTS

List of Figures xxiii
List of Tables xxv
List of Acronyms xxvii
Structure of this Book xxxiii

1 POWER SYSTEM AUTOMATION 1
   1.1   Introduction 1
   1.2   Evolution of Automation Systems 1
   1.3   Paradox of Automation 3
   1.4   Supervisory Control and Data Acquisition (SCADA) 4
        1.4.1   Evolution of SCADA 4
        1.4.2   Components of SCADA Systems 6
        1.4.3   Communication Infrastructure And Methods 9
        1.4.4   Security Concerns of Power System Automation .. 10
   1.5   Scada Applications in Power Systems 11
        1.5.1   Computerized Control 12
        1.5.2   SCADA/EMS/GMS AND SCADA/DMS 13
        1.5.3   Benefits of SCADA in Power Systems 13
   1.6   Changing Scenario of Power System Field 15
        1.6.1   Transmission and Distribution Systems 15
        1.6.2   Customer Premises 16
   1.7   Industrial Sectors and Their Interdependencies 17
   1.8   Data Flow in Power System Scada 18
   1.9   The Need for Secure Data Transfer 19
        1.9.1   Need For Better Security Technology 20
   1.10 Changing Scada Culture 21

2 SCADA BASICS AND COMPONENTS 23
   2.1   Introduction 23
   2.2   Data Acquisition Systems (DAS) 24
        2.2.1   Objectives and Advantages 24
        2.2.2   Single Channel Data Acquisition System 24
        2.2.3   Multi-Channel Data Acquisition System 25
        2.2.4   Sensors 27

            2.2.5   Signal Conditioning 27
            2.2.6   Sample And Hold Circuit 28
            2.2.7   A to D Converters (ADC) 29
            2.2.8   Storage and Display 33
            2.2.9   Data Forwarding And Communication 34
            2.2.10   Communication in SCADA System 34
            2.2.11   Selection Criteria of DAS 34
     2.3   Power System Scada Components 37
            2.3.1   Remote Terminal Unit (RTU) 37
            2.3.2   Intelligent Electronic Devices (IEDs) 46
            2.3.3   Instrument Transformers with Digital Interfaces. 48
            2.3.4   Programmable Logic Controller (PLC) 49
            2.3.5   Data Concentrators and Merging Units 50
            2.3.6   Bay Control Unit 52
            2.3.7   Master Control Centres (MCC) 53
            2.3.8   System SCADA Software 54
            2.3.9   Master Station Hardware 54
            2.3.10   Servers in the Master Station 54
            2.3.11   GLOBAL POSITIONING SYSTEMS (GPS)-RELEVANCE TO SCADA 58
            2.3.12   HUMAN MACHINE INTERFACE (HMI) 58

3   **SYSTEM DESIGN CONSIDERATIONS** 65
     3.1   Introduction 65
     3.2   Communication Architecture 65
            3.2.1   Point-To-Point between Two Stations 65
            3.2.2   Multipoint or Multiple Stations 65
            3.2.3   Talk through Repeaters 67
     3.3   Communication Philosophies 68
            3.3.1   Polled or Master Slave 68
            3.3.2   CSMA/CD System (Peer-To-Peer) 70
                     3.3.2.1   RTU to RTU Communication 70
                     3.3.2.2   Exception Reporting (Event Reporting) 71
                     3.3.2.3   Polling plus CSMA/CD With Exception Reporting 72
     3.4   System Reliability and Availability 73
            3.4.1   Fail Safe Systems 73
            3.4.2   Fault Tolerant Systems (FTS) 74
            3.4.3   Graceful Degradation Systems 75
            3.4.4   Design Considerations for Fault Tolerant Systems 75
            3.4.5   High Availability 75
            3.4.6   Critical Functions 76
            3.4.7   System Redundancy 79
            3.4.8   Channel Redundancy 80
     3.5   Design and Configuration Considerations of MCC 81

4 TRANSMISSION AND DISTRIBUTION AUTOMATION 83
4.1 Introduction 83
4.2 Substation Automation-technology Drivers 83
4.3 Smart Devices for Substation Automation 84
4.4 Smart Transmission 84
    4.4.1 Design of Digital Substations 85
    4.4.2 Energy Management Systems (EMS) 85
    4.4.3 HVDC and FACTS -Enhancing Power System Performance 89
    4.4.4 Hybrid Transmission System 93
    4.4.5 Phasor Measurement Units (PMU) 93
    4.4.6 Wide-Area Monitoring System (WAMS) 97
    4.4.7 Wide Area Monitoring, Protection and Control (WAMPAC) 98
4.5 Distribution Systems 99
    4.5.1 Feeder Automation 99
        4.5.1.1 Automated Feeder Switching 100
    4.5.2 Distribution Management System (DMS) 101
        4.5.2.1 Distribution Network Model 102
        4.5.2.2 Network Connectivity Analysis (NCA) 102
        4.5.2.3 State Estimation (SE) 103
        4.5.2.4 VOLT /VAR Control (VVC) 104
        4.5.2.5 Load Flow Studies (LFS) 104
        4.5.2.6 Load Flow Application (LFA) 105
        4.5.2.7 Load Shed Application (LSA) 105
        4.5.2.8 Fault Management And System Restoration (FMSR) 106
        4.5.2.9 Loss Minimization Via Feeder Reconfiguration (LMFR) 106
        4.5.2.10 Load Balancing Via Feeder Reconfiguration (LBFR) 107
        4.5.2.11 Distribution Load Forecast (DLF) 107
        4.5.2.12 Outage Management System (OMS) 108
    4.5.3 Geographical Information Systems (GIS) 109
    4.5.4 Customer Information Systems (CIS) 110
    4.5.5 High-Efficiency Distribution Transformers (HEDT) 111
    4.5.6 Phase Shifting Transformers (PST) 111
    4.5.7 Solid State Transformer (SST) 112

5 INDUSTRIAL COMMUNICATION BASICS 113
5.1 Introduction 113
5.2 Types of Transmission 113
    5.2.1 Analog and Digital 114
    5.2.2 Synchronous And Asynchronous 115
    5.2.3 Broadcast, Multicast, and Unicast 115
    5.2.4 Simplex, Half Duplex and Full Duplex Communication Channels 117
    5.2.5 Baseband And Broadband 118
5.3 Guided Media 119
    5.3.1 Twisted Pair Copper 119
    5.3.2 Copper Co-Axial Cables 121

- 5.3.3 Fiber Optic Cables ... 122
- 5.3.4 Cabling Considerations... 124
  - 5.3.4.1 Noise... 124
  - 5.3.4.2 Cabling Connection Types ... 124
  - 5.3.4.3 Attenuation ... 125
  - 5.3.4.4 Crosstalk ... 125
  - 5.3.4.5 Fire Rating of Cables ... 126
- 5.4 Unguided Media ... 126
  - 5.4.1 Microwaves Communication... 127
  - 5.4.2 Terrestrial Communication ... 127
  - 5.4.3 Satellite Communication ... 128
  - 5.4.4 Mobile Communication ... 129
- 5.5 Scada Communication Technologies ... 134
  - 5.5.1 WIRED OR GUIDED MEDIA TECHNOLOGIES ... 134
    - 5.5.1.1 Copper Utp... 134
    - 5.5.1.2 Optical Fiber... 135
    - 5.5.1.3 Fiber To The Home (FTTH) ... 135
    - 5.5.1.4 Hybrid Fiber Coax (HFC)... 135
    - 5.5.1.5 Power Line Carrier Communication (PLCC)... 136
    - 5.5.1.6 Broadband Over Power Line (BPL) ... 136
    - 5.5.1.7 Homeplug... 137
  - 5.5.2 WIRELESS OR UNGUIDED MEDIA TECHNOLOGIES ... 137
    - 5.5.2.1 IEEE and Wireless Standards... 138
    - 5.5.2.2 Frequency Hopping Spread Spectrum (FHSS)... 138
    - 5.5.2.3 4G Cellular ... 140
    - 5.5.2.4 WiFi ... 141
    - 5.5.2.5 WiMax... 141
    - 5.5.2.6 ZigBee ... 142
    - 5.5.2.7 ZWave... 143
    - 5.5.2.8 VSAT ... 145
- 5.6 Security in Wireless Communications ... 145
  - 5.6.1 Endpoint Threat Detection and Response (ETDR) ... 146
  - 5.6.2 Transparency ... 147
  - 5.6.3 Redundancy ... 147

6 INDUSTRIAL NETWORKING BASICS... 149
- 6.1 Introduction ... 149
- 6.2 Protocols-the Rules That Govern Communications... 149
  - 6.2.1 Structure Of Communication Protocols... 150
  - 6.2.2 Interoperability ... 150
  - 6.2.3 Open and Closed Architecture ... 151
    - 6.2.3.1 Open Systems ... 151
    - 6.2.3.2 Closed Systems... 151
  - 6.2.4 Connection Oriented and Connectionless Protocols... 152
    - 6.2.4.1 Connection Oriented Communication... 152

```
            6.2.4.2   Connectionless Communication ............................. 152
    6.2.6   ISO Open Systems Interconnection Reference Model............... 153
            6.2.6.1   Protocol ...................................................... 153
            6.2.6.2   OSI Reference Model ........................................ 154
    6.6.7   TCP/IP MODEL ..................................................... 170
    6.6.8   PORTS AND PORT NUMBERS......................................... 173
    6.6.9   TCP/IP VULNERABILITIES............................................ 174
    6.6.10  ENHANCED PERFORMANCE ARCHITECTURE (EPA) ................... 174
6.3 Network Devices ............................................................. 176
6.4 De Militarized Zone (DMZ) .................................................. 195
6.5 Network Topologies ......................................................... 196
6.6 Media Access Methods ...................................................... 202
6.7 Pros and Cons of Computer Networks........................................ 207
6.8 Computer Networks: Present Day Technologies.............................. 208
    6.8.1   Local Area Networks (LANs) ........................................ 208
    6.8.2   Metropolitan Area Networks (MAN) ................................. 209
    6.8.3   WAN Technologies .................................................. 210
    6.8.4   Internet ............................................................. 211
    6.8.5   Intranets and Extranets............................................. 212
    6.8.6   Value Added Networks (VAN) ....................................... 212

7   PROTOCOLS: THE NERVES AND VEINS ....................................... 215
    7.1 Introduction ............................................................. 215
    7.2 Evolution of Scada Communication Protocols ............................. 215
    7.3 Scada Communication Protocols.......................................... 216
        7.3.1   Distributed Network Protocol 3 (DNP 3) ......................... 216
                7.3.1.1   Protocol Architecture of DNP3 .......................... 217
        7.3.2   Modbus ......................................................... 218
                7.3.2.1   Modbus Limitations ..................................... 221
                7.3.2.2   Attacks on DNP3 and Modbus............................ 221
        7.3.3   Profibus ........................................................ 222
                7.3.3.1   Profibus Process Automation (PA) ...................... 223
                7.3.3.2   Profibus Factory Automation (Decentralized Peripherals-
                          DP)................................................... 223
                7.3.3.3   Profibus Fieldbus Message Specification (FMS).............. 223
                7.3.3.4   Communication Architecture of Profibus .................... 224
        7.3.4   IEC 60870-5-101/103/104....................................... 225
        7.3.5   IEC 60870-5-101 [T-101] ........................................ 225
        7.3.6   IEC 60870-5-103 [T-103] ........................................ 226
        7.3.7   IEC 60870-5-104 [T-104] ........................................ 227
                7.3.7.1   Protocol Architecture of IEC 60870-5 .................... 227
                7.3.7.2   Attacks on IEC 60870-5................................. 228
        7.3.8   IEC 61850 ...................................................... 228
                7.3.8.1   Comparison of DNP3 and IEC-61850 GOOSE................ 230
                7.3.8.2   Attacks on IEC 61850 Protocol........................... 231
```

			7.3.9	ICCP TASE 2(IEC 60870-6) 231
				7.3.9.1	ICCP Functionalities. 232
				7.3.9.2	Protocol Architecture of ICCP TASE 2. 232
				7.3.9.3	Implementation Issues and Interoperability. 232
				7.3.9.4	ICCP-Product Differentiation 233
				7.3.9.5	ICCP- Product Configurations 234
	7.4	Other Sg Pertinent Standards. 234
		7.4.1	IEEE C37.118.1synchrophasor Standard 234
		7.4.2	IEC 61968 Standard 235
		7.4.3	IEC 61970 Standard 235
		7.4.4	IEC 62325 Standard 236
		7.4.5	IEC 61508 Standard 236
		7.4.6	IEC 62351 Security Standard 237
		7.4.7	IEC 62056 Electricity Metering Data Exchange Standard 238
		7.4.8	IEC 62056-21 Standard 239
	7.5	Secure Communication (scommunication) 239
	7.6	Selecting the Right Protocol for Scada. 239

8	**PHYSICAL-CYBER SECURITY OF SCADA** 243
	8.1	Introduction 243
	8.2	It Security and Scada Security. 244
	8.3	Remote Access and Open Communication Systems 246
	8.4	VPN and MPLS in Power System Automation 248
		8.4.1	VPN Architecture 249
		8.4.2	Security Issues of VPN 249
		8.4.3	Remote Access VPN 250
		8.4.4	VPN Termination in Remote Access. 250
		8.4.5	Site-To-Site VPN. 251
		8.4.6	Difference between IPSec VPN and SSL VPN 253
		8.4.7	Deploying VPN 255
		8.4.8	MultiProtocol Label Switching (MPLS). 256
		8.4.9	Choosing MPLS VPN Services 257
		8.4.10	Advantages and Disadvantages of MPLS VPNS. 258
	8.5	Critical Infrastructure Protection and NERC CIP 260
		8.5.1	Critical Infrastructure. 260
		8.5.2	Critical Infrastructure Protection (CIP) 260
		8.5.3	Critical Information Infrastructure Protection 261
		8.5.4	NERC CIP and Bulk Electric System (BES) 261
	8.6	Security Concerns in Substation Automation 262
		8.6.1	Attack Vector through Substation HMI 264
		8.6.2	Security Concerns of SCADA Control Center 266
		8.6.3	Defense-in-Depth Architecture 267
		8.6.4	Firewall deployment and Policies 268
		8.6.5	Design Considerations of PSS Security 271
		8.6.6	PSS Potential Risks 272

9 CYBER RISKS AND MITIGATION STRATEGIES ... 275
- 9.1 Introduction ... 275
- 9.2 Common Scada Network Security Attacks ... 276
  - 9.2.1 Threat Sources to PSS ... 279
  - 9.2.2 Power System SCADA Vulnerabilities, Threats and Attacks ... 280
  - 9.2.3 Alarming Power System SCADA Threats ... 282
    - 9.2.3.1 Zero Day Vulnerabilities ... 282
    - 9.2.3.2 Non-Prioritization of Tasks ... 282
    - 9.2.3.3 Database Injection ... 283
    - 9.2.3.4 Communication Protocol Issues ... 283
    - 9.2.3.5 Stealthy Integrity Attack ... 283
    - 9.2.3.6 Replay Attack ... 283
    - 9.2.3.7 False Data Injection Attack ... 284
    - 9.2.3.8 Zero-Dynamics Attack ... 284
    - 9.2.3.9 Covert Attack ... 285
    - 9.2.3.10 Surge Attack, Bias Attack, and Geometric Attack ... 285
  - 9.2.4 Dreadful PSS Malwares ... 285
    - 9.2.4.1 BlackEnergy ... 286
    - 9.2.4.2 Ukraine Incident ... 286
    - 9.2.4.3 Stuxnet ... 287
    - 9.2.4.4 Iranian Experience ... 288
    - 9.2.4.5 Spreading of Stuxnet ... 289
    - 9.2.4.6 Havex ... 290
    - 9.2.4.7 Sandworm ... 291
    - 9.2.4.8 Duqu and Flame ... 291
  - 9.2.5 Flash Drive Usage and End Node Security (ENS) ... 292
    - 9.2.5.1 BadUSB ... 292
    - 9.2.5.2 Cyber Incidents Using USB ... 293
- 9.3 Purdue Reference Architecture for ICS ... 293
- 9.4 Physical Security ... 298
- 9.5 Network Security ... 301
- 9.6 Computer/server Security ... 304
- 9.7 Application Security ... 304
- 9.8 Device Security ... 308
- 9.9 Modern Approach to IT/OT Integration ... 310
- 9.10 Increasing Resiliency By Segmenting the OT Network ... 311

10 SMART GRID CONCEPTS AND APPLICATIONS ... 313
- 10.1 Introduction ... 313
- 10.2 Smart Grid Definition and Development ... 314
- 10.3 Constraints of Old Grid ... 315
- 10.4 Benefits of Smart Grid ... 316
- 10.5 Scada: Heart and Brain of Smart Grid ... 317
- 10.6 Stakeholders of Smart Grid ... 318
- 10.7 Transforming to Smart Grid ... 319

10.8 Smart Grid Interoperability Framework...320
    10.8.1 Asset Optimization..320
    10.8.2 Customer Side Optimization...320
    10.8.3 Distribution Optimization 321
    10.8.4 Customer Automation And Communications 321
    10.8.5 Transmission Optimization 322
    10.8.6 Human Resource Development And Optimization 322
10.9 Smart Grid Road Map...323
10.10 Standards and Interoperability 323
10.11 Smart Distribution 324
    10.11.1 Demand Side Energy Management..............................324
    10.11.2 Renewable Energy Sources (Res) 325
    10.11.3 Distributed Generation And Technologies 326
    10.11.4 Grid Energy Storage 327
    10.11.5 Advanced Metering Infrastructure (Ami).......................327
    10.11.6 Energy Efficient Smart Homes (EESHS).........................328
    10.11.7 E-Mobility 328
10.12 Multiple Distributed Microgrids 329
    10.12.1 Characteristics of A Microgrid.....................................329
    10.12.2 Microgrid Components..330
    10.12.3 Benefits of Microgrid 331
    10.12.4 Multiple Microgrids And Hierarchical Control 332
    10.12.5 DC Microgrids..334
    10.12.6 Microgrid Challenges...335
10.13 Smart Transmission 336
10.14 Sg Big Data Analytics and Challenges 336
    10.14.1 AMI Data Analytics-Present Status 337
10.15 Communication - the Key Enabler 337
    10.15.1 Smart Grid Communication Requirements.......................338
    10.15.2 Networking the Smart Grid 339
    10.15.3 Smart Communication for Smart Grid 340
    10.15.4 Communication Technologies For Smart Grid 340
10.16 Smart Grid Security-the Foremost Challenge....................................340
    10.16.1 Cyber-Physical Security 341
    10.16.2 System Security 341
    10.16.3 Cyber-Security 341
    10.16.4 Smart Grid Vulnerabilities.......................................342
    10.16.5 Security Protocols..343
10.17 Other SG Implementation Challenges and Considerations..................345

11 SCADA SECURITY STANDARDS 347
  11.1 Introduction 347
  11.2 Selection of Standards...348
  11.3 ISA 99 / IEC 62443 Standard...348

11.4 NERC CIP Standard for BES 350
       11.4.1 Achieving NERC CIP Compliance 354
11.5 ISO/IEC 27001 Standard 355
       10.5.1 Audit Controls For the ISO 27001 357
       10.5.2 Maintaining ISO 27001 Compliance 358
11.6 NIST SP 800-53 360

12   DOCUMENTED SCADA CYBER ATTACKS 361

12.1 Introduction 361
12.2 Documented Power System Cyber Incidents 361
       12.1.1 ICS Cyber Attacks On US 363
       12.1.2 Cyber Threats to Indian ICS 364
12.3 Common ICS Security Issues 366
12.4 Mitigation Strategies 367

Index *371*

LIST OF FIGURES

Figure 1.1:	Fourth generation SCADA or IoT	6
Figure 1.2:	Single Line Diagram of a typical substation	16
Figure 1.3:	Data flow in a typical power system SCADA environment	19
Figure 2.1:	Single Channel Data Acquisition System	25
Figure 2.2:	Multi-Channel DAS	26
Figure 2.3:	Multi-Channel DAS for simultaneous measurement	26
Figure 2.4:	Sample and Hold circuit	29
Figure 2.5:	Integrating or dual slope ADC	30
Figure 2.6:	Successive Approximation ADC	30
Figure 2.7:	Parallel Comparator (Flash) type ADC	31
Figure 2.8:	Counter type ADC	32
Figure 2.9:	RTU Architecture	38
Figure 2.10:	Block diagram of an IED	46
Figure 2.11:	Block Diagram of a Typical Master Control Centre	53
Figure 3.1:	Point to Point Master-Slave Communication	66
Figure 3.2:	Master RTU communicating with multiple station RTUs	66
Figure 3.3:	Store and forward station	67
Figure 3.4:	Talk through repeaters	68
Figure 3.5:	Illustration of master-slave polling with RTUs	69
Figure 3.6:	Polling cycle as per the polling table	70
Figure 3.7:	Block diagram of a typical DCS MCC with Fail Safe HA connectivity	77
Figure 3.8:	De Militarized Zone (DMZ)	78
Figure 3.9:	SCADA Control System LAN	78
Figure 3.10:	Dispatch Training Simulator	78
Figure 3.11:	Fail Safe connectivity to DCS MCC using HA	79
Figure 3.12:	Channel redundancy achieved with a resilient and reliable cloud	80
Figure 3.13:	Channel redundancy achieved with two different communication clouds	81
Figure 4.1:	Digital substation with process and station LAN architecture	86
Figure 4.2:	Block Diagram of an EMS	86
Figure 4.3:	Components of a PMU	95
Figure 4.4:	Building blocks of a synchrophasor based system	96
Figure 4.5:	Configuration of the WAMPAC	99
Figure 4.6:	(a). Initial Outage and Fault Location; (b). Supply Restoration by Feeder Automation	100
Figure 4.7:	Data exchange between GIS and DMS	110

Figure	Description	Page
Figure 5.1:	Twisted Pair Copper	120
Figure 5.2:	Co-axial copper cable	121
Figure 5.3:	Fiber Optic Cable	123
Figure 5.4:	Terrestrial communication	128
Figure 5.5:	Satellite communication	129
Figure 5.6:	Nonadjoint cells can use the same frequency ranges	130
Figure 5.7:	changing carrier frequency at random	139
Figure 6.1:	OSI Layer Protocol Data Encapsulation	155
Figure 6.2:	Application request through API	158
Figure 6.3:	Data into standard formats by the Presentation Layer	160
Figure 6.4:	The three phases of a Session Layer	161
Figure 6.5:	TCP formats data into a stream suitable for transmission	163
Figure 6.6:	The most efficient path by the Network Layer	165
Figure 6.7:	Two sub-layers of the Data Link Layer	167
Figure 6.8:	Comparison of the OSI model with the TCP/IP model	170
Figure 6.9:	Four layers of TCP/IP and its component protocols	171
Figure 6.10:	Comparison of the EPA model with the OSI model	175
Figure 6.11:	Data packets go to all workstations reaching the HUB	176
Figure 6.12:	Data packets sent to only destination workstations through Switch	177
Figure 6.13:	A Bridge connects two LAN segments	180
Figure 6.14:	Repeater showing its functionality	185
Figure 6.15:	VLANs enable administrators to manage logical networks	187
Figure 6.16:	VLANs showing the existence in a higher level than the physical network	188
Figure 6.17:	Functional steps of a Proxy Firewall	193
Figure 6.18:	Ring Topology	197
Figure 6.19:	(a) Linear bus topology (b) Distributed (tree) bus topology	199
Figure 6.20:	Star Topology	200
Figure 6.21:	Mesh topology	201
Figure 7.1:	A Master-Slave Networking Relationship	218
Figure 7.2:	Communication layers of IEC 60870-5	228
Figure 7.3:	IEC 61850 Layered Architecture	229
Figure 8.1:	IPSec and SSL VPN	254
Figure 8.2:	Block diagram of an automated substation	263
Figure 8.3:	Attack points in an automated substation with DCU and HMI	264
Figure 8.4:	Firewall with DMZ between Corporate and Control Networks	266
Figure 8.5:	Defense-in-Depth Architecture	268
Figure 9.1:	Enterprise zone	295
Figure 9.2:	Industrial Demilitarized Zone (IDMZ)	296
Figure 9.3:	Manufacturing Zone	297
Figure 9.4:	Modern IT/OT DMZ Scenario No.1	310
Figure 9.5:	Modern IT/OT DMZ Scenario No.2	310
Figure 10.1:	Components of a Microgrid	330
Figure 10.2:	Hierarchical control scheme for Microgrid	333
Figure 12.1:	ICS Security Key Reminders	369

LIST OF TABLES

Table 1.1:	Sensors with signal conditioning requirements	28
Table 3.1:	Polling table with the master station	70
Table 5.1:	Characteristics of baseband and broadband	119
Table 5.2:	UTP cable ratings	120
Table 5.3:	Different characteristics of Mobile Technology	134
Table 5.4:	Characteristic Comparison of the various Gs	140
Table 6.1:	Network Devices and Functionalities	183
Table 6.2:	Comparison of Different Types of Firewalls	195
Table 6.3:	Comparison of Network Topologies	201
Table 7.1:	DNP3 and OSI model layers	217
Table 7.2:	Profibus Versions	224
Table 7.3:	Profibus PA, DP, and FMS layered protocols	224

LIST OF ACRONYMS

AAM	Advanced Assets Management
ADO	Advanced Distribution Operation
AH	Authentication Header
AMI	Advanced Metering Infrastructure
ANSI	American National Standard Institute
API	Application Programming Interface
APT	Advanced Persistent Threat
AR	Auto Re-closure
ASD	Adjustable Speed Drives
ATM	Asynchronous Transfer Mode
ATO	Advanced Transmission Operation
BES	Bulk Electricity System
BGP	Border Gateway Protocol
BMS	Building Management System
BPLC	Broadband Power Line Communication
CENELEC	European Committee for Electrotechnical Standardization
CFE	Communication Front End
CIM	Common Information Model
CIP	Critical Infrastructure Protection
CIS	Customer Information System
COSEM	Companion Specification for Energy Metering
DA	Distribution Automation
DCU	Data Control Unit
DER	Distributed Energy Resources
DG	Distributed Generation
DLC	Direct Load Control
DLF	Distribution Load Forecast
DLMS	Device Language Message Specification
DMD	Distribution Network Model or Dynamic Mimic Diagram
DMS	Distributed Management System

DNP	Distribution Network Protocol
DOS	Denial Of Service
DRMS	Demand Response Management System
DSM	Demand Side Management
DST	Decision Support Tools
EAP	External Access Point
ECC	Energy Control System
ECM	Equipment Condition Monitor
EDGE	Enhanced Data Rates For GSM Evaluation
EMS	Energy Management System
ENS	End Node Security
EPA	Enhanced Power Architecture
ESP	Electronic Security Perimeter
EU	European Union
EUC	Equipment Under Control
EV	Electric Vehicle
EVT	Electronic Voltage Transformer
FACTS	Flexible AC Transmission System
FCC	Federal Communication Commission
FEP	Front End Processor
FES	Flywheel Energy Storage
FMSR	Fault Management and System Restoration
FOCS	Fibre Optical Current Sensor
FPI	Fault Passage Indication
FTTH	Fibre To The Home
FTTP	Fibre Transmission Transfer Protocol
GIS	Geographical Information System
GMPLS	Generalizes Multi-Protocol Label Switching
GOOSE	General Object Oriented Substation Event
GPS	Global Positioning System
GRE	General Routing Encapsulation
GSM	Global System For Mobile Communication
HAN	Home Area Network
HASR	High Availability Seamless Redundancy
HDLC	High Level Data Link Control
HEDT	High Efficiency Distribution Transformer
HEV	Hybrid Electronic Volt
HHU	Hand-Held Unit
HMI	Human Machine Interfaces

HVDC	High Voltage Direct Current
ICCP	Intercontrol Centre Communication Protocol
ICE	Internal Combustion Engine
ICT	Information and Communication
IEC	International Electrotechnical Commission
IED	Intelligent Electronic Device
IEEE	Institute Of Electrical and Electronics Engineering
IHEM	In-Home Energy Management
IP	Internet Protocol
IP FRR	IP Fast Reroute
IPFC	Interline Power Flow Controller
IPSec	Internet Protocol Security
ISA	International Society for Automation
ISO	Independent System Operation
IT	Information Technologies
LAN	Local Area Network
LBFR	Load Balancing via Feeder Reconfiguration
LDMS	Local Data Monitoring System
LDP	Label Distribution Protocol
LFA	Load Flow Application
LFP	Lithium Ferro Phosphate
LIB	Lithium Iron Battery
LMFR	Loss Minimization via Feeder Reconfiguration
LNMC	Lithium Nickel Manganese Cobalt
LSA	Load Shed Application
LSP	Label Switched Path
LTE	Long Term Evaluation
MAC	Media Access Control
MBP	Manchester Bus Powered
MCC	Master Control Centre
MDMS	Meter Data Management
MITM	Man In The Middle
MMS	Manufacturing Messaging Specification
MPLS	Multi-Protocol Label Switching
MU	Merging Unit
NAS	Network Access Server
NCA	Network Connectivity Analysis
NERC	North American Electric Reliability Corporation
NIST	National Institute For Standard and Technology

O&M	Operation and Maintenance
OMS	Outage Management System
OSI	Open System Interconnection
PCMCIA	Personal Computer Memory Card International Association
PDC	Phasor Data Concentrator
PDU	Protocol Data Unit
PHEV	Plug In Hybrid Electric Vehicle
PLCC	Power Line Carrier Communication
PMU	Phasor Measurement Unit
PPP	Point-To-Point Protocol
PPTP	Point To Point Tunnelling Protocol
PSS	Power System SCADA
PST	Phase Shifting Transformer
PV	Photo Voltaic
RES	Renewable Energy Sources
RSVP	Resource Reservation Protocol
RTOS	Real Time Operating System
RTU	Remote Terminal Unit
SCADA	Supervisory Control and Data Acquisition
SE	State Estimation
SGN	Smart Grid Communication Network
SHEMS	Smart Home Energy Management System
SIL	Safety Integrity Level
SM	Smart Meter
SMES	Super Conducting Magnetic Energy Sources
SOC	State Of Charge
SPS	Special Protection System
SSH	Secured Shell
SSL	Secured Socket Layer
SST	Social State Transformer
TA	Topology Analyser
TC	Technical Committee
TCP	Transmission Control Protocol
TDM	Time Division Multiplexing
THD	Total Harmonic Distortion
TLS	Transport Layer Security
ToU	Time Of Use
TSO	Transmission System Operation
UDP	User Data Protocol

UN	United Nation
UPFC	Unified Power Flow Controllers
USB	Universal Serial Bus
V Ring	Virtual Ring
V2G	Vehicle To Grid
VAR	Voltage Ampere Reactive
VFD	Variable Frequency Drives
VPDN	Virtual Private Dial-Up Network
VPN	Virtual Private Network
VVC	Volt VAR Control
VVO	Volt VAR Optimization
WAAPCA	Wide Area Adaptive Protection Control and Automation
WAMPAC	Wide Area Monitoring Protection and Control
WAMS	Wide-Area Monitoring System
WAN	Wide Area Network
WASA	Wide Area Situational Awareness
WG	Working Group
ZDA	Zero Dynamic Attack
ZDV	Zero Day Vulnerability

STRUCTURE OF THIS BOOK

This book mainly intended for practicing power system engineers and senior undergraduate and post graduate students of Electronics, Electrical, ICT and Instrumentation engineering students who wish to be the part of implementation of Power System Automation, the evolving Smart Grid and Microgrid, etc. Every action is a depended variable of energy and it is an accepted fact that there is a revolutionary change taking place in the energy sector and one of them is the automation of Power Systems. Hence the book starts with the chapter Power System Automation which explains its need and necessities.

Chapter 1 Power System Automation: This chapter begins with the concept of automation and its evolution. The innovation of SCADA into the automation and its various stages are also explained. Then moves on to explaining the advantages and benefits of implementing SCADA based automation in power system. The chapter concludes with explaining the need of secure data transfer and the changing SCADA based automation work culture of power system professionals.

Chapter 2 SCADA Basics and Components: This chapter gives an introduction to SCADA with an emphasis on Data Acquisition Systems (DAS) and its components. The objectives and advantages as well as the evolution of SCADA are briefly discussed but with clarity. A brief discussion of the various components of DAS such as Sensors, Signal conditioners, Sample and Hold circuits, Analog to Digital Converters, etc. are also given in this chapter. A brief discussion regarding the selection criteria of the DAS is then given before moving to give a comprehensive description of Remote Terminal Units (RTU), Programmable Logic Controller (PLC) and the different modern components of Power System SCADA such as Intelligent Electronic Devices, Data Concentrator Units, Merging Units, Human Machine Interface, etc. The brief introduction of Data Concentrators and Merging Units are presented in such a manner that how digital substations can be designed. This chapter then gives an introduction of architecture of SCADA Master Stations and its hardware and software components. It concludes with a brief description of Geographical Positioning System (GPS), Situational Awareness and Alarm Processing.

Chapter 3 System Design Considerations: This chapter begins with describing the communication architecture of basic SCADA. Then moves on to describing the common communication philosophies adopted in DCS. As the reliability and availability of DCS functions are the most important, they are briefly introduced, but cater the necessary understanding to power system professionals and engineering students who are engaged or intend to embark into the DCS, Smart Grid or Microgrid domain. It then explains the concepts of Fault Tolerant Systems, Fail Safe and redundant systems, High Availability, etc. Based on these concepts, this chapter then elaborates the design of a typical SCADA MCC architecture having redundant and HA connectivity.

Chapter 4 Transmission and Distribution Automation: Transformation of substations to digital substations is one of the major areas of power system modernization to adopt Smart Grid. Hence the fourth chapter is dedicated for describing the Substation Automation, the modern components required for realizing a digital substation, the advanced transmission components such as, High Voltage Direct Current (HVDC), Flexible AC Transmission System (FACTS), Phasor Measurement Unit (PMU), Wide Area Monitoring System (WAMS) the smart distribution systems such as Feeder Automation, Distribution Management System (DMS), Outage Management System (OMS), Customer Information Systems (CIS), Geographical Information Systems (GIS), High-Efficiency Distribution Transformers (HEDT), Phase Shifting Transformers (PST), etc.

Chapter 5 Industrial Communication Basics: This chapter has focused on various communication aspects of the industrial SCADA with an emphasis on DCS and Smart Grid. It begins with discussing various types of transmission technology in very modest way so that it is very apt and most essential for a power engineer who engaged in the design and implementation of SCADA and Smart Grid. The chapter then discusses the guided and unguided media used today for communication in such a manner that it is very useful for a practicing communication professional, which includes the various cabling issues as well.

The various but most relevant communication technologies which find space not only in industrial SCADA but also in other smart automation technologies today are discussed comprehensively. Finally the chapter focused on the security issues of the wireless communication technology.

Chapter 6 Industrial Networking Basics: The advancement of ICT innovates the four major areas in industry viz. Data Acquisition Systems(DAS), computer and communication, cyber-security and the pertinent ICS applications. The present domain of the ICS does not require much knowledge of secure M2M communication. But engineers who wish to associate with the development and implementation of the advanced and distributed SCADA and ICS, a basic understanding of computer and communication, is most essential. Hence this chapter Industrial Networking Basics, which describes the basic terminologies and the most fundamentals of computer and communication needed for an industrial security engineer is included.

Chapter 7 Protocols: The Nerves and Veins: In Smart Grid whatever data is gathered in the field, has to be send to the control room for processing and decision making. Being an industrial process, which is to be controlled in real-time mostly from a remote control room, the data and the control information have to be exchanged in a most secure manner. In computer science, this is achieved by encrypting the data and using proper protocols. Hence an essential and appropriate knowledge of various relevant protocols adopted and developed for Smart Grid and Power System SCADA are required for an electrical engineer to get a proper control over the automation process. The seventh chapter briefly described the various SG Protocols.

Chapter 8 Physical-Cyber Security of SCADA: It gives a description about difference between the OT (SCADA) security and the IT security. The requirement of open communication system and standardization are discussed with emphasis on security. The VPN and MPLS technology with their advantage and disadvantages, selection criterion are also discussed. The Critical Infrastructure Protection requirements as well as the NERC CIP standards are also described. Security concerns of the Substation Automation and control center architecture are deliberated with solutions. Malware threats especially the attack of Stuxnet which is a nightmare for Power System SCADA are discussed in this chapter. The attack vectors, the Zero Day Vulnerabilities and proposed solutions are also briefly described. The chapter concludes with describing the threats and vulnerabilities of ICS and Power System SCADA with various types of attacks and mitigating techniques.

Chapter 9 Cyber Risks and Mitigation Strategies: In recent years power sector is one of the main target of cyber-attackers and a number of cyber incidents have occurred to the power systems. This chapter describes different types of attacks and incidents to make the power system engineers aware of the severity. It also briefly describes the different types of threats that PSS may encounter from the perspective of the utility and implementation agencies. It then explains the lethal malware threats which mainly exploit the Zero Day Vulnerabilities of the Windows OS. This chapter then move on to explaining the Purdue Reference Model which is generally accepted as the basic security architecture model in industry. Then explains Physical and Environmental Security, Physical Security Threats, Attacks and Mitigation Strategies. Then it explains Network Security, Network Security Threats and Mitigation Strategies. Also explains Computer Security, Device security and Application Security. Briefly describes the software development security and attacking the air gapped systems tactics.

Chapter 10 Smart Grid: Fundamentals and Applications: This chapter begins with explaining the need, necessities and definition of Smart Grid and then moves to describe the benefits and road map for implementating the Smart Grid. It further elucidates how the ICT enabled Smart Grid will enhance bi-directional communications, bi-directional power flow, overall efficiency, reliability, and reduce the costs of electricity services. A brief introduction of Microgrids also has been given. The major challenges in implementing Smart Grid are AMI Big Data management, Smart Grid networking and communication, cyber-security, etc. The ICT innovations to the present power grid make it a massive efficient system but unfortunately is prone to cyber-attacks. Since the Smart Grid is considered as a Critical Infrastructure of a Nation, any vulnerability, should be identified and addressed immediately with utmost importance and appropriate solutions must be implemented to reduce the risks to an acceptable level. The risk and challenges of the same has also been examined with proposed solutions.

Chapter 11 ICS Security Standards: This chapter briefly explains the various security standards presently existing in the industrial sector and their selection requirements. IEC 62443, NIST 800, NERC CIP and ISO 27001 standards are briefly explained.

Chapter 12 Documented SCADA Cyber Attacks: In recent years ICS is one of the main target of cyber-attackers and a number of cyber incidents have occurred to the ICS. This chapter describes some of the cyber-incidents as well as different sources of attacks to make the ICS engineers aware of the severity of the attacks. It also briefly describes the different types of threats that ICS may encounter from the perspective of the utility and implementation agencies. It then explains the lethal malware threats like Stuxnet which is a nightmare for Power System SCADA and ICS as it mainly exploits the Zero Day Vulnerabilities of the Windows OS.

CHAPTER 01

POWER SYSTEM AUTOMATION

1.1 INTRODUCTION

As the term suggests, Power System Automation is the execution of various tasks in substations, generating stations and power distribution without periodic human interference. It aims to minimize and gradually keep away the human intervention by simplifying the complicated tasks by reducing them to a single instance. Today, automation has the effect of making human resources available for more engaging tasks, leading to greater efficiency and improved job satisfaction. In other words, power system automation is the act of automatically controlling the power system via instrumentation and control devices. Substation automation refers to using data from Intelligent Electronic Devices (IED), control and automation capabilities within the substation, and control commands from remote users to control power-system devices.

Since full substation automation relies on substation integration, the terms are often used interchangeably. Power system automation includes processes associated with generation, transmission and distribution of power. Monitoring and control of power delivery systems in the substation and on the pole reduce the occurrence of outages and shorten the duration of outages. The IEDs, communications protocols, and communications methods, work together as a system to perform power-system automation. The term *power system* describes the collection of devices that make up the physical systems that generate, transmit, and distribute power. The term instrumentation and control (I&C) system refers to the collection of devices that monitor, control, and protect the power system.

1.2 EVOLUTION OF AUTOMATION SYSTEMS

Automation or *automatic control*, is the use of various control systems for operating equipment such as machinery, processes in factories, boilers and heat treating ovens, switching on telephone networks, steering and stabilization of ships, aircraft and other applications with minimal or reduced human intervention. Some processes have been completely automated.

The fundamental benefit of automation is that it saves human labor. It also saves energy and materials in addition to the improvement in quality, accuracy and precision. The term *automation,* derived from the Latin word *automaton,* was not widely used before 1947, until General Motors established the automation department. It was during this time that industry was rapidly adopting feedback controllers, which were introduced in the 1930s.

Automation has been achieved by various means including mechanical, hydraulic, pneumatic, electrical, electronic devices and computers, usually in combination. Complicated systems, such as modern factories, airplanes and ships typically use all these combined techniques. The main advantages of automation are,
1. Increased throughput or productivity,
2. Improved quality or increased predictability of quality,
3. Improved robustness of processes or product,
4. Increased consistency of output, and
5. Reduced direct human labor costs and expenses.

The main disadvantages of automation are,
1. Security Threats/Vulnerability: An automated system may have a limited level of intelligence, and is therefore more susceptible to committing errors outside of its immediate scope of knowledge,
2. Unpredictable/excessive development costs: The research and development cost of automating a process may exceed the cost saved by the automation itself, and
3. High initial cost: The automation of a new product or plant typically requires a very large initial investment in comparison with the unit cost of the product, although the cost of automation may be spread among many products and over time.

There are certain limitations to automation and they are,
1. Current technology is unable to automate all the desired tasks,
2. Many operations using automation have large amounts of invested capital and produce high volumes of product, making malfunctions extremely costly and potentially hazardous. Therefore, some personnel are needed

to insure that the entire system functions properly and that safety and product quality are maintained,
3. As a process becomes increasingly automated, there is less and less labor to be saved or quality improvement to be gained. This is an example of both diminishing returns and the logisticHYPERLINK "https://en.wikipedia.org/wiki/Logistic_function" function, and
4. As more and more processes become automated, there are fewer remaining non-automated processes. This is an example of exhaustion of opportunities. New technological paradigms may however set new limits that surpass the previous limits.

The current limitations existing in the field of automation are explained below.

Many roles for humans in industrial processes presently lie beyond the scope of automation. Human-level pattern recognition, language comprehension, and language production ability are well beyond the capabilities of modern mechanical and computer systems. Tasks requiring subjective assessment or synthesis of complex sensory data, such as scents and sounds, as well as high-level tasks such as strategic planning, currently require human expertise. In many cases, the use of humans is more cost-effective than mechanical approaches even where automation of industrial tasks is possible. Overcoming these obstacles is a theorized path to post-scarcity economics.

1.3 PARADOX OF AUTOMATION

The paradox of automation says that the more sophisticated the automated system, the more crucial the human contribution of the operators. Automation makes human involvement less, but their involvement becomes more critical. If an automated system has an error, it will multiply that error until it's fixed or shut down. This is where human operators come in.

A fatal example of this was Air France Flight 447 having the automated fly-by-wire system, which is designed to reduce human error by letting computers control many aspects of the flight and was considered as the safest mode of operation. But due to a failure in automation, the flight entered into an aerodynamic stall put the pilots into a manual mode but the crew lacked practical training in manually handling the aircraft both at high altitude and in the event of anomalies of speed indication. As a result it leads to a fatal accident, disappeared from the sky claiming 228 lives.

1.4 SUPERVISORY CONTROL AND DATA ACQUISITION (SCADA)

Supervisory Control and Data Acquisition (SCADA) is a system that operates with coded signals over communication channels so as to provide control of remote equipment (using typically at least one communication channel per remote station). In SCADA, the control system is combined with the data acquisition system by fetching the coded signals over communication channels to acquire information about the status of the remote equipment's for displaying or for recording or controlling equipment functions. It is a typical type of industrial control system (ICS) which uses computer-based systems to monitor and control industrial processes which exist in the physical world. SCADA systems historically distinguish themselves from other ICS systems by being large-scale processes that can include multiple sites, spread out over large areas. These processes can be industrial, infrastructure, and facility-based processes, as described below.

Industrial processes include those of manufacturing, production, power generation, fabrication, and refining. The process can be run in continuous, batch, repetitive, or discrete modes. Infrastructure processes may be public or private, and include water treatment and distribution, wastewater collection and treatment, oil and gas pipelines, electrical power transmission and distribution, wind farms, civil defense siren systems, and large communication systems.

Facility processes occur both in public facilities and private ones, including buildings, airports, ships, and space stations. They monitor and control heating, ventilation, and air conditioning systems (HVAC), access, and energy consumption.

1.4.1 Evolution of SCADA

SCADA systems have been evolved through the following four generations.
- *Monolithic SCADA:* These SCADA systems were in use before the revolution of computer networking. They were stand-alone systems having no connectivity to other systems and developed as local SCADA. These mostly used guided communication with proprietary protocols. Further these systems did not envisage the fail safe and fault tolerant design aspects seriously. Hence these SCADA systems faced reliability issues considerably. The hardware used was mainly minicomputers capable of large computing capabilities.
- *Distributed SCADA:* This generation came into existence when the computer networking technology has been developed and incorporated

into the SCADA systems. Here the data gathered and the data processing across the multiple stations are connected through computer networking. The latency has been reduced considerably to an extent that system as a whole functioned in near real time. These SCADA system were economical when compared with first generation, as each station has been assigned with particular tasks. However the protocols used was still proprietary and was not interoperable. But the proprietary protocols had an advantage that beyond the developers, the extent of security or the security flaws are unknown. In other way these SCADA systems were secured through *security through obscurity.*

- *SCADA with standard protocols:* The advancement of communication technology and with the introduction of interoperable SCADA protocols, SCADA broke the geographical barriers and spread across more than one LAN network called Process Control Network. The master station may have several servers running parallel to handle various tasks, such as historian, SCADA, NMS, Development, etc. This makes the system very economical and real-time. However the physical-security is a major concern and must be addressed while designing and implementing.
- *Internet of Things:* By appropriately integrating the advancement of cloud computing, SCADA has taken a new shape and adopted the name Internet of Things.

The advantages of amalgamating the various technologies are:
1. Capable of being flexible and affordable with the ability to go private.
2. Capable of ingest massive amount of machine data
3. Capable of connecting different machines and systems such as SCADA, DCS, and historians.
4. Capable of connecting various machines having net connectivity and process data across all these sources together.
5. Capable of real time complex and real time processing of data from multiple sources.
6. Capable of Big data processing and apply supervised and unsupervised machine learning algorithms to predict outcomes.

Another advantage of incorporating the cloud computing technology is that it significantly reduces the infrastructure costs and increases the ease of maintenance. Further the SCADA operations become near real time and the use of open protocols with TLS security improves the security boundary considerably.

Certain ICT professionals envisage IoT as an appropriate amalgamation of Machine to Machine (M2M) communication, Wireless Sensor Networks (WSN), Radio Frequency Identification (RFID), and SCADA as shown in Figure 1.1.

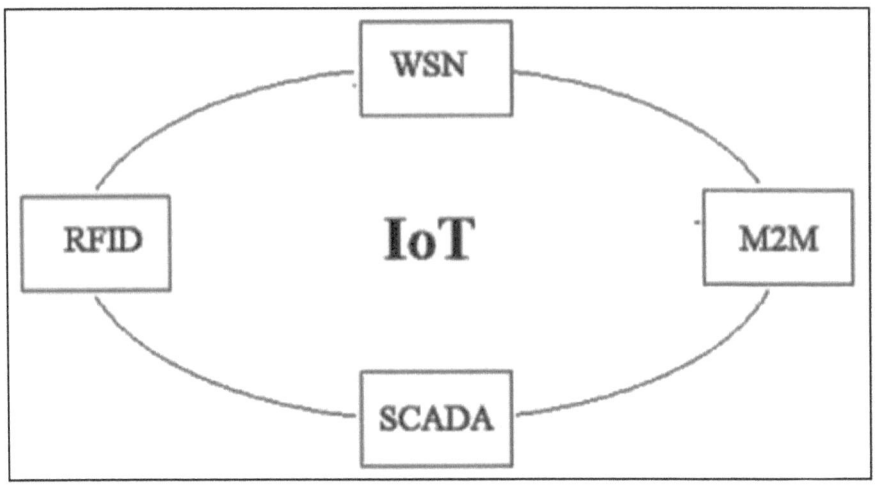

▲ **Figure 1.1:** Fourth generation SCADA or IoT.

In fact, the fourth generation SCADA or IoT transforms the human centric internet into objects or things centric. It is expected that about 70 billion things may be hooked into internet in the near future while people hooked into the internet may be only 7 billion. However at present, IoT is not a tangible reality but it is a prospective vision of a number of technologies. But once materialized, it can drastically change the way of functioning of our society. Realizing the potential and the opportunity of business, IoT has become a buzzword in many countries. Further it is anticipated that the whole world will be soon under the influence of IoT wave.

1.4.2 Components of SCADA Systems

A SCADA system mainly consists of the following subsystems which are briefly explained below.

Sensors
Sensors are sophisticated devices that are frequently used to detect and respond to electrical or optical signals. It converts the physical parameter into a signal which can be measured electrically. Transducer characteristics

define many of the signal conditioning requirements of the measurement system.

Signal Conditioners
The main function of a signal conditioner is to pick up the signal and convert it into a higher level of electrical signal. Signal conversion is common industrial applications that use a wide range of sensors to perform measurements. They protect personnel and equipment from dangerous voltages. In case of failure, high AC voltages or voltage pulses can enter the measuring circuit. Signal conditioners prevent the passage of high voltages into the control-side loop by galvanic isolation. Signal conditioning is more elaborated in succeeding chapter.

A to D Converters (ADC)
An analog-to-digital converter, or ADC, is a device or peripheral that converts analog signals into digital signals. In the real world, signals mostly exist in analog form. An ADC with a S/H circuit can be used to sample such signals and the signals can be converted to the digital values. There are four types of A/D converters generally used, and they are,
1. Integrating or dual slope ADC,
2. Successive approximation ADC,
3. Parallel Comparator (Flash) type ADC, and
4. Counting type ADC.

Remote terminal units (RTUs)
Remote terminal units connect to sensors in the process and convert sensor signals to digital data. They have communication hardware module capable of sending digital data to the supervisory system, as well as receiving digital commands from the supervisory system. RTUs often have embedded control capabilities such as ladder logic in order to accomplish Boolean logic operations.

Programmable Logic Controllers (PLCs)
Programmable logic controller connects to sensors in the process and converts sensor signals to digital data. PLCs have more sophisticated embedded control capabilities (typically one or more IEC 61131-3 programming languages) than RTUs. PLCs do not have communication hardware module, although this functionality is typically installed alongside them. PLCs are sometimes used in place of RTUs as field devices because they are more economical, versatile, flexible, and configurable.

Communication Hardware Module
Communication hardware module is typically used to connect PLCs and RTUs with control centers, data warehouses, and the enterprise. Examples of wired telemetry media used in SCADA systems include leased telephone lines and WAN circuits. Examples of wireless telemetry media used in SCADA systems include satellite (VSAT), licensed and unlicensed radio, cellular and microwave.

Data Acquisition Server
Data Acquisition Server is a software service which uses industrial protocols to connect software services, via communication channels, with field devices such as RTUs and PLCs. It allows clients to access data from these field devices using standard protocols.

Human Machine Interface (HMI)
Human Machine Interface or HMI is the apparatus or device which presents the processed data to a human operator, and through this, the human operator monitors and interacts with the process. The HMI is a client that requests data from a data acquisition server and through which the human operator controls the process.

HMI is usually linked to the SCADA system's databases and software programs, to provide trending, diagnostic data, and management information such as scheduled maintenance procedures, logistic information, detailed schematics for a particular sensor or machine, and expert-system troubleshooting guides.

The HMI system usually presents the information to the operating personnel graphically, in the form of a mimic diagram. This means that the operator can see a schematic representation of the plant being controlled. For example, a picture of a pump connected to a pipe can show the operator that the pump is running and how much fluid it is pumping through the pipe at the moment. The operator can then switch the pump off. The HMI software will show the flow rate of the fluid in the pipe decrease in real time. Mimic diagrams may consist of line graphics and schematic symbols to represent process elements, or may consist of digital photographs of the process equipment overlain with animated symbols.

The HMI package for the SCADA system typically includes a drawing program that the operators or system maintenance personnel use to change the way these points are represented in the interface. These representations can be as simple as an on-screen traffic light, which represents the state of an actual traffic light in the field, or as complex as a multi-projector display representing the position of all of the elevators in a skyscraper or all of the trains on a railway.

Historian Server
Historian is a Software Service which accumulates time-stamped data, Boolean events, and Boolean alarms in a database which can be queried or used to populate graphic trends in the HMI. The historian is a client that requests data from a data acquisition server.

Supervisory Station
Supervisory Station refers to the servers and software responsible for communicating with the field equipment (RTUs, PLCs, SENSORS etc.), and then to the HMI software running on workstations in the control room, or elsewhere. In the simplest sense it gathers (acquires) data on the process and sends commands (control) to the various SCADA system components. In smaller SCADA systems, the master station may be composed of a single PC. In larger SCADA systems, the master station may include multiple servers, distributed software applications, and disaster recovery sites. To increase the integrity of the system the multiple servers will often be configured in a dual-redundant or hot-standby formation providing continuous control and monitoring in the event of a server malfunction or breakdown.

1.4.3 Communication Infrastructure And Methods

Communication infrastructure connects the supervisory system to the remote terminal units or Data Concentrators. SCADA systems have traditionally used combinations of radio and direct wired connections. SONET/SDH is also frequently used for large systems such as railways and power stations. The remote management or monitoring function of a SCADA system are achieved through wired connectivity often referred to as telemetry. Hence RTU sometimes referred as Remote Telemetry Unit. In certain SCADA systems data needs to flow to the corporate networks which are already established with the legacy of the low-bandwidth protocols.

SCADA protocols are designed to be very compact. Many are designed to send information only when the master station polls the RTU. Typical legacy SCADA protocols include Modbus RTU, RP-570, Profibus and Conitel. These communication protocols are all SCADA-vendor specific but are widely adopted and used. Standard protocols are IEC 60870-5-101 or 104, IEC 61850 and DNP3. Today these communication protocols are standardized and recognized by all major SCADA vendors. Many of these protocols now contain extensions to operate over TCP/IP. Although the use of conventional networking specifications, such as TCP/IP, blurs the line between traditional and industrial networking, they fulfill fundamentally differing requirements.

As the Power system SCADA becoming exposed to cyber space, the security standards such as North American Electric Reliability Corporation (NERC) and Critical Infrastructure Protection (CIP), there is increased use of satellite-based communication. This has the key advantages that the infrastructure can be self-contained by not using circuits from the public telephone system, can have built-in encryption, and can be engineered to the availability and reliability required by the SCADA system operator. Earlier experiences using consumer-grade VSAT were poor. Modern carrier-class systems provide the quality of service required for SCADA.

RTUs and other automatic controller devices were developed before the advent of industry wide standards for interoperability. The result is that developers and their management created a multitude of control protocols. Among the larger vendors, there was also the incentive to create their own protocol to *lock in* their customer base. A list of automation protocols is compiled here.

Recently, OLE for Process Control (OPC) has become a widely accepted solution for intercommunicating different hardware and software, allowing communication even between devices originally not intended to be part of an industrial network.

OPC is a series of standards specifications. The current OPC specifications form a set of standard OLE/COM interface protocols based upon the functional requirements of Microsoft's OLE/COM technology. Such technology defines standard objects, methods and properties for servers of real-time information like distributed process systems, programmable logic controllers, smart field devices and analyzers in order to communicate the information that such servers contain to standard OLE/COM compliant technologies enabled devices. The OPC Foundation is the body that maintains the standard.

1.4.4 Security Concerns of Power System Automation

With the advancement of power system automation new intelligent technologies utilizing two-way communications and other digital advantages are being optimized by Internet connectivity. Modernization of Industrial

Control Systems (ICS) which employs, Presently SCADA systems also has dependency with internet many ways including data transfer, IDS updating with patch up release, etc. These developments enhanced the efficiency and performance of the power grid, but it also increased the vulnerability of the grid to latent cyber-attacks. Stuxnet, Black Energy, Havex, and Sandworm are all recent examples of lethal malware targeting SCADA systems and nightmares of SCADA engineers. The IEDs like smart meters, network devices with improper end node security and increasing External Access Points (EAP) like integration of renewable energy sources, introduce new additional areas through which a potential cyber-attack may be launched at the grid.

SCADA systems that tie together decentralized facilities such as power, oil, and gas pipelines and water distribution and wastewater collection systems were designed to be open, robust, and easily operated and repaired, but not necessarily secure. The move from proprietary technologies to more standardized and open solutions together with the increased number of connections between SCADA systems, office networks, and the Internet has made them more vulnerable to types of network attacks that are relatively common in computer security. It is not a secret, and confirmed by the United States Computer Emergency Readiness Team (US-CERT) that the cyber jihad and many unauthenticated users are downloading sensitive configuration information including password hashes of many SCADA systems which control the critical infrastructure.

1.5 SCADA APPLICATIONS IN POWER SYSTEMS

Power networks are complex systems, and today the grid has grown to an extent that it cannot be efficiently and securely operated without Energy Management Systems (EMS). SCADA/EMS/GMS (supervisory control and data acquisition/ Energy Management System/Generation Management System) supervises, controls, optimizes and manages transmission and generation systems. SCADA/ DMS (Distribution Management System) performs the same functions for power distribution networks.

Both systems enable utilities to collect, store and analyze data from hundreds of thousands of data points in national or regional networks, perform network modeling, simulate power operation, pinpoint faults, anticipate outages, and participate in energy trading markets. The systems are a vital part of modern power networks and are enabling the development of smart grids, the highly automated energy systems of the future. Smart grids will need to handle large quantities of renewable power from both large and small-scale generators. To maintain grid stability in spite of this potentially disruptive source of power and the two-way flow of power in what is currently a one-way system, advanced monitoring systems will be very important.

1.5.1 Computerized Control

Although the roots of power control go back to the 1920s when power equipment manufacturers supplied its first remote control system for a power plant, it was not until the 1960s and the advent of computerized process control that modern power network control systems as we know them today became possible.

Most SCADA/EMS/GMS systems at that time were designed exclusively for a single customer. Power systems were vulnerable, and there was a need to develop applications and tools for preventing faults from developing into large-scale outages like the New York blackout of 1977. In the 1980s it became possible to model large-scale distribution networks in a standardized way. A key project that mirrored this achievement was the integration of the generation, transmission and distribution networks.

As the deregulation and privatization of the power industry began in the 1990s, the biggest structural change has been occurred in the power industry. Specialization became increasingly common, with many utilities focusing on generation, transmission or distribution. At the same time the need to interconnect national or regional power systems brought new requirements for cross-border control systems. Energy trading systems were required to enable independent system operators (ISO), such as those for California and New York, to operate real-time markets for energy trading.

With specialization, today the norms and the needs of power operators were beginning to differ. Generation companies needed an interface with the energy trading markets, and the capability to plan and optimize supply to meet spot market demand. Transmission companies required advanced systems to manage their high voltage networks and prevent a fault in one part of the system from cascading across the entire network. Distribution operators needed to take network management down to the level of individual customer connection points (often numbering several million) to minimize customer outage times. All these real time operations are possible only with fail-safe and fault tolerant SCADA systems which can monitor and control power system operations within the accepted latency.

1.5.2 SCADA/EMS/GMS AND SCADA/DMS

Supervisory Control And Data Acquisition/Energy Management Systems/ Generation Management System (SCADA/EMS/GMS) supervises, controls, optimizes and manages generation and transmission systems. A SCADA/DMS distribution management system performs the functions which are required for the power distribution networks. Both systems enable utilities to collect, store and analyze data from thousands of data points of national or regional networks, perform network modeling, simulate power operation, identify the exact fault locations, prevent outages, and participate in energy trading markets.1960's SCADA/EMS/GMS systems at that time were designed exclusively for a single customer. Power systems were vulnerable, and there was a need to develop applications and tools for preventing faults from developing into large-scale outages. In the 1980s it became possible to model large-scale distribution networks in a standardized way.

The deregulation and privatization of the power industry that began in the 1990s was the biggest structural change in the industry's history. Specialization became increasingly common, with many utilities focusing on generation, transmission or distribution. Smart Metering and Smart Grid technologies impacts, Network management is a prerequisite and vital for any smart grid of the future. These grids will have to incorporate and manage centralized and distributed power generation, intermittent sources of renewable energy like wind and solar power, allow consumers to become producers and export their excess power, enable multi-directional power flow from many different sources, and integrate real-time pricing and load management data.

1.5.3 Benefits of SCADA in Power Systems

The main benefits of the SCADA in power system are Deferred Capital Expenditure, Optimized Operation and Maintenance Costs, Equipment Condition Monitoring (ECM), Sequence of Events (SOE) Recording, Power Quality Improvement, and Data Warehousing for Power Utilities. A brief description of these benefits is described below.

- A SCADA system gives a real-time picture of loading on various transmission lines, feeders, transformers, circuit breakers, and other equipment. Further supervisory control has been done at the Master Control Center. A proper load distribution and load management can be performed with a SCADA system. A proper load distribution prevents unwanted overloading of equipment. This guarantees a longer service life for the power system components. Hence an appropriate SCADA system with better equipment monitoring and load management extends the economic life of the very expensive power equipment. This defers capital expenditures on assets to a considerable extent. Further it helps to defer the additional expansion and investment for a while as the increase in load can be properly distributed and balanced.

- Power utilities can save considerably in Operation and Maintenance (O&M) expenditure through automation with SCADA. Functions such as predictive and preventive maintenance, Volt-Var Control, self-diagnostic programs, and access to the automation data help the utility to optimize their costs by allowing it to make better and optimum decisions on O&M strategies that are based on comprehensive and accurate operational data rather based on assumptions and thumb rules.
- Equipment Condition Unit (ECU) is an IED which automatically keep a track of the functionalities, abnormalities, malfunctions, healthiness, and other important parameters. These parameters can be used for proper maintenance and care so that the life can be extended. When ECUs are properly linked to the SCADA system and operated, it can prevent many hence major equipment failure and service disruption. Power transformers, bushings, tap changers, and substation batteries are some of the power system equipment monitored by ECUIEDs.
- In power system many functions are critical and need to be recorded as a time stamped event. This will be very much beneficial for post-event analysis as it provides relevant information regarding the system loading patterns, malfunctioning of any specific devices, etc. Further, these events can be recreated in the same sequence as they occurred in the system which benefits the engineers to design the new transmission lines, feeders, and networks for the future.
- Power Quality (PQ) monitoring devices can be connected to the network and monitored centrally to achieve PQ by reducing voltage fluctuation such as sag, swell, spikes, unbalances, etc. and minimize the harmonic effects in the network, automatically. Remedial actions such as switching of capacitor banks and voltage regulators can be implemented to improve the power quality, so that the customer is benefitted with 24x7 quality power supply.

Thus, automation brings in a new set of solutions for managing the assets better for the benefit of customer satisfaction and reliable operation of the system. Hence, utilities across the world prefer automation with SCADA systems and are acquiring its advantages.

1.6 CHANGING SCENARIO OF POWER SYSTEM FIELD

Decade back electricity was generated at the generating stations and transmitted over the transmission system to the distribution substation, from where electric power was distributed to the consumers. But presently, to this traditional system, renewable generation like Wind, Solar, Biomass, etc. are added to the grid of transmission and distribution, including from the customer premises. The SCADA systems acquire data from all these components, monitor, and issues necessary control commands. A brief discussion of these components follows.

1.6.1 Transmission and Distribution Systems

The centrally generated electricity reaches the customer premises passing through a variety of substations which are classified as follows.
1. Switchyard or generating substations,
2. Bulk power substations or grid substations,
3. Distribution substations, and
4. Special-purpose substations such as traction substation, mining substation, mobile substations, etc.

A transmission substation (generating or grid substation) usually has the following components.
1. Transformers (with or without tap changers)
2. Station buses and insulators
3. Current transformers(CT)
4. Potential transformers(PT)
5. Circuit Breakers(CB)
6. Disconnecting switches or isolators
7. Reactors, series or shunt
8. Capacitors, series or shunt
9. Relays/IEDs
10. Storage batteries and battery chargers
11. Wave trap and coupling capacitors for Power Line Carrier Communication.

The present day distribution substations have similar equipment with reactive and capacitive compensation equipment in place, however with all the equipment at a lower rating. Figure 1.2 shows Single Line Diagram (SLD) of a typical substation.

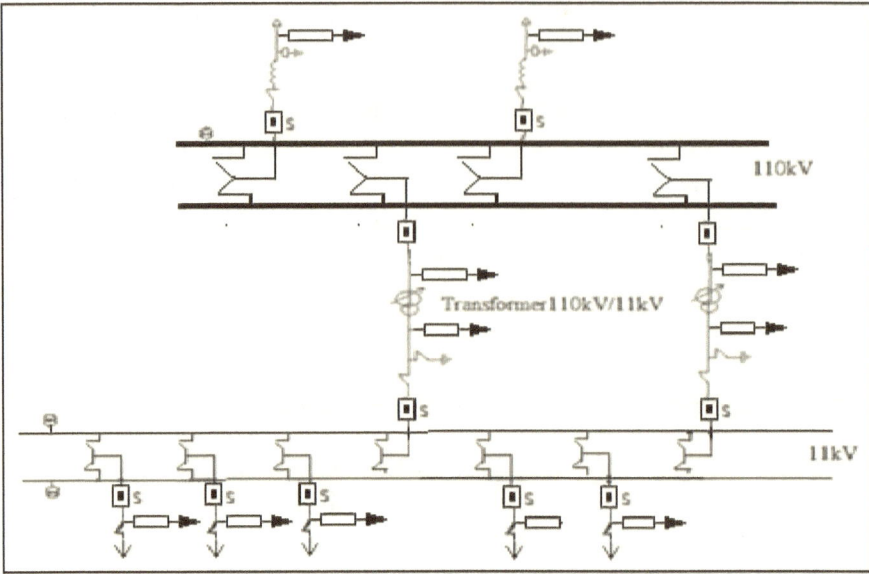

▲ **Figure 1.2:** Single Line Diagram of a typical substation

As far as the SCADA systems are concerned, analog data are acquired from the transformers and station buses via the current and voltage transducers and are further processed for transmission to the control room. Status data or digital data are acquired from the circuit breakers, isolators, and on/off positions of the shunt and series compensation devices and are conveyed to the master station. Environmental data such as temperature, pressure, humidity, and weather conditions are also collected by the appropriate sensors and are processed for onward transmission to the control center.

1.6.2 Customer Premises

With the customer taking center stage in an automated distribution system, the devices in the customer premises hold the key to successful implementation of the future smart grid. The smart energy meter capable of two-way communication, the smart appliances in the house, and also the smart plugs with communication facility are inevitable for customer automation. The main challenge here will be the integration of existing plugs and devices with the new smart meters at the customer premises. The integration of data from a variety of customer meters communicating in different protocols to the collecting hubs and further communication and processing of the data at the substation are challenges to be addressed.

1.7 INDUSTRIAL SECTORS AND THEIR INTERDEPENDENCIES

Today in the modern power grid, both the electrical power transmission and distribution industries use geographically distributed SCADA control technology to operate highly interconnected and dynamic systems consisting of many different public and private utilities and rural initiatives for supplying electricity. SCADA systems monitor and control electricity distribution by collecting data from and issuing commands to geographically remote field control stations from a centralized location. SCADA systems. These central control stations are also used to monitor and control water, oil and gas distribution including pipelines, movement of ships, trucks, and rail systems, as well as wastewater collection systems. In fact SCADA systems and DCSs are often tied together. This is the case for electric power control centers and electric power generation facilities. Although the electric power generation facility operation is controlled by a DCS, the DCS must communicate with the SCADA system to coordinate production output with transmission and distribution demands. Critical infrastructure is often referred to as a *system of systems* because of the interdependencies that exist between its various industrial sectors as well as interconnections between business partners.

Critical infrastructures are highly interconnected and mutually dependent in complex ways, both physically and through a host of information and communications technologies. An incident in one infrastructure can directly and indirectly affect other infrastructures through cascading and escalating failures. Electric power is often thought to be one of the most prevalent sources of disruptions of interdependent critical infrastructures. As an example, a cascading failure can be initiated by a disruption of the wireless communications network used for an electric power transmission SCADA system. The lack of monitoring and control capabilities could cause a large generating unit to be taken offline, an event that would lead to loss of power at a transmission substation. This loss could cause a major imbalance, triggering a cascading failure across the power grid. This could result in large area blackouts that affect oil and natural gas production, refinery operations, water treatment systems, wastewater collection systems, and transport systems that rely on the electric grid for power. So utmost importance must be given to secure the power system SCADA, else the impact will be overwhelming due to the interdependencies of the various utilities.

1.8 DATA FLOW IN POWER SYSTEM SCADA

The data flow in SCADA system and various technologies used are very important from the secured communication point of view. Today, as the industrial networking highly dependent on computer and communication technology, a fair understanding of various communication protocols are most essential as far as a SCADA system expert is concerned especially during the design session. The data flow in a SCADA system can be traced by analyzing the flow of an analog signal from the field to the display screen of a dispatcher and from the flow of control command from control station to the actuators.

In power system SCADA normally monitoring of variables starts from the substation bus bar, the potential transformer connected to the bus converts the 220 kV into 110 V. An appropriate transducer converts this 110V into a 4 to 20 mA analog signal. An analog to digital converter convert this analog signal to digital signal. This is achieved through the AI module of the RTU. The RTU further converts the digital data into another format as required by the communication protocol existing between the RTU and the master station. The packets are then transmitted to the master station through a secured communication media and received by the front-end processor/ communication front end (FEP/CFE) at the control center. At CFE the payload (data) is decrypted, decoded and retrieved. This data is then scaled (normalized) up to the 220 kV range and displayed at the appropriate bus bar in the mimic diagram of the operator console. This completes the monitoring *cycle* of the bus bar voltage. In a similar way, a control command can be sent to a field device from the control station and the control operation can be executed through the actuator.

The revolutionary advancement of computer and communication technology, an RTU with hardwired I/O modules, integrated with IEDs with data pass through capability and decision making, and capability of advanced communications, transforms a substation to a digital substation. Today this is a technical passion of power engineers which is also expedient and beneficial in many ways.

In a typical power system SCADA, bidirectional data flow starts between the IEDs and RTU. The RTU, which also acts as a Data Concentrator Units (DCU) normally have IED pass through capabilities. RTUs which are installed in substation environment also have the capability of communicating with the field devices directly. The RTU communication with the SCADA master control center either through utilities dedicated communication link or through a third party communication link which must be secured. Preferably VPNs are preferred for this purpose with proper VPN termination.

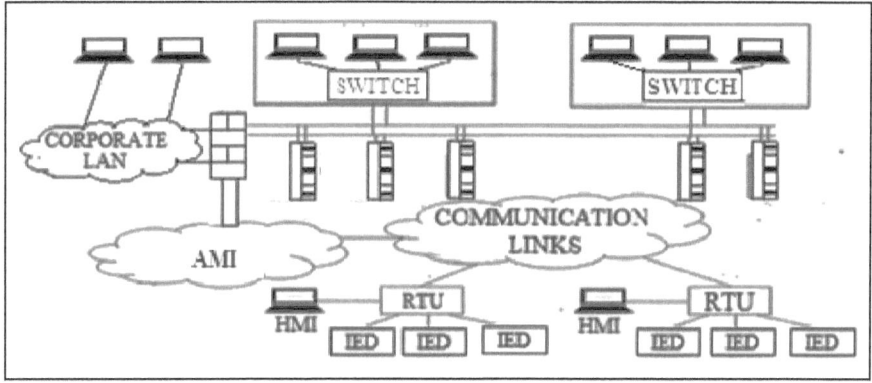

▲ **Figure 1.3:** Data flow in a typical power system SCADA environment.

In the power distribution sector, an efficient and authentic power supply to the consumer is the primary function of any distribution system. In distribution systems certain measures are taken for supervision, control, operation, measurement and protection. These are highly burdensome works that take lot of manpower. So, the need of advanced automatic control systems to reach the required destination is becoming mandatory, to supersede antiquated ways that are persisting in the present distribution system. EMS (Energy Management System) of power system SCADA monitors the Energy consumption of receiving stations and also energy delivered to interface point at Distribution Boundary meters. Hence the data flow between the border meters and the control center also takes place in a power system SCADA. Another area where the data flow required is from the smart meters installed as a part of AMI. A bidirectional data flow is inevitable here also. In fact a secured data flow is the benchmark which maintains a successful load control process in a power system SCADA. Figure 1.3 describes the data flow in a typical power system SCADA environment.

1.9 THE NEED FOR SECURE DATA TRANSFER

The world's manufacturing, energy and transportation infrastructures are currently facing a serious security crisis. These critical systems are largely based on legacy SCADA and Industrial Control System (ICS) products and protocols. Many of these products are decades old and were never designed with security in mind. Yet industry has also embraced new network technologies like Ethernet and TCP/IP, which have enabled instant access to data throughout the organization, including the plant floor. While this interlinking improves efficiency, it also significantly increases the exposure of these control systems to external forces such as worms, viruses and hackers. Given the 20 year life cycle common for industrial systems, it will be many years before more secure ICS and SCADA devices and protocols are in widespread use. This leaves millions of legacy control systems open to attack from even the most inexperienced hacker. If a hacker or worm can get any control system access, it can exploit the protocol to disable or destroy most industrial controllers. The good news is that there is an effective and easy-to-deploy solution to this security crisis. Using an advanced technology called *Deep Packet Inspection* (DPI), SCADA-aware firewalls can offer fine-grained control of control system traffic. This white paper explains what DPI is and how it compares to traditional IT firewalls. It then outlines how engineers can use DPI to block the malicious or inappropriate traffic, while avoiding needless impact on the control system. A case history illustrates how a seaway management company used Modbus DPI firewalls to secure a mission critical canal system.

1.9.1 Need For Better Security Technology

Over the past decade, industry has embraced network technologies like Ethernet and TCP/IP for SCADA and process control systems. This has enabled companies to operate cost effectively and implement more agile business practices through instant access to data throughout the organization, including the plant floor. While companies reap the benefits of these new technologies, many are also discovering the inherent dangers that result from making control systems more accessible to a wider range of users. Linking corporate systems together to provide access to managers, customers and suppliers significantly increases the exposure of these systems to external forces such as worms, viruses and hackers. To make matters worse, network protocols used by SCADA and Industrial Control Systems (ICS) were never designed with security in mind. If they offer any capability to restrict what users can do over the network, it is primitive and easy to subvert. If an individual is allowed to read data from a controller, then they can also shut down or reprogram the controller. These issues are likely to remain for at least the next decade. Industrial control systems are rarely replaced as their expected lives may be 10, 20 or more years. Similarly, the security limitations of the SCADA and ICS protocols cannot be addressed through patches, as their functionality is defined in established standards that take years to change. It will be years before newer, more secure ICS and SCADA devices are in widespread use. This leaves millions of legacy control systems open to attack from even the most inexperienced hacker. If a hacker or worm can get any control system access, it can exploit the protocol to disable or destroy most industrial controllers. The solution is a technology called *Deep Packet Inspection* (DPI) and it offers fine-grained control of SCADA network traffic.

1.10 CHANGING SCADA CULTURE

SCADA and distributed control systems are traditionally the domain of Electrical Engineers. But with the integration of Information Communication Technology (ICT) to SCADA systems broke the geographical barriers which requires expertise in advanced hardware and software platforms, internet communication and security protocols, flawless connectivity to corporate networks, physical cyber security, etc. Thus involvement of ICT and cyber security experts becomes unavoidable with modern SCADA systems. This introduced conflicting cultures and priorities and differing stances especially on implementing intrusion detection systems, firewalls, authentication, and encryption.

Until recently, there was a view that the very nature of most SCADA systems, they are less vulnerable than IT systems. It is true that control systems are less visible than IT systems and many were not connected to external networks or cyber space. Further their components required detailed technological knowledge to implement and operate. Hence the myth of security through obscurity was a fact until the end of 1999. But the Y2K issue exposed the information of SCADA systems and potential security problems especially to the critical infrastructure. Today manufactures of SCADA components provide detailed product information in brochures and Web sites. Further media and Web articles continuously report attacks, threats and vulnerabilities, which make public as well as the attackers aware of SCADA technology.

Open communication systems become a necessity as they bring costs down, but as the name implies these systems are more open to cyber-attack than their proprietary and more closed alternatives. Proprietary systems not only have fewer connections to other systems, they are also less familiar to professional hackers, creating a possible *security through obscurity* defense. On the other hand, communication systems based on Ethernet, TCP/IP protocols, the Internet and widely used operating systems such as Windows invite attack from literally millions of hackers worldwide.

In fact, presently the power system automation is a combined and cooperative effort of power engineers, ICT professionals, and cyber security experts. With the introduction of more and more IEDs with advanced resilience and communication technology, the role of ICT professionals become more predominant in SCADA systems.

SUMMARY

In this chapter SCADA systems are briefly introduced with the history of power system automation, and the increasing requirement of SCADA system in the power sector. The advantages and uses of SCADA system in the power sector are also discussed. Further the present day security concerns especially the increasing cyber incidents to the SCADA systems deployed on critical infrastructure like power sector are briefly described for reader's awareness along with a fleeting narration of the changing SCADA culture.

CHAPTER 02

SCADA BASICS AND COMPONENTS

2.1 INTRODUCTION

SCADA (Supervisory Control and Data Acquisition) has been around us since there have been control systems and extensively used for monitoring and controlling geographically distributed processes in a variety of industrial processes. The first SCADA systems have been employed only for data acquisition by means of panels of meters, lights and strip chart recorders. Until recently many of the SCADA related products are proprietary, and the knowledge of the components has been acquired by the personnel while operating the system. Now the SCADA component manufactures started to follow standards which support interoperability. These help SCADA professionals to understand and design the SCADA systems in a systematic and structured manner. All SCADA or Distributed Control System (DCS) starts with the field equipment depending on the process/plant to be monitored and controlled. The appropriate sensors pick up the process parameters and convert into proportional electrical voltage or current signal. These electrical signals are then conditioned and converted as per the requirement into digital form by means of an ADC. This electrical signal in digital format is then communicated to the SCADA server through different devices and protocols depending upon the SCADA architecture and the communication technologies employed. In the succeeding sessions an attempt has been made to elaborate on the essential components of the SCADA starting with Data Acquisition Systems (DAS) with an emphasis on Power System SCADA (PSS). Then moves on to elaborate various components such as Remote Terminal Unit (RTU), Programmable Logic Controller (PLC), Bay Control Unit (BCU), Merging Units (MU), Data Concentrators (DC), Master Control Station and SCADA Servers, HMI, etc. which are employed in the modern Power System SCADA environment.

2.2 DATA ACQUISITION SYSTEMS (DAS)

A DAS is a system which gathers the input data in digital form accurately, quickly and economically as required. This consists of sensors with suitable signal conditioning, data conversion, data processing, multiplexing, data handling, associated transmission, and storage and display systems.

2.2.1 Objectives and Advantages

The basic objectives of the Data Acquisition System are briefly described below.
1. Acquiring the required data reliably at required intervals.
2. The acquired data has to be appropriately processed and inform the status of the process/plant to the operator from time to time for controlling and decision making.
3. In general, DAS are designed to conceive a complete picture of the process/plant under monitoring intend to keep the plant/process operation safe and optimal.
4. With an effective HMI system or SCADA software, identify problem areas and minimize unit unavailability and maximize productivity at minimum cost.
5. DAS must be able to prepare sequence of events (SOE), summaries and store data for diagnosis, forecasting, etc.
6. Generally DAS are designed to compute unit performance indices using real-time data.
7. DAS must be flexible and capable of being expanded to accommodate future requirements.
8. DAS architecture design must be fail-safe, fault tolerant and reliable and down time must be less than 0.1%.

The major advantages of using DAS are increased reliability, lower operation and maintenance cost, faster restoration of the industrial process, reduction in human intervention and errors, accelerated decision making, with better accuracy, etc.

2.2.2 Single Channel Data Acquisition System

A Block Diagram of Single Channel Data Acquisition System is shown in Figure 2.1. The various components of the system are briefly explained below.

It consists of Sensors, Signal Conditioner, Sample and Hold circuit, Analog Digital Converter and a storage device/printer/PC. Followed by an Analog to Digital Converter (ADC), performing repetitive conversions at a free running, internally determined rate. The outputs are in digital code words including over range indication, polarity information and a status output to indicate when the output digits are valid.

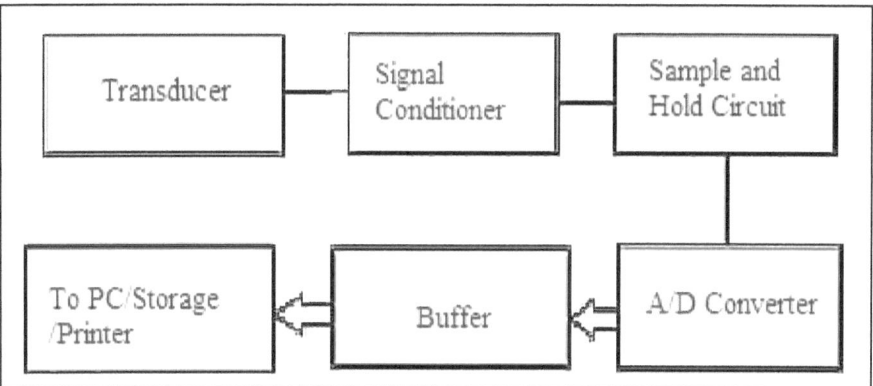

▲ **Figure 2.1:** Single Channel Data Acquisition System

The digital outputs are further fed to a storage or printout device, or to a digital computer device for analysis. The popular Digital Panel Meter (DPM) is a well-known example of this. However, there are two major drawbacks in using it as a DAS. It is slow and the BCD has to be changed into binary coding, if the output is to be processed by digital equipment. While it is free running, the data from the A/D converter is transferred to the interface register at a rate determined by the DPM itself, rather than commands beginning from the external interface.

2.2.3 Multi-Channel Data Acquisition System

By suitably incorporating a Multiplexer, a multi-channel DAS can be designed. By suitably selecting the address of the MUX, the input channels can be polled depending upon the priority. A block diagram of a multi-channel DAS is shown in Figure 2.2. Multi-channel DAS is elaborated in subsequent chapters.

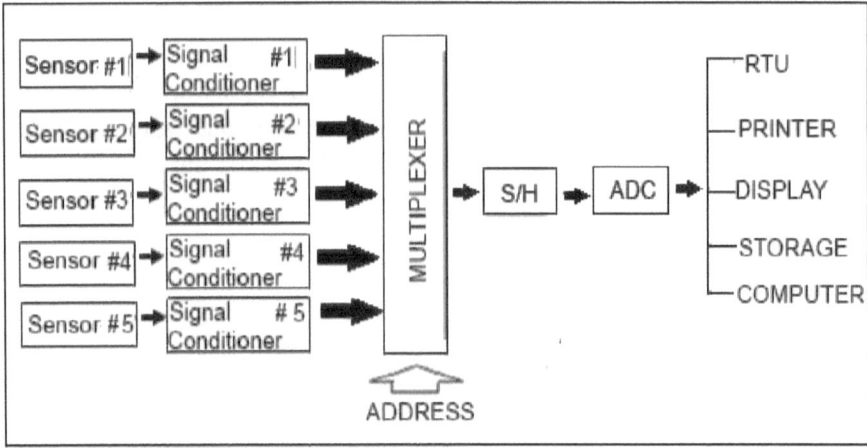

▲ **Figure 2.2:** Multi-Channel DAS

When simultaneous measurements of the physical quantities are to be taken, the S/H circuit generally placed before MUX as shown in Figure 2.3. All the required parameters which are to be captured simultaneously are sampled and hold at the same instance and then digitized one after the other with software polling.

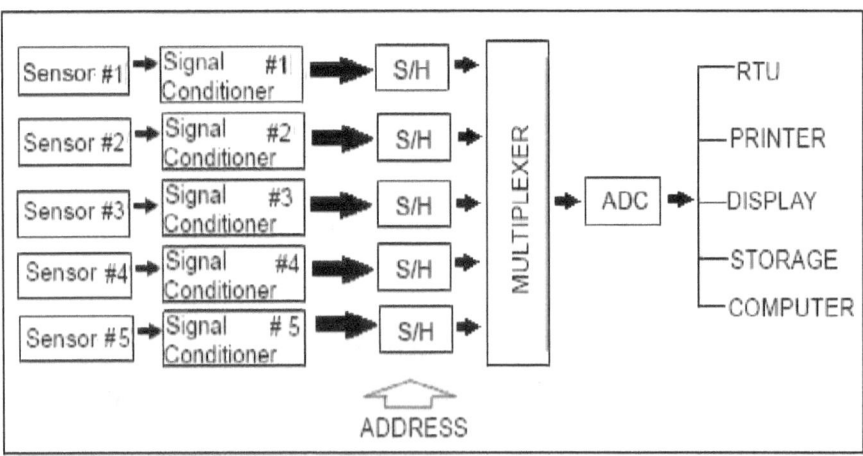

▲ **Figure 2.3:** Multi-Channel DAS for simultaneous measurement

2.2.4 Sensors

Sensors or Transducers are devices that detects and measure a physical quantity such as pressure, force, temperature, acceleration, etc. and provide a corresponding output in the form of an electrical signal like voltage, current, resistance or frequency. Transducer characteristics define many of the signal conditioning requirements of the measurement system.

2.2.5 Signal Conditioning

In general the quality of the signals obtained from the sensors has to be enhanced appropriately to bring it to an acceptable level to the Analog to Digital Converters (ADC). This includes signal scaling, amplification, attenuation, linearization, filtering, anti-aliasing, excitation etc. Further signal conditioners have an additional responsibility to protect from unintentional or accidental high voltage inputs or surges, etc. Direct digital conversion carried out near the signal source is very advantageous in cases where data needs to be transmitted through a noisy environment. Even with a high level signal of 10 V, an 8 bit converter having a 1/256 resolution can produce 1 bit ambiguity when affected by noise of the order of 40 mV. Presently transducers being developed which is combined with ADC capable of converting to encrypt digital data.

Excitation: Not all sensors are active. In such cases, where the sensing devices are passive, it requires a voltage or current excitation and is supplied by the signal conditioners.

Amplification and Attenuation: If the signal acquired is large, then a simple attenuator, is used to scale down the input gains, in order to make it acceptable to the input signal range of the ADC. However most of the sensors generate the signals of low amplitude of voltage, current, or resistance, etc. In this case an amplifier circuit of suitable gain is employed to bring them to the acceptable level. If the sensor output is in the form of change in resistance, then a bridge circuit is most ideal to detect the change in resistance. A bridge amplifier is most suitable for amplifying the bridge outputs and improving the sensitivity of detection.

Isolation: When involving with high voltage it has to be ensured that the signals are physically isolated between the sensor output and the rest of the system. It is achieved by breaking the conductor paths by magnetic coupling, optical coupling, or by capacitive coupling. Optical coupling uses an LED at the transmitting side and a photo diode at the receiving side. Capacitive coupling uses a capacitor to isolate input and output signals. Magnetic coupling uses a transformer to isolate the input and output. These techniques are advantageous in handling signals from high voltage sources and transmission towers. In biomedical applications such isolation is unavoidable.

Linearization: Sensors which gives non-linearization response can be corrected by proper signal conditioning techniques. **Linearization of the data, can be performed by analog techniques using either linear-approximation, or smooth series approximation using a low cost IC amplifier. Alternately linear approximation can be performed digitally after data acquisition and conversion by the use of ROMs. by storing a suitable linearization table or programme initially.**

Filtering and Anti-Aliasing: As most of the sensor output level is very low, they are very prone to Electro Magnetic Interference. Appropriate filters are used to eliminate the noise from the signals. Table 1.1 summarizes the basic characteristics and signal conditioning requirements of typical transducers.

▼ **Table 1.1:** Sensors with signal conditioning requirements

1	Thermocouple	Amplification, Linearization and Reference temperature for cold junction compensation.
2	Strain Gauge	Required excitation voltage or current, Bridge formation, amplification, and linearization.
3	LVDT	Excitation and linearization.
4	RTD	Current excitation and linearization
5	Thermistor	Current or voltage exciter and linearization

2.2.6 Sample And Hold Circuit

To achieve a reliable and accurate analog to digital conversion, ADCs require a fixed time during which the input signal remains constant called aperture time. This is a requirement of the conversion algorithm used by the A/D converter. If the input changes during this time, the ADC output will be inaccurate. This situation can be managed by suitably incorporating a sample-and-hold device. It samples the output signal from the multiplexer or gain amplifier very quickly and holds it constant for the ADCs aperture time. Usually sample and hold circuit is placed between multiplexer and ADC.

Sample-and-hold circuit can be approximated by a capacitor and a high gain opamp. A typical sample-and-hold using an opamp is shown in Figure 2.4. When the switch is closed, the capacitor charges to the input voltage. When the switch is opened, the capacitor holds the voltage level until the next sampling time. The opamp provides large input impedance so that the capacitor is not discharged appreciably and at the same time offers the gain to drive external circuit.

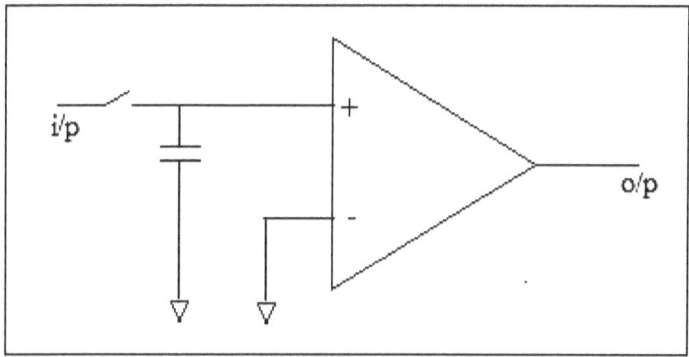

▲ **Figure 2.4:** Sample and Hold circuit

2.2.7 A to D Converters (ADC)

An analog-to-digital converter, or ADC, is a device or peripheral that converts analog signals into digital signals. In the real world, signals mostly exist in analog form. An ADC with a S/H circuit can be used to sample such signals and the signals can be converted to the digital values. There are four types of A/D converters generally used, and they are:
1. Integrating or dual slope ADC,
2. Successive approximation ADC,
3. Parallel Comparator (Flash) type ADC, and
4. Counting type ADC.

Integrating or dual slope ADC
These are used for very low frequency and may have very high accuracy and precision is shown in Figure 2.5. They are found in thermocouple and RTD modules. Other advantages include very low cost, relatively less noise and mains pickup tend to be reduced by the integrating and dual slope nature of the A/D converter. The A/D procedure essentially requires a capacitor to be charged with the input signal for a fixed time, and then uses a counter to calculate how long it takes for the capacitor to discharge. This length of time is proportional to the input voltage. It is more accurate ADC type among all. It has greater noise immunity compare to other ADC types. However, it is the slowest ADC among all.

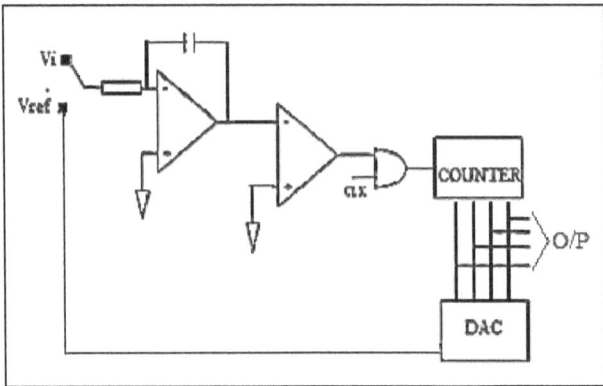

▲ **Figure 2.5:** Integrating or dual slope ADC

Successive approximation ADC

Successive approximation A/Ds allow much higher sampling rates (up to a few hundred kHz with 12 bits is possible) while still being reasonable in cost. The conversion algorithm is similar to that of a binary search, where the ADC starts by comparing the input with a voltage (generated by an internal DAC converter), corresponding to half of the full-scale range. If the input is in the lower half, the first digit is zero and the ADC repeats this comparison using the lower half of the input range. If the voltage had been in the upper half, the first digit would have been 1. This dividing of the remaining fraction of the input range in half and comparing to the input voltage continues until the specified number of bits of accuracy has been obtained. It is obviously important that the input signal does not change when the conversion process is underway. Figure 2.6 shows the block diagram of a typical Successive Approximation ADC.

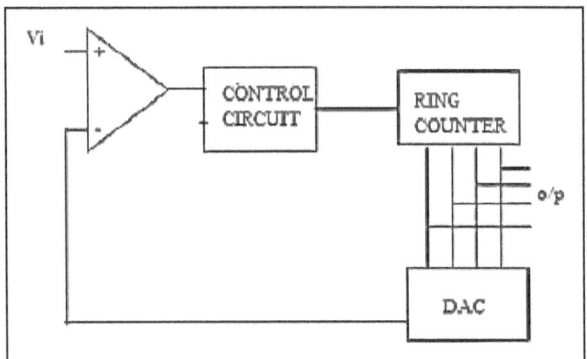

▲ **Figure 2.6:** Successive Approximation ADC

Parallel Comparator (Flash) ADC

It is the fastest ADC among all the ADC types. In flash ADC of n-bit in size, (2n - 1) comparators and 2n registers are needed. Each comparator compares Vin to a different reference voltage, starting with Vref = 1/2 (LSB). Op-amp is used as comparator here. Figure-2 depicts block diagram of parallel comparator ADC. It is very fast as mentioned. But hardware requirements are very high. As an example, 255 comparators are required for an 8-bit ADC. Also it has low resolution and large power consumption is other disadvantages. Figure 2.7 shows the block diagram of typical Parallel Comparator (Flash) type ADC.

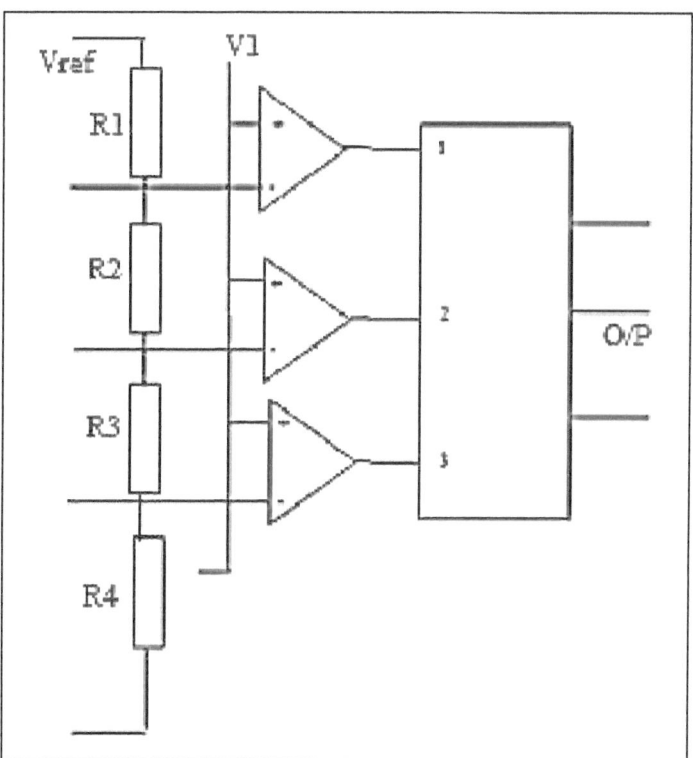

▲ **Figure 2.7:** Parallel Comparator (Flash) type ADC

Counting type ADC

Counting type ADC uses counter and DAC. It compares DAC output with analog voltage and does the same till both are equal in magnitude. At this moment counter will stop. The conversion time depends on analog input voltage. Figure2.8 depicts the block diagram of a counter type ADC.

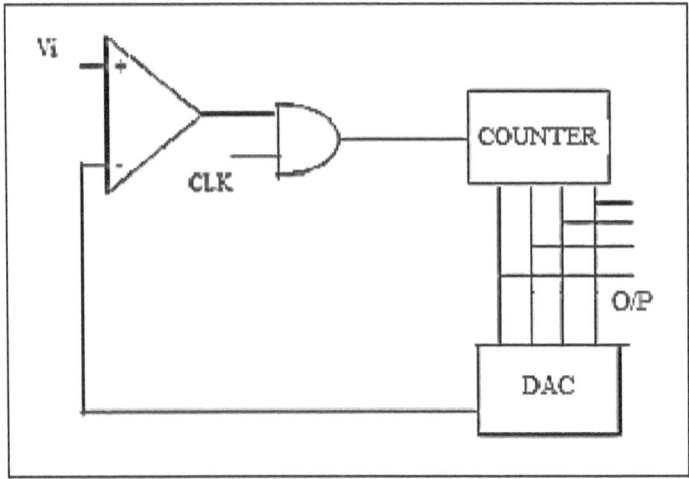

▲ **Figure 2.8:** Counter type ADC.

ADC specifications

Certain important specifications of A to D converters are briefly explained below.

- *Accuracy:* It includes quantization error, system noise, linearity, etc. Absolute accuracy refers to the maximum analog error.
- *Common Mode Rejection Ratio (CMMR):* This is the ratio of the resulting output signal to a changing input common-mode signal. It is the degree of the rejection of a common mode signal across the differential output stage.
- *Conversion time:* The conversion time is the time required for an ADC to complete the single conversion. This time does not include the acquisition time and MUX time. The conversion time for a given ADC is preferably less than the throughput time.
- *Crosstalk:* In the multi-channel DAS, coupling between the adjacent channels, and sharing the transmission path, results in crosstalk. This interference appears as a noise in the digital output and quite unwanted.
- *Input Range:* The specified range of the peak to peak, input signal of an A/D converter.
- *Latency:* Latency is the time required for an ideal step input to converge, within an error margin to a final digital output value. The error-band is expressed as a predefined percentage of the total output voltage step. The latency of a conversion is that period between the time where the signal acquisition begins to the time to the next conversion starts.

- *Linearity errors:* With most ADCs gain, offset and zero errors are not critical as they may be calibrated out. Linearity errors, differential non-linearity (DNL) and integral non-linearity INL) are more important because they cannot be removed.
- *Differential non-linearity(DNL):* **It is** the difference between the actual code width from the ideal width of 1 LSB. If DNL errors are large, the output code widths may represent excessively large and small input voltage ranges. If the magnitude of a DNL is greater than 1 LSB, then at least one code width will vanish, yielding a missing code.
- *Integral Non-Linearity (INL):* INL describes the non-linearity of ADC. It is considered as an important parameter because it is a measure of an ADC non-linearity error. However, as in any Analog or Mixed-Signal Design project, some specifications are important, some are not. It all depends on the project requirements regarding accuracy and precision. Understanding INL enables the circuit designer to avoid surprises in his or her project. INL is defined as the maximum deviation of the ADC transfer function from the best-fit line.
- *Resolution:* Resolution can be used to describe the general performance of an ADC. It is the smallest change that can be distinguished by an ADC converter. For example, for a 12-bit ADC converter resolution would be 1/4096 = 0.0244%.. It's an important ADC specification because it determines the smallest analog input signal an ADC can resolve.
- *Monotonicity:* This requires a continuously increasing output for a continuously increasing input over the full range of the converter. This term implies that an increase (decrease) in the analog voltage input will always produce no change or an increase in the digital code.
- *Quantizing uncertainty:* Because the ADC can only resolve an input voltage to a finite resolution of 1 LSB, the actual real-world voltage may be up to ½ LSB below the voltage corresponding to the output code or up to ½ LSB above it. An ADC's quantizing uncertainty is therefore always ±½ LSB.
- *Relative accuracy:* This refers to the input to output error as a fraction of full scale with gain and offset error adjusted to zero.

2.2.8 Storage and Display

Once the analog data is converted into digital format, it can be suitably stored, displayed, processed, or send to remote master stations through an appropriate communication medium.

2.2.9 Data Forwarding And Communication

In certain occasions, the information acquired has to be send to remote locations, where other SCADA components or master stations are situated. This is general requirement in the case of Distributed SCADA such as Power System SCADA (Energy Management System and Distribution Management System). This arises the requirement for an appropriate communication technology which can provide a secure communication. Today a variety of communication technologies suitable are available which will be discussed in the subsequent chapters.

2.2.10 Communication in SCADA System

The innovations of ICT in SCADA in fact helped to overcome the geographical barriers and made the remote monitoring and control an easy task. The main communication requirements are between field devices and the RTU/PLC/DCU, and the communication between the RTU and the Master Control Center. Many proprietary and open communication protocols suitable for industrial SCADA have been developed. Today many open protocols developed are under modification to incorporate security features. A detailed description of the SCADA communication and communication protocols are given in the subsequent chapters.

2.2.11 Selection Criteria of DAS

Today many companies are manufacturing the SCADA components using different technologies. Choosing the right SCADA system components from this wide array of SCADA products is a daunting task. As already explained SCADA component starts with data acquisition components which translate the physical world analog and discrete signals to digital data suitable for processing by the digital computers. Hence the fundamental and prime component of the DAS is the ADC. Many DAS products today available in the market are with communication and control capabilities and having a range of I/O modules. Hence with an in-depth knowledge of the DAS specifications, one can fully optimize their requirements and select the most suitable and economical solution so that one can avoid paying for features which one don't require.

Input Range: Selection of input range is very important as it depends the nature of the physical quantity to be measured and the type of the transducer used. Usually DAS provide a range which matches the maximum output range of the transducer. This results in providing the highest possible resolution. Many DAS provide multiple input ranges by using software-programmable-gain amplifiers.

Number of input channels: The number of input channels on data acquisition boards typically ranges from 4 to 64. Input channels can be single-ended (SE) or differential (DI). Differential inputs offer noise immunity, and can improve accuracy when long cables, low-level input voltages (less than 1 V full-scale), or high-resolution converters are used.

Accuracy: The ADC of the DAS converts the analog value of the physical quantity to digital data. Accuracy has been defined by how closely the binary code matches the true value of the monitoring analog signal. The major component of accuracy, and its limiting factor, is resolution. Stated in bits, resolution determines the number of counts or binary numbers used to represent the analog signal. When evaluating a data acquisition board, it is critical to understand the relationship between resolution and accuracy. Resolution is simply one factor affecting accuracy.

Total harmonic distortion (THD): It is the ratio of the sum of the harmonics of the fundamental frequency of the input signal to the fundamental frequency itself. THD is a good indicator of the quality of the circuit design. A high THD measurement indicates a flawed analog-input design.

Speed: The throughput of a board, generally specified in mega samples per second or kilosamples per second is a crucial measurement for high-speed applications. If multiple ADCs are used on a single board, the specified throughput represents the sum total of the individual converter throughputs. ADC throughput is mainly determined by three elements.
1. conversion time which is the time needed to do actual conversion,
2. acquisition time which is the time needed by associated acquisition circuitry-the multiplexer, signal conditioning, and sample and hold-to acquire a signal accurately, and
3. transfer time which is the time needed to transfer data from the board to system memory.

Some high-speed DAS boards increase throughput by overlapping the acquisition time on one sample with the analog-to-digital conversion time of the previous sample, in effect handling two signals at one time. Most analog output circuits have a separate data buffer for each channel. Settling time varies proportionately with the size of the output change, and is specified in microseconds.

Clocks, Triggers, etc. With multiple pacer clock circuits, multiple conversions with precision timings suitable for time stamped data acquisition are possible. Pacer clock triggering is achieved by a software instruction or with a hardware digital pulse or with an analog voltage. Most of the RTUs and PLCs contain general purpose counter/ timer circuits for time stamped data acquisition. These consist of several counters and a frequency source. While confirming the pacer clock circuits in DAS, one should not be confused with the counter/ timers available on the board, which are dedicated to the clocking and triggering of the analog-to-digital and digital-to-analog subsystems.

In simple local SCADA systems, PCI data acquisition boards can feed acquired data directly to the PC's memory, eliminating the need for on-board memory and the resulting gaps in data. Furthermore, PCI boards are auto-configured upon installation to match the system's resources. Manual configuration of jumpers or DIP switches is a thing of the past. When frequency-domain performance is characterized, one can be sure that the acquired data will be as accurate as required it to be. The Effective Number of Bits (ENOB) specification is the way to clearly convey a data acquisition board's ac performance. Combining all critical, real-world performance considerations such as accuracy, settling time, and dynamic performance in one easy-to-comprehend specification, ENOB specifies the overall accuracy of the analog-to-digital transfer function.

Standards and Certifications: The FCC and CE certification standards are two good indicators of quality and reliability. These certifications guarantee the buyers that DAS components are meeting certain standards and will perform robustly in a real-world situation.

Planning For Future Changes: A data acquisition system should meet the present needs and provide flexibility for the future, while justifying the utilities investment in hardware and software. Common upgrades to a DAS is changing especially the hardware and hence with new add-on boards. If the application software is not designed with an open systems approach, adding new boards can cause expensive, time-consuming reprogramming. Making sure that the Application Programming Interface (API) is hardware-independent will allow changing boards with significantly no or little reprogramming. Also, make sure the data acquisition software support multiple operating systems. Ensure that DAS software is supported for the upgraded version of Operating System. This will reveal the fact of difficulty of migration from one of the early versions of OS to the upgrades.

Another hardware issue involves the use of interrupts. Traditionally, computer peripherals request the host CPU's attention via the hardware interrupts on the CPU. However, the number of peripheral components on a typical system (modem, scanner, CD-ROM drive, etc.) has increased to the point where it frequently exceeds the fifteen interrupts available. To address this problem, Data Translation and some other board manufacturers have stopped using hardware interrupts on new boards, achieving the same function with a software feature.

Hence while planning a data acquisition system or SCADA for an utility, keep the application in mind, comprehend the requirements clearly and choose wisely the data acquisition system, the software and the communication technology which is relevant and most appropriate. Also look for simple solutions rather complex one but with accuracy and robustness. Also ensure that software designed for future upgradability.

2.3 POWER SYSTEM SCADA COMPONENTS

The modern Power System SCADA uses Remote Terminal Unit (RTU), Intelligent Electronic Devices (IEDs), IED Architecture and Components, Instrument Transformers with Digital Interfaces Programmable Logic Controller (PLC), Data Concentrators and Merging Units and Bay Control Unit (BCU) depending upon the need necessity, adaptability and economic factors. The Master Control Center (MCC) is made up of SCADA System Software, Master Station Hardware, Servers, Global Positioning Systems (GPS) and Human Machine interface (HMI). The following sections elaborates these components.

2.3.1 Remote Terminal Unit (RTU)

Remote Terminal Unit (RTU) is a microprocessor based device connected to appropriate sensors, transmitters and process equipment for the purpose of remote telemetry and control of plant or process. RTUs find applications in oil and gas remote instrumentation monitoring, networks of remote pump stations, environmental monitoring systems, air traffic equipment, power utilities, etc. RTUs with the aid of appropriate sensors, monitors production processes at remote site and transmits all data to a central station where it is gathered and monitored. An RTU can be interfaced using serial ports such as RS232, RS485, etc. or Ethernet to communicate with the central stations. They also support various protocol standards such as Modbus, IEC 60870, DNP3 making it possible to interface with 3rd party software.

RTU Architecture

The RTU architecture comprises of a CPU, volatile memory and non-volatile memory for processing and storing programs and data. It communicates with other devices via either serial ports or an on-board modem with I/O interfaces. It has a power supply module with a backup battery, surge protection against spikes, real-time clock and a watchdog timer to ensure that it restarts when operating in the sleep mode. Figure 2.9 shows the block diagram of an RTU configuration. A typical RTU hardware module includes a control processor and associated memory, analog inputs, analog outputs, counter inputs, digital inputs, digital outputs, communication interfaces and power supply. Centralized RTU design where all I/O modules are housed in RTU panels and communicating with master station through communication port.

Distributed RTU design in which distributed I/O modules or processor with I/O modules are housed in respective RTU panel. All these distributed I/O modules or I/O modules with processor shall be connected to a central processor for further communication with master station. The customer shall asses the requirement of RTU panels for such design and supply panels accordingly.

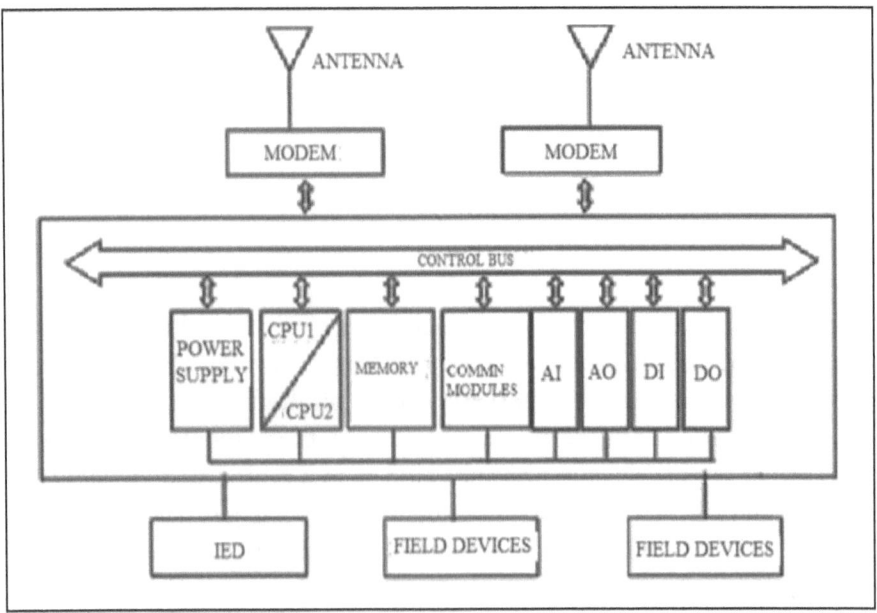

▲ **Figure 2.9:** RTU Architecture

RTU Components

Central Processing Unit (CPU): Most of the available RTU designs utilize a 16 bit or 32 bits microprocessor with a total memory capacity of 256kbytes expandable to 4 Mbytes. It also has two or three communication ports or multiple Ethernet links. This system is controlled by a firmware and a real-time clock with full calendar used for accurate time stamping of events. A watchdog timer provides a check that the RTU program is executing regularly. The RTU program regularly resets the watchdog timer and if this is not done within a certain time-out period the watchdog timer flags an error. Depending on the requirements of safety, reliability, high availability, and functionality, dual CPU with dual power supplies are often preferred. Thus the system is capable of automatic monitoring without fail with primary and hot standby CPUs. Failures are detected automatically and trigger a switch over from the primary CPU to hot standby CPU.

Analog Input Modules (AI): An analog input signal is generally a voltage or current that varies over a defined value range, in direct proportion to a physical process measurement. 4-10 milliamp signals are most commonly used to represent physical measurements like pressure, flow and temperature. Analog inputs of different types including 0-1 mA, 0–10 V., ±1.5 V, ±5.0 V etc.are also common and acceptable to RTU. Five main components of the analoginput(AI) module are described below.

1. *Multiplexer:* This samples several analog inputs in turn and switches each to the output in sequence. The output goes to the analog digital converter.
2. *Signal Conditioner:* This amplifies and transforms the low-level voltages to match the input range of the board's A/D converter.
3. *Sample and hold circuit:* An analog device that samples the voltage of a continuously varying analog signal and holds its value at a constant level for a specified minimum period of time.
4. *A/D converter:* A system that converts an analog signal into a digital format or a digital code corresponding to the input voltages.
5. Bus interface and board timing system.

Typical analog input modules have the following features.
1. 8, 16, or 32 analog inputs.
2. Resolution of 8 to 12 bits.
3. Range of 4-10 mA.
4. Input resistance typically 140 kΩ to 1 MΩ.
5. Conversion rates typically 10 microseconds to 30 milliseconds.

RTU can also receive analog data via a communication system from a master or Intelligent Electronic Device (IED) which sends data values to it.

Analog Output Module (AO): Though it is not commonly used, analog outputs (AO) modules are included in control devices to deal with varying quantities, such as graphic recording instruments or strip charts. The function of an Analog Output module is to convert a digital value supplied by the CPU to an analog value, by means of a Digital ToAnalog Converter (DAC). This analog representation can be used for variable control of actuators. The basic features of the Analog output modules are as follows.

1. 8, 16 or 32 analog outputs.
2. Resolution of 8 or 12 bits.
3. Conversion rate from 10 micro seconds to 30 milliseconds.
4. Outputs ranging from 4-10 mA or 0 to 10 volts.

Digital or status inputs (DI): These are used to indicate status and alarm signals. Most RTUs incorporate an input section or input status cards to acquire two state real world information. This is usually accomplished by using an isolated voltage or current source to sense the position of a remote contact (open or closed) at the RTU site. This contact position may represent many different devices, including electrical breakers, liquid valve positions, alarm conditions, and mechanical positions of devices.

Digital Output Modules (DO): RTUs may drive high current capacity relays to switch power on and off to devices in the field. The DO modules are used to drive an output voltage at each of the appropriate output channels with three approaches possible.

1. Triac Switching,
2. Read Relay Switching, and
3. TTL voltage outputs

Power Supply Module: RTUs need a continuous power supply to function, but there are situations where RTUs are located at quite a distance from an electric power supply. In these cases, RTUs are equipped with alternate power source and battery backup facilities in case of power losses. Solar panels are commonly used to power low-powered RTUs, due to the general availability of sunlight. Thermo electric generators can also be used to supply power to the RTUs where gas is easily available like in pipelines. Normally RTU is expected to operate from 110/140 V AC ± 10% 50 Hz or 11/14/48 V DC± 10% typically. Batteries that should be provided are lead acid or nickel cadmium. Typical backup requirements are for 10-hour standby operation and a recharging time of 11 hours for a fully discharged battery at 15°C. The power supply, battery and associated charger are normally contained in the RTU housing. The monitoring parameters of the battery system of RTU which should be transmitted back to the central site/master station are analog battery reading and alarm for battery voltage outside normal range

Communication interfaces: Modern RTU are designed to be flexible enough to handle multiple communication media such as

1. RS 232/RS442/RS 485 etc.
2. Ethernet.
3. Dial up telephone lines/dedicated landlines.
4. Microwave, and Satellite.
5. X.15 packet protocols, and
6. Radio via trunked/VHF/UHF.

An RTU may be interfaced to Multiple Control Stations and Intelligent Electronic Device (IEDs) with different communication media such as RS232, RS485, etc. An RTU may support standard protocols (Modbus, IEC 60870-5-101/103/104, DNP3, IEC 60870-6-ICCP, IEC 61850 etc.) to interface any third party software.

Data transfer may be initiated from either end using various techniques to insure synchronization with minimal data traffic. There are two methods in general viz. polling and report by exception. In polling method, the master may poll its subordinate unit (Master polls RTU or RTU polls IED) for changes of data on a periodic basis. The report by exception method is used where a subordinate unit initiates an update of data upon a predetermined change in analog or digital data. Periodic complete data transmission must be used periodically, with either method, to insure full synchronization and eliminate stale data. Most communication protocols support both methods, programmable by the installer. Analog value changes will only be reported, usually on changes outside a set limit from the last transmitted value. Digital Status values observe a similar technique and only transmit groups (bytes) when one included point (bit) changes.

Multiple RTUs or multiple IEDs may share a communications line, in a multi-drop scheme, as units are addressed uniquely and only respond to their own polls and commands. IED communications transfer data between the RTU and an IED. This can eliminate the need for many hardware status inputs, analog inputs, and relay outputs in the RTU. Communications are accomplished by copper or fiber optics lines. Multiple units may share communication lines.

Communications to a Master Control Center are generally envisaged and suitably incorporated in larger systems. The communication media used to transfer data may be copper, optical fiber or wireless media communication system. Multiple units usually share communication channels.

RTU environmental enclosures

Typically, the printed circuit boards of DI, AI, AO, DO, etc. are plugged into a backplane in the RTU cabinet. The RTU cabinet usually accommodates inside an environmental enclosure which protects it from extremes of temperature, humidity, etc. The following factors are the typical considerations while selecting the enclosures.

1. Circulating air fans and filters: This should be installed at the base of the RTU enclosure to avoid heat build-up. Hot spot areas on the electronic circuitry should be avoided by uniform air circulation. It is important to have a heat soak test too.
2. Hazardous areas: RTUs must be installed in explosion proof enclosures. In substations it is recommended to keep a safe distance from the circuit breakers.
3. Operating temperatures:Operating temperatures of RTUs are variable when the remote monitoring is located outside the building in a weather-proof enclosure. These temperature specifications can be relaxed if the RTU is situated inside a building, where the temperature variations are not as extreme (provided consideration is given to the situation, where there may be failure of the ventilators or air-conditioning systems).
4. Humidity: Typical humidity ranges are 10–95%. Ensure at the high humidity level that there is no possibility of condensation on the circuit boards or there may be contact corrosion or short-circuiting. Lacquering of the printed circuit boards may be an option in these cases. Be aware of the other extreme, where low humidity air (5%) can generate static electricity on the circuit boards due to stray capacitance. CMOS based electronics is particularly susceptible to problems in these circumstances. Only screening and grounding the affected electronic areas can reduce static voltages. All maintenance personnel should wear a ground strap on the wrist to minimize the risk of creating and transferring static voltages.
5. Electromagnetic interference (EMI): If excessive electromagnetic interference (EMI) and radio frequency interference (RFI) is anticipated in the vicinity of the RTU, special screening and earthling should be used. Some manufacturers warn against using handheld transceivers in the neighbourhood of their RTUs. Continuous vibration from vibrating plant and equipment can also have an unfavourable impact on an RTU, in

some cases. Hence vibration shock mounts should be specified for such RTUs. Other areas which should be considered with RTUs are lightning (or protection from electrical surge).

RTU Design Standards

The RTUs available in the market are generally designed in accordance with applicable International Electro technical Commission (IEC), Institute of Electrical and Electronics Engineer (IEEE), American National Standards Institute (ANSI), and National Equipment Manufacturers association (NEMA) standards. For easy maintenance the architecture design generally support pluggable modules on backplane. The field wiring are terminated such that these are easily detachable from the I/O module.

Selection Criteria of RTUs

The SCADA RTUs need to communicate with all on-site equipments and survive under the harsh conditions of an industrial environment. Hence the following points may be generally ensured while selecting the RTU.
1. Sufficient capacity to support the equipment at the site with the scope for the future expansion. Hence at every site, the RTU (Remote Telemetry Unit) that can support expected growth over a reasonable period of time, but within the budget may be selected.
2. Rugged construction and ability to withstand extremes of temperature and humidity. Depending upon the environmental factors where the RTU is intend to be kept, a rugged design which can keep the RTU system as the most reliable element.
3. Secure, redundant power supply. The SCADA system has to be up and working 24x7 without any excuses. RTU should support battery power ideally with redundant power inputs.
4. Redundant communication ports. Network connectivity is as important to SCADA operations as a power supply. A secondary serial port or internal modem may be kept in the RTU online to counter the LAN failure. Further RTU with multiple communication ports can easily support a LAN migration strategy.
5. Non-volatile memory (NVRAM) for storing software and/or firmware. NVRAM retains data even when power is lost. New firmware can be easily downloaded to NVRAM storage, often over LAN - so that capabilities of RTUs may be kept up to date without excessive site visits.
6. Intelligent control. Sophisticated SCADA remotes can control local systems by themselves according to programmed responses to sensor

inputs. Though it is not necessary for every application, but it does come in handy for certain cases.
7. Real-time clock for accurate date/time stamping of reports.
8. Watchdog timer to ensure that the RTU restarts after a power failure.

Securing Remote Terminal Units

As RTU be the one of the critical component of the DCS and ICS, physical-cyber security of the RTU is most important especially when deployed in critical infrastructure SCADA such as Power System SCADA(PSS), Oil and Gas Industry, etc. Today the Electric utilities are most concerned of statutory security requirements of North American Electric Reliability Corporation Critical Infrastructure Protection (NERC CIP). The flexibility and range of RTU technologies necessitates a proper incorporation of its security and safety procedures.

Physical security of an RTU can be ensured by keeping restricted and authenticated access. Door opening of RTU enclosure can be properly monitored by a status signal, or by placing a surveillance camera which communicates the MCC in real-time RTU to alert system operators regarding the physical security breach of a remote RTU. One of the main standards that define procedures for implementing electronically secure Industrial Control Systems (ICS) is ISA/IEC-62443. This guidance applies to end-users, system integrators, security practitioners, and control systems manufacturers who design, implement, and manage the ICS. While deal with Power System SCADA (PSS), most of the utilities are following the NERC CIP standards as they are exclusively developed for Bulk Electric System (BES).

NERC CIP defines a Critical Asset, as the facility, or system which, if destroyed or degraded, would affect the smooth and proper operation of the Bulk Electric System (BES). In other way, they are the programmable electronic devices and communication networks which comprise hardware, software, and the data. The Critical Cyber Assets (CCA) is defined as Cyber Assets, essential to the reliable operation of Critical Assets. Bulk Electric System(BES) are defined as, the electrical generation resources, transmission lines, sub stations, interconnections with neighbouring systems, and associated equipment, generally operated at voltages of 100 kV or higher. Obviously the RTUs installed at substations which deals with voltages more than 100kV or higher is a Critical Cyber Assets and need NERC CIP compliance and protection to ensure safety and security.

In order to frame the security policies and procedures, it is critical for utilities to understand the necessary RTU functionality and how it can be incorporated to achieve goals. Majority of the RTUs which are Critical Cyber Asset may not be kept in a completely protected area, and may be accessible to people who are not explicitly authorized to access them. This necessitates the key requirement of physical security of RTUs. Most RTU equipment includes, digital input points as a standard feature, and adding a door alarm is a simple way to meet NERC CIP requirements. Including the door alarm input as a sequence of events (SOE) point provides a means of meeting another aspect of the NERC CIP procedures for logging data and maintaining historical records.

Cyber security cannot be mentioned without encryption. Encrypted data transmission and reception makes RTUs more secure. It is obviously harder for an unauthorized user to manipulate the data sent and received. Encrypted data, however, has a disadvantage. Most notably, it will impact the scan rate of the device. Further encrypting and decrypting every message received and transmitted by the RTU amounts to a considerable increase in the processing time, which in turn affects the latency. This demands high computing power and processing capability of RTU CPU. Many RTUs being used in the field are already pushed to the upper limits without consideration for encryption. Another disadvantage is that the data is encrypted and the user will need special tools to view it, as it is often important to examine the data exchanged between devices to ensure they operate properly.

An important feature of RTU is its ability to provide logs of important events. Many of the alarms need to be logged for time of occurrence and the relevant information. Maintenance engineers who are making connections and configuration changes to the unit can be properly evaluated with this chain of events. Sequence of Event (SOE) logs, user logs and system logs can all be used to document events from the safe and secure operations of RTU.

The formation of procedures and plans, which need to be written with current and future capabilities in mind, is most important part of the SCADA security standards such as NERC CIP, IEC 62443, etc. Due to the wide technology range it will not be practical to replace or upgrade every device as quickly or easily. Securing an ICS or DCS is not a job for the faint hearts; rather it is a job that must be done with courage, intelligence, and imagination. A work which requires many hours of hard, thoughtful work for developing a security policy in tune with the utilities general safety and security policy. Today, it is considered that providing cyber security to critical infrastructure is no way inferior to the job of defending a nation by the army generals.

2.3.2 Intelligent Electronic Devices (IEDs)

IEDs are the main components of substation integration and automation. IEDs which integrate metering, protection, and control functions reduce the number of equipment installations, reduce panel and control room space and in turn reduces the cost of the capital expenditure. It also eliminates redundant equipment and databases. Present day IEDs are capable of directly communicating with the SCADA systems in the required standard and interoperable protocols. A single IED is capable of monitoring and providing a number of parameters and status signals such as barker status, switch positions, etc. They are capable of event recording and provide SoE. In fact IEDs revolutionize the substation by Substation.

IEDs facilitate the exchange of operational and non-operational data. Operational data are instantaneous values of power system analog and status points such as volts, amps, MW, MVAR, circuit breaker status, switch position, etc. They are also called SCADA data. Non-operational data consists of files and waveforms such as even summaries, oscillographic event reports, SoE records, SCADA-like points (status and analog points) which have a logical state or a numerical value. These data are not needed in SCADA to monitor and control the power system. The main functions of IEDs are Protection, Control, Monitoring, Metering, and Communication.

IED Architecture and Components

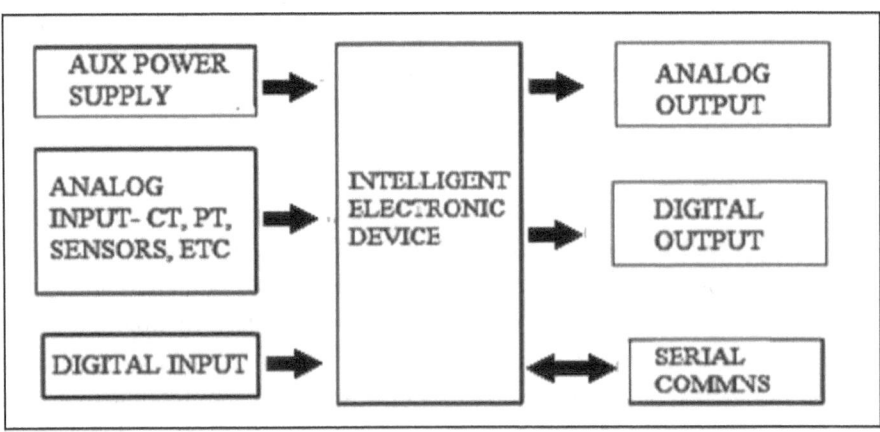

▲ **Figure 2.10:** Block diagram of an IED

Figure 2.10 depicts the organizational block diagram of a typical Intelligent Electronic Device. The modern IED architecture ensures that the device is multipurpose, modular in nature, flexible and adaptable, and has robust communication capabilities which include multiple selectable protocols, multi-drop facilities with multiple ports, and rapid response for real-time data. Current IEDs coming with tremendous computing power which is capable to carry out a variety of functions, various applications like protection and metering. IEDs have Sequence of Event (SOE) recording capability that can be very useful for post-event analysis, for fault waveform recording, and for power quality measurements. The advantage of this is, it eliminates the requirement of additional digital fault recorders and power quality monitors. IEDs can also accept and send out Analog and Digital signals with selectable settings, thus making the IEDs versatile. The details of the IED building blocks are briefly explained below.

- *Auxiliary power supply:* Unlike older protection relays which may not need an auxiliary supply, IEDs always require an auxiliary power supply. Most IEDs accept an extended range of power supply, usually ranging from 14 -150 V DC or 110 -140 V AC.
- *Analog inputs:* Protection relays are always provided with current transformer (CT) and potential transformer (PT) inputs. In addition, IEDs may be provided with sensor inputs. Hence it is important to specify the rated secondary current and frequency before ordering.
- *Digital inputs:* Some IEDs require potential-free contacts for digital inputs(DI), while others recognize the positive power supply voltage (source) or negative power supply voltage (sink) as a logical '1'. Digital inputs may be commands or status information.
- *Analog outputs:* Some IEDs are provided with transducer outputs. Mostly these outputs are programmable. These outputs can be active or passive. The passive type requires external power supply.
- *Digital outputs:* Digital outputs can be potential free normally open (NO), normally closed (NC) or solid state contacts. It is important to check the switching capability of the output contacts, as the differences can be significant. Digital outputs may be commands or status information.

IEDs have the real time and rapid data exchange capabilities with multi ports. Many IEDs which are available in the market are capable of handling Analog Inputs, Discrete Outputs, Analog Outputs and Discrete Inputs. SoE recording, fault recording, metering and protection are the common features with all IEDs.

The IED integrates many single-function electromechanically relays, control switches, extensive wiring, and much more into a single unit. In addition the IED handles additional features like self and external circuit monitoring, real-time synchronization of the event monitoring, local and substation data access, programmable logic controller functionality, and an entire range of software tools for commissioning, testing, event reporting, fault analysis, etc.

2.3.3 Instrument Transformers with Digital Interfaces

A range of digital instrument transformer solutions enabling full IEC 61850 implementation and contributing to transmission architectures for the Smart Grid networks of present and future have been developed by various manufactures. The innovative design with smart sensor digital instrument transformers is very accurate, intelligent, safe, and cost-effective and very importantly core-less.

These modern product range serves both AC (up to 1200kV) and DC (up to 800kV) transmission systems as well as high current DC applications and provide increased safety, intelligence, flexibility, availability and global savings through a variety of advanced characteristics such as those described below.
1. Simplified and reduced cabling,
2. Compact solution and lightweight,
3. Smaller substation footprint,
4. Near zero maintenance, and
5. Reduced inventories.

A new range of Optical Sensor Intelligence (OSI) based Instrument Transformers, are developed which are capable of serving both AC and DC transmission systems. These Cost-effective OSI range brings key technological features and benefits such as wider dynamic range from 1 A to 4800 A. It is very sophisticated and Smart Grid ready, delivering direct digital output for metering applications. It also has measurement improvements resulting from the fact that there is no saturation of the network and a high bandwidth. These products offer greater flexibility due to the ratio being modifiable at any time via a computer connection. There are also considerable economical features compared to conventional instrument transformers. These products are more compact and lighter, uses less wiring, as all signals are transmitted through a fiber optic cable, requires less maintenance, is easy to install with greater flexibility and can be easily upgraded. Further these are more environmentally friendly and more secure with the risk of explosion minimized, reduced leakage, SF6 free, and no issue of end-life-disposal. Typically an optical sensor represents only 10% of the weight of an oil filled transformer, reducing transportation costs and the quantities of materials used. State-of-art Fiber Optical Current Sensor (FOCS) solutions are presented together with Rogowski coil and Electronic

Voltage Transformer (EVT) in innovative solutions that deliver significant operational performance, environmental, safety and substation engineering benefits supporting future Smart Grid substation investments.

2.3.4 Programmable Logic Controller (PLC)

A Programmable Logic Controller (PLC) is a computer based solid state device that controls industrial equipment and processes. It was initially designed to perform the logic functions executed by relays, drum switches and mechanical timer/counters. Advanced PLC with Analog control capability are also available today, but RTU based SCADA systems are preferred by power utilities as the RTU can also function as Data Concentrating Unit(DCU) and offers more security especially when unguided remote communication is used. The advantage of a PLC over the RTUs which are available in the market can be used in a general purpose role and can easily be set up for a variety of different functions. The actual construction of a PLC can vary widely and are popular for the following reasons.

1. PLC solution is more economical when compared to general purpose RTU solution.
2. The logic of the PLC can easily be modified to cope with new situations and requirements.
3. Design and installations are relatively easier as many hardware requirements can be suitably substituted with software.
4. If properly installed, PLCs are a far more reliable solution than a conventional hardwired relay solution or RTU based solution.
5. PLCs has more sophisticated control than RTU, mainly due to the software capability.
6. PLCs are very compact and occupy less space when compared to alternative solutions.
7. Trouble shooting is simple and easy to diagnose and fix hardware/firmware/software problems.

Ladder Logic: Ladder diagrams or ladder logic are a type of electrical notation and symbology frequently used to illustrate how electromechanical switches and relays are interconnected. It is a rule-based programming language rather than a procedure language, which creates and represents a program through ladder diagrams. It is mainly used in developing programs or software for PLCs.

Ladder logic is widely used in industrial settings for programming PLCs where sequential control of manufacturing processes and operations is required.

The programming language is quite useful for programming simple yet critical systems or for reworking old hard-wired systems into newer programmable ones. This programming language is also used considerably in advanced automation systems such as electronics and automobile manufacturing companies.

The idea behind ladder logic is that even employees without programming backgrounds can quickly program since it makes use of conventional and familiar engineering symbols for programming. But this advantage is quickly negated since manufacturers of PLCs often also provide ladder logic programming systems with their products, which sometimes do not use the same symbols and conventions as those made for other models of PLCs from other manufacturers. These proprietary symbols and conventions is usually meant only for specific make and models. Hence the programs cannot be ported easily to other PLC models or must be completely rewritten.

2.3.5 Data Concentrators and Merging Units

Another two important building blocks that are very useful in modular design of DCS are Data Concentrators and Merging Units. A brief description is given below.

Data Concentration Units (DCU)

In certain situations the DCS design, especially in PSS, is an intimidating task because of the extremely large number of input and output data. The variety of large number of field devices and IEDs with different protocols makes the DCS design very complex. Considering the fact that most substations are automated and intended to operate unmanned, these data has to be transmitted to MCC for control and analysis. The RTU in a substation can serve as a data concentrator by gathering or concentrating the data from the field devices. It utilizes a data concentrating device that has lots of I/O ports on it. When connected to all the IEDs and field devices which can send their details as data in a star topology, this devices are polled to gather the data and transmits it to a remote master and/or client server stations. I/O ports on modern data concentrators available with Ethernet ports such as fibre ports and RJ45 ports suitable for CAT5 cable. These ports can enable a TCP/IP communication and establish a LAN or WAN connectivity. Data concentrators offer several additional capabilities. One of them is to serve as a gateway device. That is, any connection to a WAN system is made at only one point in the substation network and is using this device alone. Modern gateway devices, suitable in DCS environment have the following capabilities.

1. It has the capability to set up a firewall, with SCADA specific IDS loaded to keep the intruders away,
2. It has the capability to provide a secure connection for secure data transmission. Many RTUs/DCUs have the NERC CIP compliance. Remote connection with DCU can be established via the internet using VPN, leased telephone lines, OFC, or unguided medias.
3. Modern RTUs support a number of protocols. They also have protocol conversion capabilities. This helps the field devices which do not support modern SCADA protocols such as DNP3 or IEC 61850, can also be connected and communicated with the RTU without replacing the legacy IEDs, and
4. Many RTUs have the capability of providing a web based HMI. In certain cases, an HMI has to be kept at remote locations with limited capability or use it as a Local Data Monitoring System (LDMS). In such situations, accommodating devices like a display monitor, keyboard, and mouse, etc. must be connected to the RTU or LDMS in a secure manner.

If the substation is small, then all station IEDs and field devices can be connected directly to the data concentrator. No Ethernet switches are required and additional I/O ports on this device can be installed if necessary.

Merging Unit (MU)

In the simplest sense, Merging Units (MU) is a device that enables the Substation Automation with IEC 61850 protocol process bus by converting analogue signals from conventional CTs and PTs into IEC sampled values. Today MU is one of the most critical elements required for the development of modern digital substations. The MU provides the suitable interface for the implementation of the process bus concept in modern substation automation systems. The IEC61850 standard specifies the communication architecture for communication systems and networks for substations. The basic function of the MU in automation systems is to convert the analogue values of current and voltages measured at the process level to digital format and transmit same to the bay level of the substation where the microprocessor relays are located. In modern substations, its function is to collect multichannel digital signals output by electronic current and electronic voltage transformers synchronously and transmit these signals with the protocol of IEC 61850 to protective, measurement and control devices.

In substations, these multichannel digital signals output by electronic current and electronic voltage transformers are collected synchronously and transmit these signals with the protocol of IEC 61850 to protective, measurement and control devices. The use of merging units for elimination of several multiple wire connections running from the switchyard to the microprocessor relays located in the control room. The conventional MU model collects current and voltage signals from various current and voltage transformers in the switchyard, converts them into digital form and sends the digital equivalents to the microprocessor relays via a single fibre optic cable known as the process bus. Today modern MUs are available with additional features of in-built overcurrent protection and bay control functions. These new models are intended to provide local over-current protection and bay control for all equipment in the bay being monitored by a particular MU.

Existing substations with an RTU can be integrated with IEDs depending on the availability of interoperable features. If the RTU is not supporting the standard protocols, the gateways for protocol conversion have to be appropriately integrated. The implementation of modern IEDs provides a rich new source of data that can benefit the entire utility organization.

2.3.6 Bay Control Unit

A Bay Control Unit (BCU), is a very adaptable panel mounted unit which provides wide range of monitoring, control and automation capabilities at the individual bay or circuit level. The main advantage of the BCU is the capability of remote monitoring. The status of the substation equipment and supporting devices are obtained through contact multiplying relay or from auxiliary contacts. Thus both the OPEN and CLOSED positions or a fault or intermediate circuit-breaker or auxiliary contact position can be detected or monitored. BCU control the power equipments in agreement with control commands issued to control equipment and monitoring the bay power apparatus

Monitoring
Remote monitoring of the system is the main advantage of using BCU. The status of primary equipment or auxiliary devices are obtained from auxiliary contacts. Therefore it is possible to detect and indicate the OPEN and CLOSED position or a status of fault circuit-breaker or auxiliary contact position.

Control
BCU also support all control functions that are required for operating substations. The main application is reliable control of switching and other

processes. The automation engineer can set specific functions for the automation of switchgear or substation equipments. Functions are activated via function keys, binary input or via communication interface. Switching authority is determined according to parameters, communication or by *LOCAL/REMOTE* switch. If it is set to LOCAL, only local switching operations are possible and all remote operations are disabled.

Command processing:
This includes all the functionality of command processing especially the processing of single and double commands with or without feedback. Sophisticated monitoring of the control hardware and software, checking of the external process, control actions using functions such as runtime monitoring and automatic command termination after output are possible. Every switching operation and change of breaker position is kept in the status indication memory.

2.3.7 Master Control Centres (MCC)

Figure 2.11 shows a large master control centre with a dual redundant LAN, with all the components, designed with full redundant hardware and software features to achieve High Availability (HA). The system redundancy and communication channel redundancy must be ensured from remote location to the servers of the master station. The master station must be fault tolerant and fail safe in every respect, so that any natural calamity affecting one station will not cause a problem with the functioning of other stations and the system can be monitored and controlled effectively. Uninterrupted power supplies (UPS) monitoring systems are also very important and must be ensured as most of the process/plants are non- stoppable and cannot be interrupted. The various SCADA components in the MCC are elaborated below.

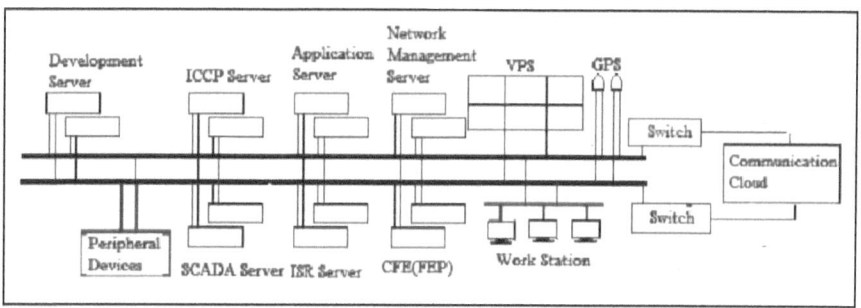

▲ **Figure 2.11:** Block Diagram of a Typical Master Control Centre

2.3.8 System SCADA Software

These are the data acquisition and control, database design and development, reporting and accounting, and the HMI software. It performs all the basic functionalities which are required in common to all SCADA application. As far as power system SCADA is concerned, all SCADA functions are considered as critical functions.

Today, substation automation is achieved with any one of the modern SCADA devices such as Remote Terminal Unit (RTU), Programmable Logic Controller (PLC), Data Concentrators(DCU), Merging Units(MU), Bay Control Unit (BCU), etc. depending on the situational requirements and economy. However most commonly the substation automation software has the following functional features.

1. Advanced protection functions having conventional and IED based,
2. Substation automation functions,
3. Advanced bus failover and automatic load restoration,
4. Feeder sectionalising and islanding,
5. Adaptive relaying, and
6. Equipment Condition Monitoring(ECM).

2.3.9 Master Station Hardware

The main hardware in a master station are the computer, server systems, display units, printers, plotters, routers, and security and protecting devices such as firewalls in addition to uninterrupted power supply ensuring the power quality. Many servers are used for executing the different tasks to be performed. Hence selection of servers is most important and selection must be based on the requirements of the master station.

2.3.10 Servers in the Master Station

As mentioned, different dedicated redundant, high availability servers are deployed in the SCADA master control station to execute specific tasks. The servers are connected through a high speed dual redundant LAN. The client server architecture ensures that the data is accessed by another server whenever it is required and properly authenticated. The dedicated server systems should have special capabilities and features depending upon the requirements of the application for which it is deployed. Some of these features are CPUs with high computing and fast processing power, high-performance RAMs, redundant and uninterrupted, quality assured power supplies, high end routers

and switches for network connections, and high performance Firewalls. The computer server systems available in a SCADA master station are as follows.
1. SCADA server,
2. Application server,
3. Information Storage and Retrieval (ISR) Server,
4. Development server,
5. Network Management Server (NMS),
6. Video Projection System (VPS),
7. Communication Front End Server (CFE),
8. Inter Control Centre Communications Protocol (ICCP) server, and
9. Dispatcher Training Simulator (DTS) server.

The main functions of each server are described below.

SCADA server

The SCADA server is assigned with the data acquisition of all inputs and presents it to other servers for displaying, further processing and decision making. It also accomplishes the control command execution as per the directions issued from the SCADA control station. In small SCADA systems the SCADA server directly gathers the data from the field devices and issue control commands directly to the devices. But when the SCADA system is large SCADA server communicates through the CFE server which is described below.

Application server

Depending upon the nature of process/plant to be controlled, the SCADA application varies. It can be EMS, GMS or DMS as far as power system is concerned. The application software modules are hosted by the application server required for the specific SCADA system.

PSS Distribution Management may have voltage reduction, load management, power factor control, two-way distribution communications, short-term load forecasting, fault identification, fault isolation, service restoration, interface to intelligent electronic devices {IEDs), three-phase unbalanced operator power flow, interface to/integration with automated mapping/facilities management (AM/FM), interface to customer information system (CIS), trouble call/outage management, and so on.

The transmission SCADA may have the Energy Management Systems (EMS) package, which includes network configuration/topology processor, state estimation, contingency analysis, three-phase balanced operator power flow, optimal power flow, etc. For the Generation Management PSS, the application software may include Automatic Generation Control (AGC), economic load dispatch, unit commitment, short-term load forecasting, etc. In case of controlling the refinery plant, necessary process monitoring and control software is loaded in the SCADA server. Same is the case with Water supply and management system.

Information Storage and Retrieval (ISR) server

An information storage and retrieval is historian server which has the following functionalities.
1. Supports the reporting accounting activities,
2. Archiving of data for the system,
3. Real-time data snapshot, and
4. Historical information recording, retrieval, and report generation.

Development server

All preliminary engineering and commissioning activities of the Power System SCADA (PSS) is done with the development server. Further the changes or developments thereafter have to be handled to keep the system real-time. The various programs such as application software development, display development, and database generation are developed also by the development server.

Network Management Server (NMS)

NMS may be used to monitor both software and hardware components in a network. It is an application or set of applications that helps the network administrator to manage network's independent components inside a larger network. It usually records data from a network's remote points to carry out central reporting to a system administrator. Modern control centres have many digital devices connected via the local network of the master station. In fact a network management system manages the following functions in a SCADA control centre.
1. Network device discovery,
2. Network device monitoring,
3. Network performance analysis,
4. Network device management, and
5. Intelligent notifications or customizable alerts.

Dispatch Training Simulator (DTS) server

DTS is also an important server in Power System SCADA. A Dispatcher Training Simulator (DTS), also known as an operator training simulator (OTS), is a computer-based training system for operators/ dispatchers of electrical power grids, is generally available at a large master station. It is used to train the dispatchers who manage the system. The DTS generally provides the power system model, hydro system model, control centre model, and instructor functions. It performs this role by simulating the behaviour of the electrical network forming the power system under various operating conditions, and its response to actions by the dispatchers. Trainees may therefore develop their skills from exposure not only to routine operations but also to adverse operational situations without compromising the security of supply on a real transmission system. Modern DTS combines or simulates the elements such as Energy Management System (EMS), Distribution Management Systems, SCADA system, load-flow study, and facilities for modeling and optimizing the economic dispatch of generating units.

Communication Front End (CFE)

Communication front end (CFE) interfaces the host computer to a network or peripheral devices. CFE is used to unburden the host computer input and output communications such as managing the peripheral devices, transmitting and receiving message packet assembly and disassembly, and error detection and error collection. This CFE, often referred to as the FEP (front-end processor) must be a high end machine with high computing capabilities so that the communication with a large number of the peripheral devices could be carried in real-time. FEP communicates with the host computer using a high-speed parallel interface. Obviously these computers are costly. FEP/CFE is synonymous with the communication controller. In power system SCADA, CFE has the role of communicating with all RTUs, other field devices which mostly connected through FRTUs, RVDU (Remote Video Display Units), etc.

ICCP server

ICCP server is an important one in Power system SCADA. This inter-control centre protocol server supports bulk data transmission between the master stations depending on the hierarchy. Typically, the ICCP server at a Regional Load Dispatch Centre (RLDC) exchange data between RLDC and National Load Dispatch Centre (NLDC) as well as with the State Load Dispatch Centre (SLDC) and sub-LDC if required but maintaining the hierarchy.

Video Projection System (VPS)

In large master control stations, especially in power industries where SCADA is employed, the distribution or transmission network is displayed using a video projection system. Master stations must be equipped with high-tech display systems that can display the area of control in a diverse manner as per the requirement of the operators. This display can be GIS information, a Single Line Diagram (SLD) of the network of the region of interest and a separate video projection system handles this function. Supporting software helps in tracing the distribution and transmission network.

2.3.11 GLOBAL POSITIONING SYSTEMS (GPS)-RELEVANCE TO SCADA

A space based radio navigation system which can provide time information and relocation to a GPS receiver which is placed on or near the Earth surface which has four or more unhindered Line of Sight(LoS) to four or more GPS satellites. There are 14 to 31 geostationary satellites in the mid orbit of the earth which are owned by United States and operated by US Air Force.

Presently time synchronization for all Industrial SCADA components especially in PSS and Smart Grid including the master control station are achieved with the GPS clocks. The GPS system does not require the user to transmit any data, rather it continuously transmit signals to the earth which contain information about the time of transmission and the satellite locations at the time of transmission. On receiving these signals, GPS receivers compute the distance to the satellite based certain equations, the location of the receiver and the time. GPS time is theoretically accurate to about 14 nanoseconds. However, most receivers lose accuracy in the interpretation of the signals and are only accurate to 100 nanoseconds. In PSS and Smart Grid, GPS plays a very vital role, as all the time stamped events, SoE, IED and PMU measurements, etc. critically needs time synchronization.

2.3.12 HUMAN MACHINE INTERFACE (HMI)

Human Machine Interface (HMI) or user interface (UI) is a combination of software and hardware that allows the interaction between the humans and the system (machine) in a SCADA environment. The main objective of this interaction is effective operation and control of the system being monitored. This aids the operator to make operational decisions and manually override automatic control operations in the event of an emergency. In local and small SCADA control centres, HMI allows a control engineer or operator to configure set points or control algorithms and parameters in the controller. In medium and large SCADA control centres, it also provides process status information, historical information, reports, and other information to operators, administrators, managers, business partners, and other authorized users.

Today the devices and instruments that an operator used in PSS have changed significantly from manual to computer-based devices. The latest expensive hardware, processors with high computing capability, dedicated software, mimic diagrams, and communication protocols with security features especially with end node security have made the system very efficient, compact and human friendly. The location, platform, and interface of HMI may vary a great deal. Further present day HMI could be a dedicated platform in the control centre, a laptop on a wireless LAN, or a browser on any system connected to the Internet.

HMI building blocks
In a SCADA system, the HMI components include operator console, operator dialogue, mimic diagram and peripheral devices, etc. are briefly explained below.

Operator console
The console where the operator monitors and controls the system is of utmost importance and includes the visual display units, alphanumeric keyboard, cursor, communication facilities, ETC. The Visual Display Unit (VDU) includes UI devices like the multiple color monitors (CRT, LCD, LED devices minimum size), with glare-reduction features (antiglare screen coatings) and should provide a display of multiple viewports (windows) on each monitor. The cursor control could be mouse, trackball, or the latest touch-screen facility. A keyboard and cursor pointing device are shared among all monitors at each console and the cursor moves across all screens without switching by user. Generally for power system SCADA each operator has three to four monitors for proper planning and multiple views and the displays should have full graphics capability with zoom facility. Audible alarms are also a prominent feature of the operator console where the operator is informed of the severity of an event in the system. The design of the operator console infrastructure including the table and chair for the operator is important and should follow ergonomics principles to make the operator comfortable during the duty period.

Operator Dialogue
Operator dialogue box is a box that pops up to enable communication between the computer and the operator. Dialogue boxes may ask you questions or give you information. The operator dialogue and commands should be

simple and easy to remember. Certain boxes may ask for an action relating to the application which is being used. Function keys of the keyboard can be programmed to incorporate major actions so that the operator can give the commands effortlessly rather than typing long dialogues.

Mimic Diagram

The mimic diagram is an essential part of any control centre or large master station where the operator and the personnel in charge get an overall view of the plant/process under control. This includes LCD/LED large-screen display with full SCADA operability with multiple screens possible. Some control centres have mosaic map board with dynamic or static tile map board and dynamic map board lamps updated by SCADA. Present trend is to use multiple video projection cubes as the dynamic map boards. The major benefit is that when the HMI is updated with system changes, the map board is also automatically updated, since the HMI drives the map board directly.

Peripheral Devices

Generally three printers are required in Master Control Centres. One of them is used to print alarms and SoEs. Another printer usually a colour printer, which is used for capturing screen shots. A black and white laser printer is used to print reports.

HMI Software Functionalities

Selecting HMI software typically starts with an analysis of product specifications and features. The key considerations can include the system architecture, performance requirements, integration and cost of procurement and operations which are described below.

1. The HMI console should have Role Based Access Control of security to protect unauthorized access to the system. As this HMI is one of the end node of the SCADA system, specific user identification (IDs) and passwordmust be used.
2. The display of power system to control may be in an effective way. There should be a provision to display all the information about the power system interconnection and the parameters of interest in the HMI, such as voltage, current, frequency, and power flow. This must be in an user friendly manner so that even a new operator could monitor and analyze the events and take corrective and control action, if needed.
3. HMI software should have the capability of preparing Logs and Reports, and Calculated Values. These reports are required many times to present

to various system hierarchies and to different departments of the utility. Today more sophisticated HMI software is structured around mobile, portable platforms. This presents a cost-saving value as the operating systems are distributed on machine-level embedded HMI, solid-state open HMI machines, distributed HMI servers and portable HMI devices.

Situational Awareness and Alarm Handling

Situational Awareness: Situational awareness is be aware of what is happening around the operator in the control room, so that the he will be able to decide how to react to the events. The operator in a control room should observe the environment, comprehend the situation, and take decisions and actions accordingly. Operators dealing with critical functions play an integral role in the safe and efficient operation of the system. Situational awareness is important when the information flow is fast and the error in judgment will lead to major consequences. In power systems, when a blackout suddenly occurs an appropriate action should be taken on time, which needs a rapid decision making and execution. Compared to other SCADA systems, operating a Power System SCADA (PSS) is more complex and mentally demanding, and thus PSS operators are more vulnerable to human errors.

HR experts consider situational awareness, as very important one, being the perception of elements in the environment within a volume of time and space, the comprehension of their meaning, and the projection of their status in the near future. Special considerations to mitigate operator errors should be taken when designing an operator-assistance system for PSS. A model developed by psychologists includes three levels viz.

1. *Perception level:* In this level the person has to perceive the status, attributes, and dynamics of the variables in the environment,
2. *Comprehension level:* In this level the data from level one have to be synthesized using interpretation, pattern recognition, and evaluation skills, and
3. *Projection level:* Here the person can extrapolate the information from the lower levels and arrive at an action plan.

Operators may make errors when they are not totally aware of the actual situation. So one of the important objectives is to equip the Master Control Centre with appropriate visual aids to meet the required perception level. Provide enough data processing and display systems for the comprehension level and finally enable the operator to take a decision and execute it. Today, SCADA software visualization and display have undergone a major transformation with new devices and tools available for increasing the perception and comprehension levels of the operators, to equip them to make better decisions at the projection level.

Alarm Handling: Once the data are communicated to the control centre, they are first processed before presentation to the operator. The processed data are then compared with the predefined values and in case of any deviation from the nominal value, an alarm is generated. The operator acknowledges the alarm and appropriate remedial action is taken. Thus, important functions of the control centre allow it to generate, annunciate, and manipulate the process and system alarms. The generation and display of the alarm and the limits are important functions of the control centre, and this information needs to be communicated to the interconnected system in case of emergency, limit violations, and malfunctions. Sometimes, the operator is confused by the series of alarms triggered by a single event, as many quantities change and unnecessary alarms of all kinds are directed to the operator. Hence, alarm filtering is important and is explained in the following section.

Intelligent Alarm Filtering

Alarm handling and processing technology ensures that the SCADA operators receive only those alerts which are most relevant to events that must be addressed immediately. The details of less critical secondary warnings are sent to databases and possibly printed for later review. With only the most important distribution system alarms presented in a prioritized fashion, SCADA operators can assess problems more easily and make better decisions to prevent a crisis.

In a plant or process which is controlled by SCADA system, if the number of alarms annunciated are too large so that it can't be managed by the human operator leading to upset the plant or process operations is generally refereed as alarm flooding. This demands an effective alarm management technique to ensure that the alarms are presented to the operator at a rate which can be assimilated by the human operator.

In PSS especially in SCADA/DMS alarms are typically triggered by faults and the events surrounding them, which occur continuously during routine operations. When a breaker on a substation feeder trips due to a transient fault, for instance, up to seven alarms may be triggered: one for the breaker trip and three each when voltages and currents on all three phases hit zero. The dispatcher needs only the breaker trip alarm and does not need any alarm information if the breaker is automatically reclosed after a transient fault since the situation resolves itself.

Today, many industrial SCADA vendors prioritise the alarms as primary and secondary alarms depending on the urgency of action by the operators. Primary alarms require urgent operator action while secondary alarms need no operator action. Certain SCADA vendors developed alarm filtering techniques, so that alarms can be configured during SCADA implementation or activated during alarm flooding.

Necessities and Requirements of Operators

The SCADA master control centre is the place from where the operator monitors the plant or process operations and issues necessary control commands. Operator on duty in industrial SCADA has to spend long hours with a good presence of mind to monitor and control the plant or process operations. An ergonomic and aesthetic design of the control centre is a must to ease the burden of the operator and to avoid stress in vital body parts. An appealing ambience has to be created and the operator console design meets the standards so that, the operator is comfortable to perform his/her responsibilities with concentration and stress free. Some of the needs and requirements of the operator in the control room, to meet the functional objectives, are summarized below.

1. A pleasant ambience in the control room must be provided so that the operator working for long durations gets facilities, such as entertainment, exercise, refreshments, time for rest, etc,
2. The operator in the SCADA control centre has to handle many I/O devices through different consoles. Hence the I/O devices must be arranged within reach, and the operator may be able to move from one device to another very effortlessly, if needed,
3. Appropriate safety slogans may be displayed in the control room for reminding the operators for carrying out safe, secure and error free operations to avoid fatal and non-fatal accidents,
4. User- friendly operation and control system must be provided for smooth and error free operation,
5. Comfortable access to the control devices to the operator must be ensured and the alarms and indicating displays must be easy to locate,
6. The Video Projection System (VPS) which display the mimic and the other related information must be easily readable and also provides accurate and proper information to the operator.
7. The operator console, the input devices, the display units, etc. must be arranged in such a manner to provide comfortable eye-hand coordination approach for handling the equipment, and

8. In order to improve the working efficiency of the operator and reduce tiredness, provide comfort in the use of devices, stability and reliability of the equipment etc.

SUMMARY

This chapter gives an introduction to SCADA with an emphasis on Data Acquisition Systems(DAS) and its components. The objectives and advantages as well as the evolution of SCADA are briefly discussed but with clarity. A brief discussion of the various components of DAS such as Sensors, Signal conditioners, Sample and Hold circuits, Analog to Digital Converters, etc. are also given in this chapter. A brief discussion regarding the selection criteria of the DAS is then given before moving to give a comprehensive description of Remote Terminal Units (RTU), Programmable Logic Controller (PLC) and the different modern components of Power System SCADA such as Intelligent Electronic Devices, Data Concentrator Units, Merging Units, Human Machine Interface, etc. The brief introduction of Data Concentrators and Merging Units are presented in such a manner that how digital substations can be designed. This chapter then gives an introduction of architecture of SCADA Master Stations and its hardware and software components. It concludes with a brief description of Geographical Positioning System (GPS), Situational Awareness and Alarm Processing.

CHAPTER 03

SYSTEM DESIGN CONSIDERATIONS

3.1 INTRODUCTION

Various SCADA architectures are adapted today depending on the requirements. It starts with the basic single channel Data Acquisition and Control architecture to the complex multi-channel, real-time, fault tolerant, fail safe system which utilizes many secure communication methodologies with proper End Node Security (ENS). The following sections elaborate the various SCADA architectures with emphasis on Distributed SCADA.

3.2 COMMUNICATION ARCHITECTURE

There are three physical communication architectures which are generally popular and deployed in SCADA systems. In certain cases they are deployed in a combined mode. They are,
1. Point to point,
2. Point to multistations, and
3. Relay Stations.

3.2.1 Point-To-Point between Two Stations

This is the simplest configuration where data is exchanged between two stations. One station can be setup as the master and one as the slave as shown in Figure 3.1. It is possible for both the stations to communicate in full duplex mode (transmitting and receiving on two separate frequencies) or simplex with only one frequency.

3.2.2 Multipoint or Multiple Stations

In this configuration, there is generally one master and multiple slaves which is shown in Figure 3.2. Generally data points are efficiently passed between the master and each of the slaves. If two slaves need to transfer data between each other they would do so through the master who would act as arbitrator or moderator. Alternatively, it is possible for all the stations to act in a peer-to-peer communications manner with each other. This is a more complex arrangement requiring sophisticated protocols to handle collisions between two different stations wanting to transmit at the same time.

Another possibility is the store and forward relay operation. This can be a component of other approaches discussed above where, one station retransmits messages onto another station out of the range of the first station which is shown in Figure 3.3.

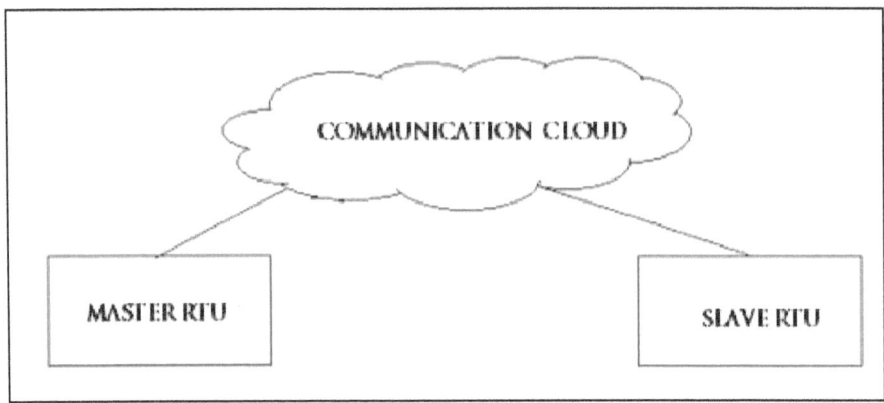

▲ **Figure 3.1:** Point to Point Master-Slave Communication

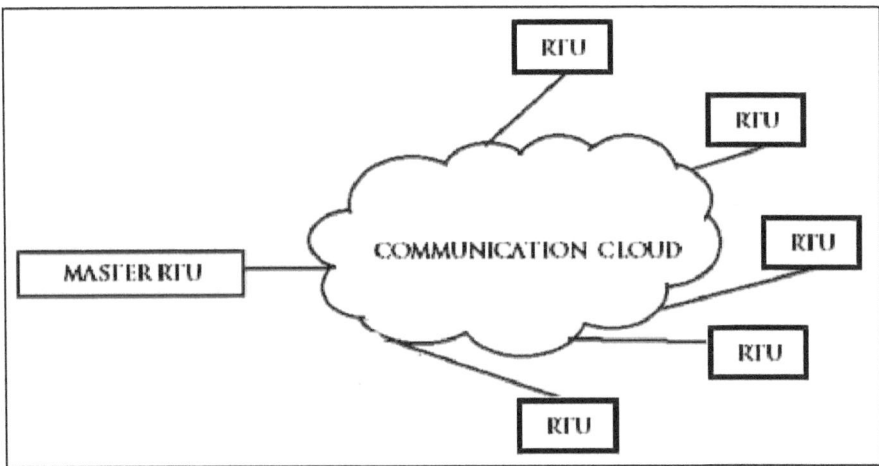

▲ **Figure 3.2:** Master RTU communicating with multiple station RTUs

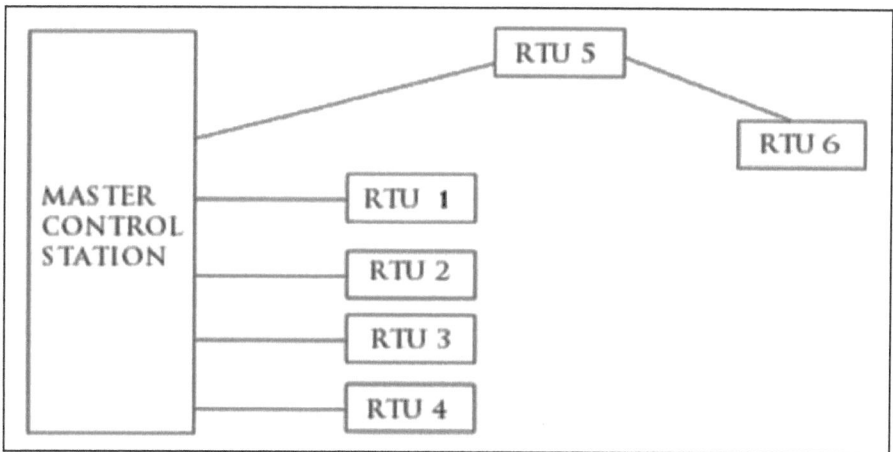

▲ **Figure 3.3:** Store and forward station

3.2.3 Talk through Repeaters

This is the generally preferred way of increasing the range of radio systems. This retransmits a radio signal received simultaneously on another frequency. It is normally situated on a geographically high point.

The repeater receives on one frequency and retransmits on another frequency simultaneously. This means that all the stations repeating the signal must receive and transmit on the opposite frequencies. It is important that all stations communicate through the talk through repeater. It must be a common link for all stations and thus have a radio mast, high enough to access all RTU sites. It is a strategic link in the communication system; failure would wreak havoc with the entire system. The antenna must receive on one frequency and transmit on a different frequency which is shown in Figure 3.4. This means that the system must be specifically designed for this application with special filters attached to the antennas. There is still a slight time delay in the transmission of data with a repeater. The protocol must be designed with this in mind with sufficient lead-time for the repeater's receiver and transmitter to commence operation.

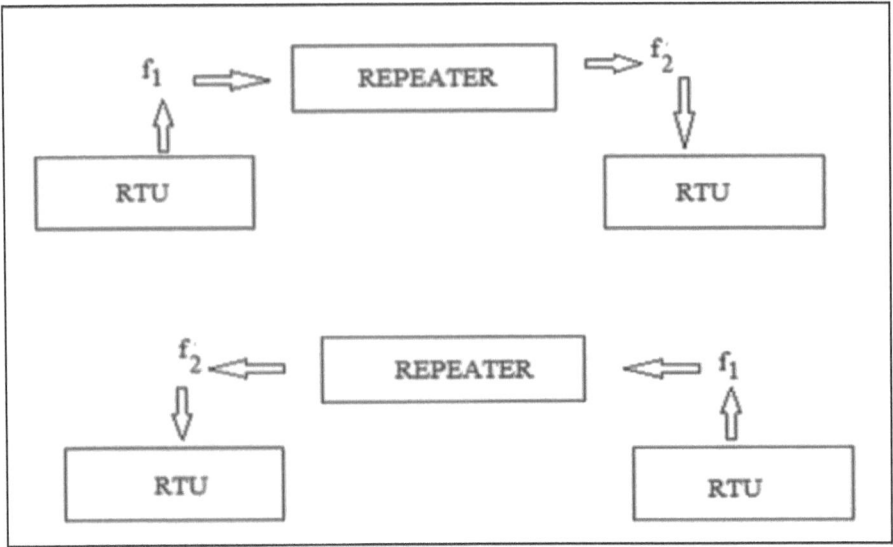

▲ **Figure 3.4:** Talk through repeaters

3.3 COMMUNICATION PHILOSOPHIES

There are two main communication philosophies in practice. These are polled (or master slave) and carrier sense multiple access/collision detection (CSMA/CD). The one of the notable methods accepted for reducing the amount of data that needs to be transferred from one point to another is to use exception reporting.

3.3.1 Polled or Master Slave

This is one of the simplest master- slave point to point or point to multipoint configuration where the master polls, the slave in predetermined regular intervals and gather the data which is shown in Figure 3.5. Here the master is in total control of the monitoring, decision making and control of the process. The slaves never initiate the data exchange rather, wait for the master's request and respond. Essentially it is a half-duplex communication. In case the slave fails to respond timely, then the master retries typically two or three times. If the slave still fails to respond, the master then moves to the next slave in the sequence with a remark that the particular slave is inactive or faulty and requires attention for rectification.

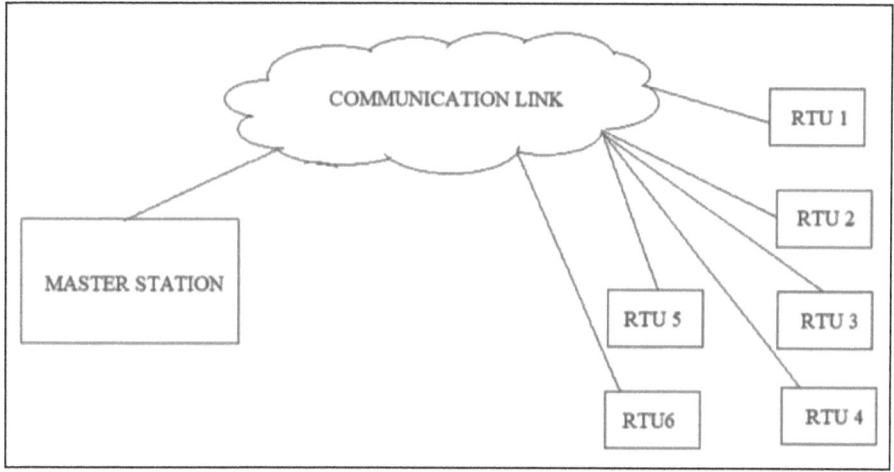

▲ **Figure 3.5**: Illustration of master-slave polling with RTUs

The advantages of this approach are:
1. Software development is fairly easy and can be made reliable due to the simplicity of the philosophy,
2. Link failure between the master and a slave node is detected fairly quickly,
3. No collisions can occur on the network, hence the data throughput is predictable and constant, and
4. For heavily loaded systems, each node has constant and bulk data transfer requirements giving a predictable and efficient system.

However it has the certain disadvantages which are described below.
1. Variations in the data transfer requirements of each slave cannot be handled,
2. Interrupt type requests from a slave cannot be entertained as the master may be either attending or processing some other slave request or data,
3. Systems, which are lightly loaded with minimum data changes from a slave, are quite inefficient and unnecessarily slow, and
4. Communication between the slaves can be achieved only through the master which adds complexity.

Two applications of the polled or master slave, approach are given in the following two implementations. This is possibly the most commonly used technique and is illustrated in the Figure 3.6 below. The master station has a polling list with details of priority and sequence. Based on this information it prepares a polling cycle as shown in Table 3.1.

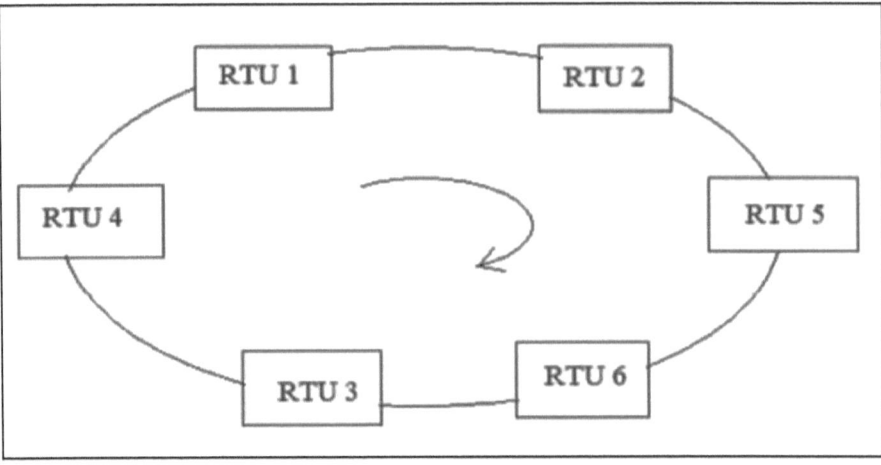

▲ **Figure 3.6:** Polling cycle as per the polling table

▼ **Table 3.1:** Polling table with the master station

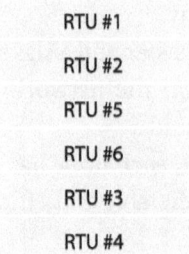

| RTU #1 |
| RTU #2 |
| RTU #5 |
| RTU #6 |
| RTU #3 |
| RTU #4 |

In the scheme of polling, in certain situations, the polling may be modified as,
1. if there is no response from a given RTU during a poll, a timeout timer has to be set and three retries (in total) initiated before flagging this station as inactive, and
2. If an RTU is to be treated as a priority station it will be polled at a greater rate than a normal priority station. It is important not to put too many RTUs on the priority list, otherwise the differentiation between high and normal priority becomes meaningless.

3.3.2 CSMA/CD System (Peer-To-Peer)

3.3.2.1 RTU to RTU Communication

In certain situations, especially in DCS an RTU in the SCADA system may need to communicate with another RTU. One of the solutions is, while responding to the master station to the poll, a message carrying a request with the destination address of the RTU can be added. The master station will then examine the destination address field of the message received from the RTU and retransmit onto the appropriate remote station. The only attempt to avoid collisions is to, listen to the medium before transmitting. If a collision occurs, the RTU wait for a random time period, and retransmit the data avoiding the collision. In this style of operation, it is possible for two nodes to try and transmit at the same time, with a resultant collision. In order to minimize the chance of a collision, the source node first listens for a carrier signal before commencing transmission. Unfortunately this does not always work where certain stations which cannot hear each other to try and transmit back to the station simultaneously.

3.3.2.2 Exception Reporting (Event Reporting)

On many occasions, the status and the RTU may be the same but the polling mechanism gathers the data from the RTU. This unnecessary transfer of data can be minimized or virtually eliminated with a technique called *exception reporting*. This approach is popular with the CSMA/CD philosophy but it could also offer a solution for the polled approach where there is a considerable amount of data to transfer from each slave.

In exception reporting, the remote station reporting devices such as RTUs monitor itself to identify a change of state or data. If there is a change of state, the remote station writes a block of data to the master station when the master station polls the remote. Typical reasons for using polled report by exception include:
1. The polling or scanning is performed with a low data rate due the communication channel constraints,
2. There is substantial data being monitored at the remote stations, and
3. The number of remote devices connected to master station is reasonably high.

Each analog or digital point that reports back to the central master station has a set of exception reporting parameters associated with it. The type of exception reporting depends on the particular environment but could be:
1. High and low alarm limits of analog value
2. Percent of change in the full span of the analog signal
3. Minimum and maximum reporting time intervals

The main advantages of this approach are quite clearly to minimize unnecessary (repetitive) traffic from the communications system.

3.3.2.3 Polling plus CSMA/CD With Exception Reporting

A practical method to combine all the approaches discussed earlier is to use the concept of a slot time and exception reporting. Here each slave station is assigned a specific time slot comprising the following sub-slot assuming that there is no requirement for communication between the slaves.
1. A slave transmitting to a master and
2. A master transmitting to a slave.

A slot time is calculated as the sum of the maximums of modem up time, plus radio transmit time, plus time for protocol message, plus muting time of transmitter. The master commences operations by polling each slave in turn. Each slave will be in synchronize with the polling and respond to the master with exception reporting if there is a status change, else respond with passive acknowledgement. As a result, the master move on to poll the next slave by overriding the remaining sub-slots. Otherwise it will complete the data exchange and then move to hear from the next slave. The master thus completes the poll cycle.

The previous and present chapter elaborated the building blocks of SCADA systems starting from the RTU, IEDs, communication systems, master stations and the HMI. Utilities have a variety of options available to mix and match the elements to building a cost-effective, efficient, and operator-friendly SCADA system as per their requirements.

Automation of the power systems started as early as the beginning of the twentieth century, and substations and control centers operate at various stages of automation all over the world. There are legacy systems with RTUs, hardwired communication from the field to the RTU, and traditional software functionalities in the control room, and it is not often financially viable to dismantle everything and purchase a completely new automation system.

Hybrid systems are a viable option, where any automation expansion project can be implemented with new devices, like IEDs, data concentrators, and merging units. The new system will coexist with the legacy RTU-based systems and the data integration and if necessary protocol conversion issues will have to be handled while commissioning the project.

If a utility decides to purchase a completely modern system, the latest building block of the SCADA system, viz, IEDs, merging units, and fiber optic communication facility with brand new HMI with situational awareness and analysis tools, can be implemented.

3.4 SYSTEM RELIABILITY AND AVAILABILITY

Real-time system operation demands high level of availability and reliability to ensure error free operations. The real-time process control systems directly control the process, and any system catastrophe or erroneous operation may lead to process damage and may affect the safety of operations. Thus these applications necessitate a high level of system reliability without compromise. Numerically system reliability is defined as the probability that the system will not fail under specified conditions.

Standards and guidelines are available for the development of safe and secure critical systems. Among these standards, the most pertinent standard for power system automation and is the NERC CIP. These development standards may be used while designing a power system SCADA starting from peripheral devices to the control center. In keeping with these guidelines, an analysis must be performed as the early stage of system design in order to assign a system integrity level which allows the utility to define the accepted failure rate of system under consideration. Before discussing the fail safe system, it is better to have a clue of the two classification of failures namely Common Cause Failures and Common Mode Failure.

Common Cause Failure (CCF): A Common cause failure occurs when two or more items fall within a specified time such that the success of the system mission would be uncertain. Item failures result from a single cause and mechanism.

Common Mode Failure(CMF): A Common-Mode Failure is the result of an event(s) which because of dependencies, causes a coincidence of failure states of components in two or more separate channels of a redundancy system, leading to the defined systems failing to perform its intended function.

3.4.1 Fail Safe Systems

A fail safe system describes a feature or a device which in the event of failure, responds in a way that will cause no harm or minimum harm to other devices or danger to personnel. Fail safe systems are used wherever the highest degree of safety needs to be guaranteed for humans, machines and the environment. A system being failsafe means, not that failure is impossible or improbable, but rather that the system design prevents or mitigates unsafe consequences of the system failure. That is, if and when a fail-safe system fails in any case, accidents and damage as a result of fault must be evaded at all costs. Thus when controlling, dangerous or critical machinery, it is necessary to device and implement a fail-safe strategy to ensure that the machine operates safely even when, the elements of the control hardware or software fail.

The types of failure are many hence a thorough analysis of the failure mode and effects are used for identifying the failure situations and design safety procedures. Some systems can never be made fail safe, as continuous availability is needed. Redundancy, fault tolerance, or recovery procedures are employed in these situations. This also makes the system, less sensitive for the reliability prediction errors or quality induced uncertainty for the separate items. On the other hand, failure detection, correction and avoidance of CCF become increasingly important to ensure system level reliability.

1. In industrial automation, usually alarm circuits are normally closed. This ensures that in case of a wire break the alarm will be triggered. If the circuit were normally open, a wire failure would go undetected, while blocking actual alarm signals.
2. In control systems, critically important signals can be carried by a complementary pair of wires. Only states where the two signals are opposite (one is high, the other low) are valid. If both are high or both are low, the control system knows that something is wrong with the sensor or connecting wiring. Simple failure modes such as dead sensor cut or unplugged wires, are thereby detected. An example would be a control system reading both the normally open (NO) and normally closed (NC) poles of a SPDT selector switch against common, and checking them for coherency before reacting to the input.

3.4.2 Fault Tolerant Systems (FTS)

Certain utility or plant operations, once started have to continue as uninterrupted and nonstop operations, even in the case of a hardware or equipment failure. In other way, these are fault tolerant systems which has the ability of preventing a catastrophic let-down, that could result from a single point of failure. A fault-tolerant system is designed from the ground up for reliability by building multiples of all critical components, such as CPUs, memories, disks and power supplies into the same computer. In the event one component fails, another should take over without skipping a single point of operation. A feasible strategy for a fault tolerant system is by anticipating exceptional conditions and design a system to cope with them. The basic aim is that the system should be able to self-stabilize after the fault and converge towards an error free state. The above strategy may not be successful in some cases. In such applications, fault tolerance is implemented by providing redundancy.

3.4.3 Graceful Degradation Systems

Fault tolerance is often used synonymously with graceful degradation, although the latter is more aligned with the more holistic discipline of fault management, which aims to detect, isolate and resolve problems pre-emptively. A fault-tolerant system swaps in backup componentry to maintain high levels of system availability and performance. But Graceful degradation allows a system to continue operations, with a reduced state of performance.

3.4.4 Design Considerations for Fault Tolerant Systems

While designing a Fault Tolerant System, one may consider the business continuity requirements, disaster recovery plan, the disaster recovery products presently available in the market, and of course the budget and available manpower for engaging. Nevertheless Fault-tolerant systems are designed to compensate for multiple failures. The failure point has to be specifically identified, and a backup component or an immediate procedure should be taken its place with no loss of service. The failure point can be computer processor unit, I/O subsystem, memory cards, motherboard, power supply, network components in the cloud, communication servers of the service provider, etc. Hence each and every stage should be thoroughly analyzed and necessary provisions may be envisaged and implemented.

In a software implementation, the operating system (OS) provides an interface that allows a programmer to checkpoint critical data at predetermined points within a transaction. In a hardware implementation, the programmer need not to be aware of the fault-tolerant capabilities of the machine.

At a hardware level, fault tolerance is achieved by duplexing each hardware component. Disks are mirrored. Multiple processors are lock-stepped together and their outputs are compared for correctness. When an anomaly occurs, the faulty component is determined automatically, and is taken out of service, but the machine continues to function as usual.

3.4.5 High Availability

Fault tolerance is closely associated with maintaining business continuity via highly available computer systems and networks. Fault-tolerant environments are defined as those that restore service instantaneously following a service outage, whereas a high-availability environment strives for setting up of independent servers coupled loosely together to guarantee system-wide sharing of critical data and resources. The loosely coupled clusters monitor each other's health and provide fault recovery, to ensure applications remain available. Conversely, a fault-tolerant cluster consists of multiple physical systems that share a single copy of a computer's OS. Software commands issued by one system are also executed on the other system. Systems with integrated fault tolerance incur a higher cost due to the inclusion of additional hardware.

3.4.6 Critical Functions

If any operation of plant or process is critical, i.e. if the downtime costs are high or cause any human fatality or any expensive hardware destruction, redundancy must be incorporated into the system to eliminate system failures, due to equipment failure. Such functionalities are categorized as mission critical functions. In fact it is the privilege of the critical function to have software and hardware redundancy to be fault tolerant in a SCADA system. In power system SCADA with remote access RTU sites, both system redundancy and channel redundancy must be ensured for critical functions. In such cases RTU with CPU redundancy and communication port redundancy are highly recommended. At the master station end, the routers, firewalls, and servers must be configured to have redundancy for a fault tolerant system. Mission-critical installations often have separate power sources in case of a power failure, and installations in areas prone to natural disasters or the threat of fire, keep the servers in different geographic locations. However, whatever type of disaster recovery is planned for, it is possible to greatly reduce lost data and downtime by planning the proper system design, and by choosing a SCADA system with built-in redundancy. While ensuring system redundancy requirement for critical functions to be fault tolerant, it is highly recommended to consider the following points.

1. Dual networks for full LAN redundancy,
2. Redundancy can be applied to specific hardware,
3. Supports primary and secondary equipment configurations,
4. Intelligent redundancy allows secondary equipment to contribute to processing load,
5. Automatic changeover and recovery,
6. Mirrored disk I/O devices,
7. Mirrored alarm servers, and
8. File server redundancy.

A typical redundant and fail safe DCS master control station configuration with HA is shown in Figure 3.7 below. The sub components such as DMZ, SCADA Control System LAN, Dispatch Training Simulator, and Fail Safe connectivity to Smart Grid MCC using HA are shown separately in Figure 3.8, Figure 3.9, Figure 3.10 respectively. Here a De Militarized Zone(DMZ) is designed carefully and kept exclusively between two Firewalls of different make. The session from the servers of DMZ to the servers of SCADA zone is strictly made forbidden by appropriately defining the Firewall ruleset.

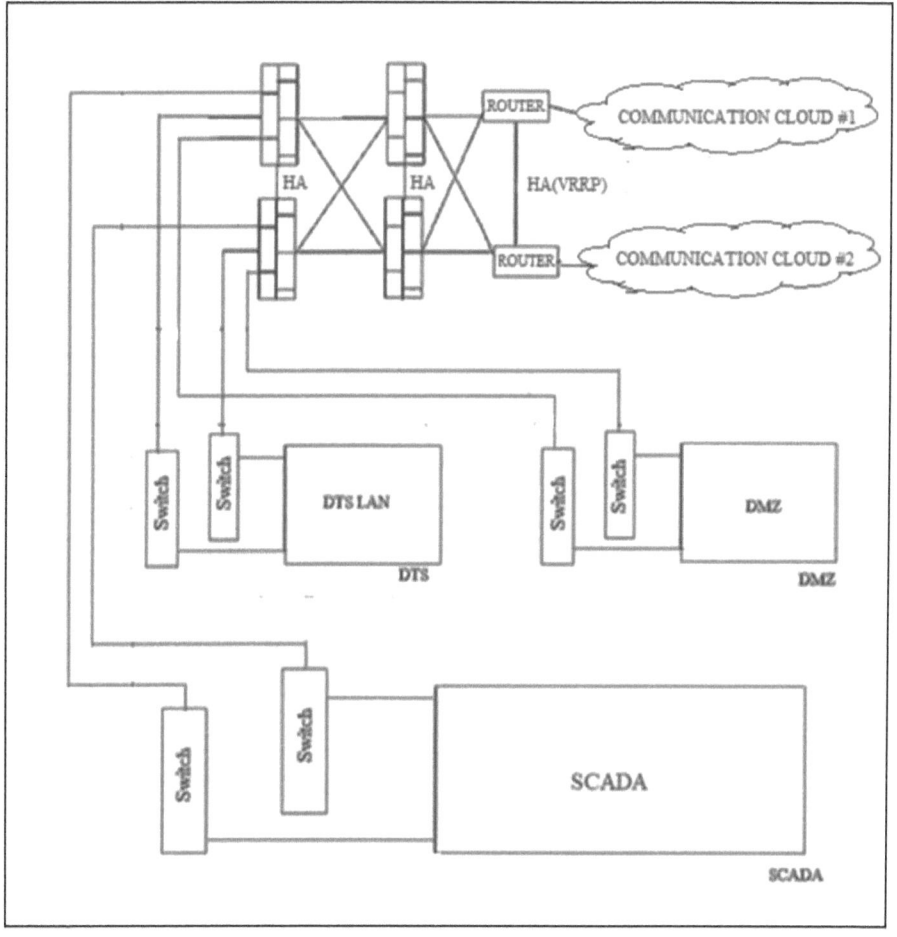

▲ **Figure 3.7:** Block diagram of a typical DCS MCC with Fail Safe HA connectivity

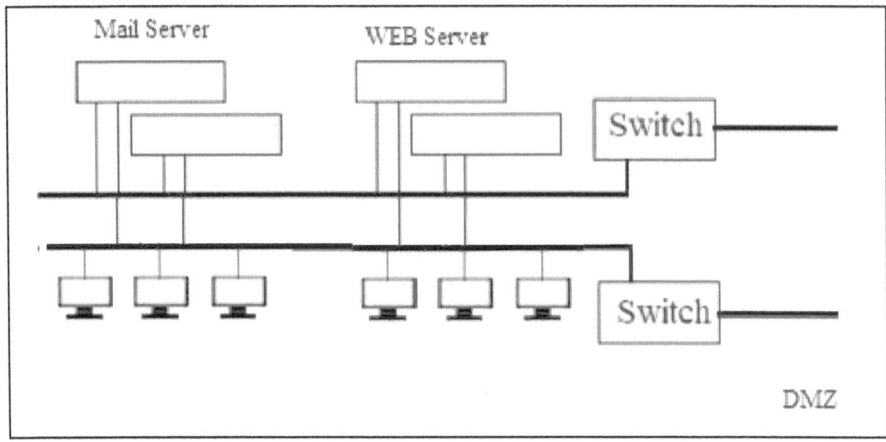

▲ **Figure 3.8:** De Militarized Zone (DMZ)

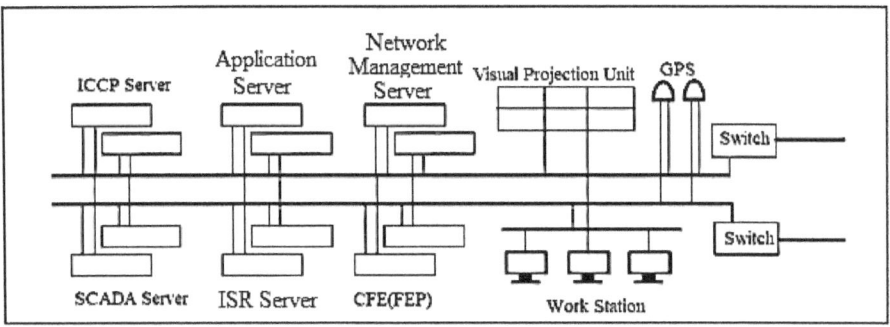

▲ **Figure 3.9:** SCADA Control System LAN

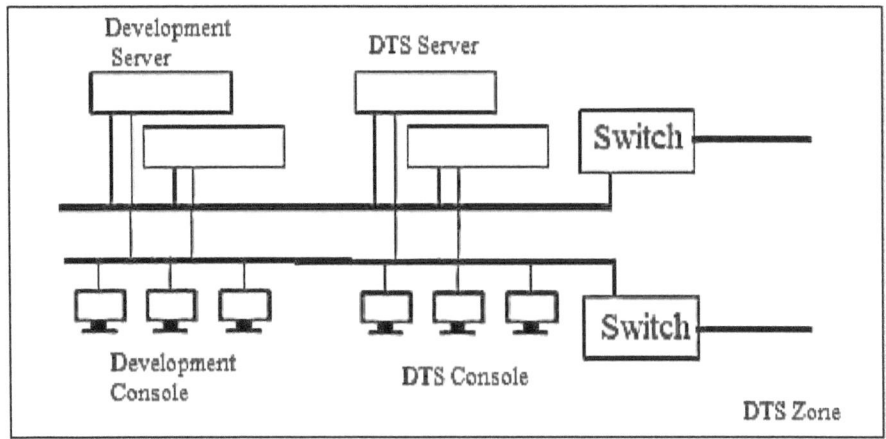

▲ **Figure 3.10:** Dispatch Training Simulator

Dispatcher Training Simulator (DTS) is a subsystem within Distribution Management System (DMS) that operates separately from the real-time system and provides a realistic environment for hands-on dispatcher training under simulated normal, emergency, and restorative operating conditions. The training is based on interactive communication between instructor and trainee. The DMS training simulator serves two main purposes.

1. Allowing personnel to become familiar with the DMS system and its user interface without impacting actual substation and feeder operations.
2. Allowing personnel to become familiar with the dynamic behavior of the electric distribution system in response to manual and automatic actions by control and protection systems during normal and emergency conditions.

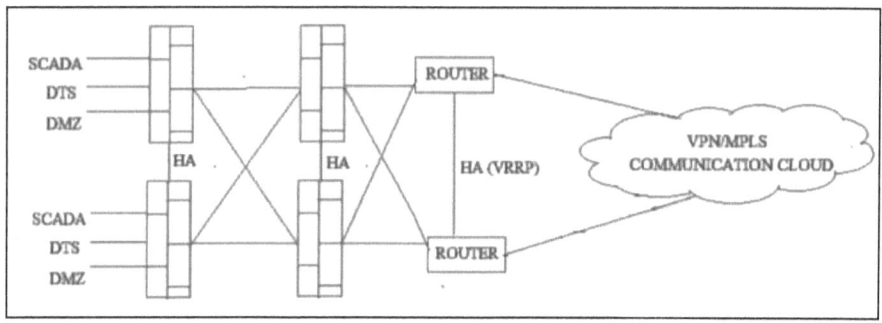

▲ **Figure 3.11:** Fail Safe connectivity to DCS MCC using HA

The Figure 3.11 describes a Fail Safe connectivity to DCS MCC using HA which is mostly preferred in the Power System SCADA and DCS which control the critical operations.

3.4.7 System Redundancy

Redundancy is the hallmark of fault tolerant systems. It is defined as additional or alternative systems, sub-systems, assets, or processes that maintain a degree of overall functionality in case of loss or failure of another system, sub-system, asset, or process. Normally power system SCADA redundancy starts from the data capturing units such as RTU or PLC. If the SCADA is a distributed system, then every system component upto the MODEM have to be redundant. Further if the communication to the master control station is by means of unguided media, then system should have antenna redundancy as well. Obviously this adds cost considerably and the implementing agencies/vendors normally find pretexts for compromise which is the general lapse observed in implementing power system SCADA. But this makes the SCADA system unreliable.

Definitely true fault tolerant SCADA systems with redundant hardware are the most costly as the additional components add to the overall system cost. However, fault tolerant systems provide the same processing capacity after a failure as before, and ensures safety.

3.4.8 Channel Redundancy

Most of the power system SCADA is distributed in nature and geographically separated. Hence remote site to site communications are required. This mainly depends on third party communication service providers.

While selecting the third party communication providers, channel redundancy has to be ensured to have true system redundancy. Mainly this is achieved in two ways which is explained below.

If the third party communication provider is highly reliable and having a true self-healing communication cloud as per the requirement of the utility, then connectivity to the two geographically different nodes is an acceptable solution on economic reasons. However in the strictest sense, true channel redundancy for critical functions ideally demands two different communication providers with active-active or active-standby mode. These are shown in Figure 3.12 and 3.13.

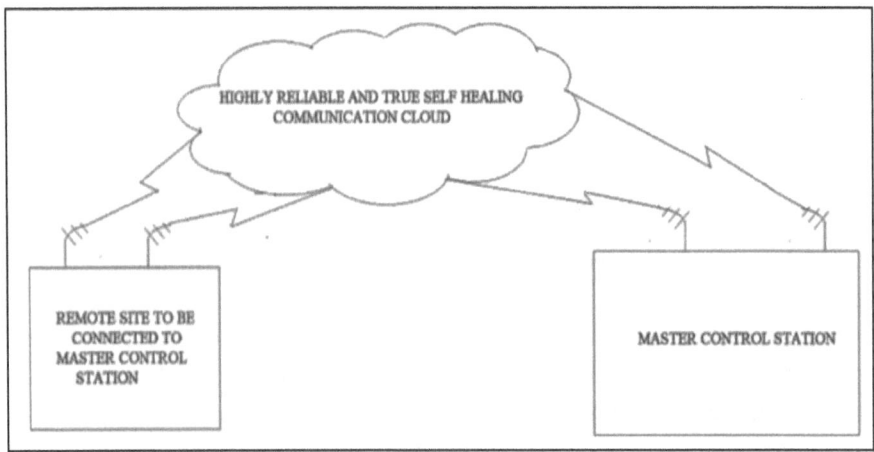

▲ **Figure 3.12:** Channel redundancy achieved with a resilient and reliable cloud

SYSTEM DESIGN CONSIDERATIONS • 81

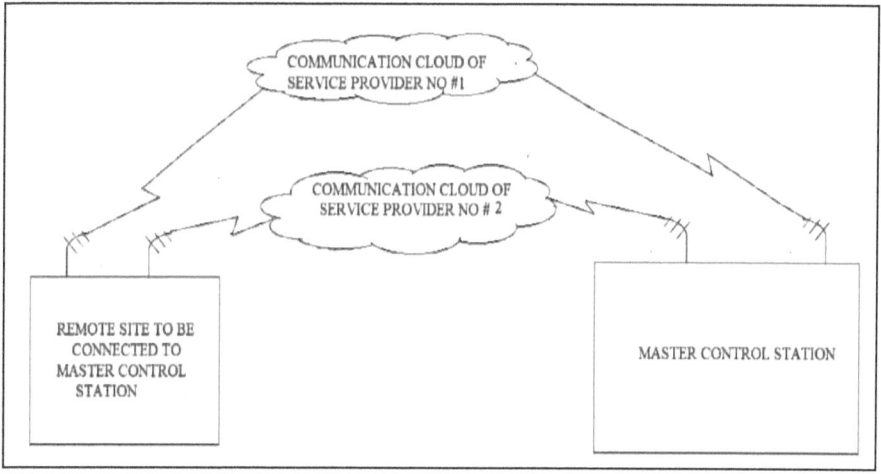

▲ **Figure 3.13:** Channel redundancy achieved with two different communication clouds

3.5 DESIGN AND CONFIGURATION CONSIDERATIONS OF MCC

The following aspects are utmost important while designing the Master Control Center of the power system SCADA.

1. Sufficient hardware and software redundancy have to be ensured, to achieve the overall system redundancy with High Availability having No Single Point of Failure(NSPF) as the SCADA functions for the power system are critical. This includes communication channel as well,
2. The firewall must be properly configured with the appropriate ruleset, after due deliberation with the security policy of the utility. Design with a two layer Firewall configuration, with different make IDS loaded are not a better option but a must, while moving for a DMZ, and
3. The CFE or the FEP of the SCADA control center must have high end capabilities with suitable IDS loaded and have cryptographic capabilities. If third party communication media are used, especially using the VPN for data transfer, utmost care must be given for proper VPN termination.

SUMMARY

This chapter begins with describing the communication architecture of basic SCADA. Then moves on to describing the common communication philosophies adopted in DCS. As the reliability and availability of DCS functions are the most important, they are briefly introduced, but cater the necessary understanding to power system professionals and engineering students who are engaged or intend to embark into the DCS, Smart Grid or Microgrid domain. It then explains the concepts of Fault Tolerant Systems, Fail Safe and redundant systems, High Availability, etc. Based on these concepts, this chapter then elaborates the design of a typical SCADA MCC architecture having redundant and HA connectivity.

CHAPTER 04

TRANSMISSION AND DISTRIBUTION AUTOMATION

4.1 INTRODUCTION

Today Substation Automation(SA) which is defined as the deployment of substation and feeder operating functions and applications ranging from Supervisory Control and Data Acquisition (SCADA) and integrating Energy Management Systems (EMS) in order to optimize the management of capital assets and enhance Operation and Maintenance (O&M) efficiencies with minimal human intervention, is gaining momentum throughout the globe. One of the reasons behind this interest behind this is to catch up with the fast changing technologies and due to the economic pressures of downsizing and reduce O&M cost. Further utilities are realizing that they must shift their focus to customer service. This compels to introduce the advanced transmission and distribution technologies to provide quality power to the customers. Following sections elaborate the need for transmission and distribution automation, relevant advanced technologies and its features.

4.2 SUBSTATION AUTOMATION-TECHNOLOGY DRIVERS

The electricity market today is deregulated and open to consumers for purchasing it directly from the power producers and suppliers. This created competition among the supplying utilities. This competition helps in improving the quality of power to be sold, and the service rendered by the utility. New energy related services and business areas allow utilities to invest more in automating the substations. The availability of various kinds of data from the system which improves decision making has also made the utilities proactive towards substation automation.

The rapid development and deployment of IEDs opened a new area of business opportunities which also increased the pace of substation automation. Many substations preferred protective relay IEDs, Equipment Condition

Monitor (ECM) IEDs, because of their unmatched advantage of data capture from the field is concerned. Another advantage is that the data can be made available to personnel working outside the control for better decision making and planning. Another technology driver is the standardization and acceptance of the protocols, especially the communication protocols by the manufacturers. The IEC 6850, IEEE 1815(DNP3), and IEC 60870 have become protocols for the substation automation. This standard based implementation and interoperability makes substation automation easier and simpler for utilities and projects are underway throughout the world. The advantages of IED to replace many electromechanical devices, reduces the construction cost and duration of construction of substation. This in turn reduces the size of the substation and hence cost considerably. All these factors lead to the acceptance of substation automation a necessity for the utilities.

4.3 SMART DEVICES FOR SUBSTATION AUTOMATION

The objective of substation automation is to minimize human intervention and improve the operating efficiency of the system. The reduction of operating costs and optimization of assets are the other benefits in the long run. Certain smart devices have been exclusively developed for power system automation which are concisely explained in the preceding chapters.

4.4 SMART TRANSMISSION

The defining feature of a Digital Substation is the implementation of a process bus. Digital Substations remove the last electrical connection between the high voltage equipment and the protection and control panels, creating a safer work environment, whilst reducing the costs for building, land, engineering, commissioning, operation and maintenance of the system.

Digital Substations enable electric power utilities to increase productivity, reduce footprint, increase functionality, improve the reliability of assets and, crucially, improve safety of service personnel. Digital Substations exploit the benefits of digital protection, control and communication technologies, mirroring the trend towards digitalization seen in many other industries.

This trend towards digitalization also applies to other areas of the substation. Within medium-voltage switchgear panels for example, the horizontal exchange of IEC 61850-8-1 GOOSE and sampled analog values reduces wiring and accelerates the testing and commissioning. Digitalized technology can continuously monitor mission-critical functions of power transformers and High Voltage Switchgear, while performing real-time simulation and diagnostics, making pro-active management of the assets lifecycle possible. As a key component towards smarter grids, where utilities continue to integrate increasing amounts of intermittent renewable energy sources, Digital Substations will also improve safety and require shorter time for restoration in case of an emergency.

4.4.1 Design of Digital Substations

A digital substation can be designed in many ways. The widely accepted substation automation protocol IEC 61850 may be implemented by starting with adaptation of existing IEDs to support the new communication standard over the station bus and process bus.

A digital substation with process bus and station bus architecture is shown in Figure 4.1. Here a Merging Units will process the sensor inputs, generate the sampled values for the three phase currents and voltages, format a communications message and multicast it on the substation LAN. Another unit called I/O device capture the status inputs, generate status data, format a communications messages, and multicast it on the substation LAN. All multifunctional IEDs will then receive sample values messages and binary status messages. The units which are destined to receive the data, and then process the data, make decision, and operate by sending a message the I/O device to perform the required operation.

4.4.2 Energy Management Systems (EMS)

With deregulation of the power industry and the development of the Smart Grid, decision-making is becoming decentralized, and coordination between different actors such as Independent System Operators (ISO), Regional Transmission Operators (RTO), and Distribution Generators (DG) including RES, energy traders, and prosumers, in various markets becomes important. All these complex changes, and the aging infrastructure which limits the operating range and increase in congestion of the networks, needs a modern management and control system to carry out all these functions with optimum efficiency. This leads to the development of Energy Management System (EMS) and a block diagram of which is shown in Figure 4.2.

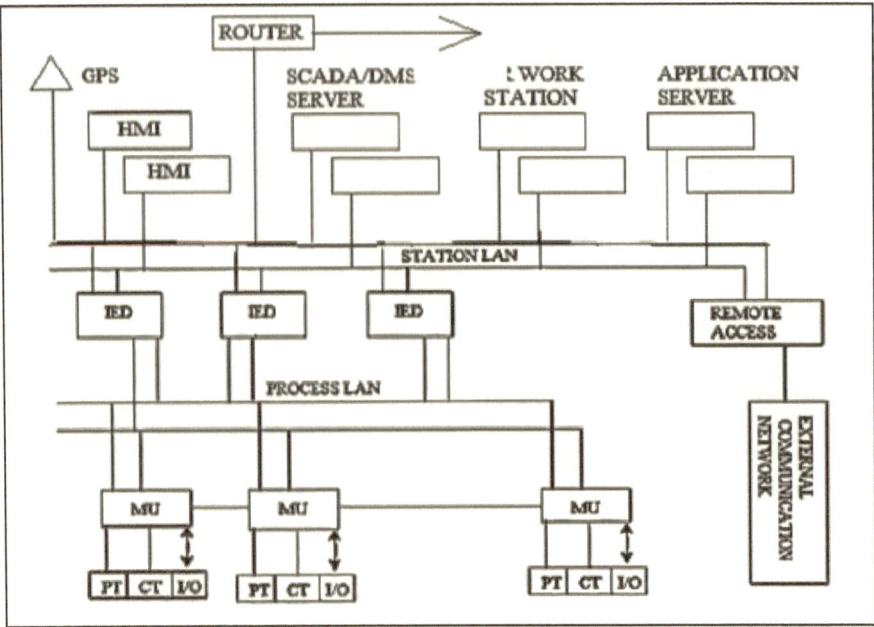

▲ **Figure 4.1:** Digital substation with process and station LAN architecture

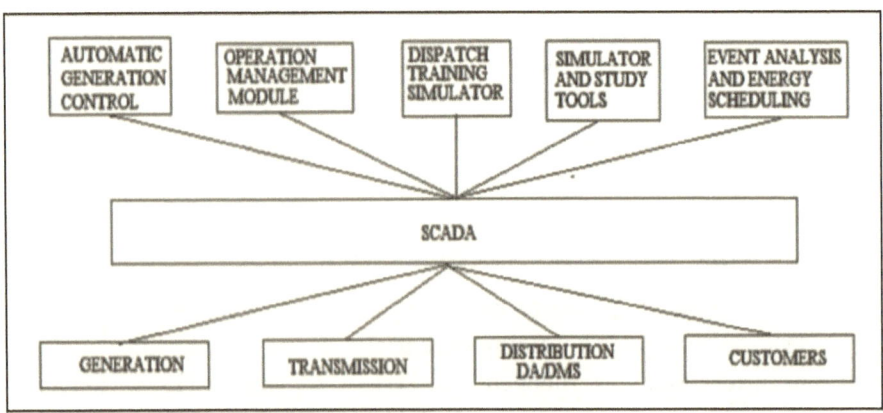

▲ **Figure 4.2:** Block Diagram of an EMS

EMS is a suitable blend of software tools and hardware preferably high end machines configured with High Availability (HA) for reliable and efficient operation of generation and transmission assets of a power utility in real time. In fact EMS monitor, control and optimize the power transmission and generation in a stringent secure manner.

The information from the power system is read through Remote Terminal Units (RTUs), which is an integral part of SCADA to an EMS or Energy Control Centre (ECC). Hardware part of EMS consists of RTU, Intelligent Electronic Device (IED), Protection, Computer networking, etc. System status and measurement information are collected by the RTUs and sent to the Control Centre through the communication infrastructure. The CFE in the EMS is responsible for communicating with the RTUs and IEDs. Different EMS Applications reside in different servers and are linked together by the LAN. Software part of EMS consists of Application programs for network analysis of power systems. In EMS, application programs are run in a real time as well as extended real time environment to keep the power system in a secure operating state. EMS Application Description has been described below.

Real-time SCADA Applications: Providing Supervisory Control and Data Acquisition including alarm/events, tagging, data historians, data links, control sequences, and load shed applications used to monitor/operate the network.

Generation Dispatch and Control: GDC provides the functions required for dispatch and closed loop digital control of multiple generators in an economic fashion while adhering to standard operating guides at the same time considering interchange schedules, dynamic schedules (load or generation in an out of the area), inadvertent interchange payback, time error correction, reserve requirements, and security constraints of the transmission network

Energy Scheduling and Accounting: ESA provides applications to monitor standard reporting criteria, production costs, interchange scheduling, inadvertent interchange accounting, and weather adaptive demand forecasting.

Transmission Security Management: TSM provides sophisticated applications to analyze and optimize the use of the transmission network in a reliable and secure manner. The important functions of an EMS can be described below.

1. Control functions: It includes real time monitoring and control functions such as Automatic Control and automation of a power system, Efficient Automatic Generation Control (EAGC) and Load Frequency Control (LFC), Reactive Power Control (RPC), Reactive Power and Voltage Control and optimal automatic generation control across multiple areas, and Tie -line control.
2. Operating functions: Ensure economic and optimal operation of the generating system, supports efficient operator decision making, improved power quality of supply and optimization functions such as,

- assists the operator for optimal utilization of the transmission network,
- assists Power scheduling interchange between areas,
- ensure optimal allocation of resources, and
- real-time overview of the power generation interchanges and reserves.

3. Planning functions: These are improved power quality of supply and system reliability, forecasting of loads and load patterns, generation scheduling based on load forecast and trading schedules, Maintaining reserves and committed transactions, and calculation of fuel consumption, production costs and emissions.
4. Protection and Security Functions: Protective relaying, Primary protection and Secondary protection or backup protection is the protection functions of EMS. The protective functions are the first which are activated in a real time operation as a protective measure. These are followed by essential control functions like LFC, AGC, RPC, etc.

The security assessment and control Application includes, security monitoring, security analysis, preventive control, emergency control, fault diagnosis and restorative control. The tools required include,
- network topology analysis,
- external system equivalent modeling,
- state estimation,
- on-line power flow,
- security monitoring, and
- contingency analysis.

When the system is insecure, security analysis informs the operator which contingency is causing insecurity and the nature and severity of the anticipated emergency. In addition to these functions, a training tool named Dispatcher Training Simulator (DTS) is embedded within the EMS. DTS were originally created as a general application to introduce operators to the electrical and dynamic behavior of a power system. Today, they model the actual power system being controlled with reasonable fidelity and are integrated within the EMS. This provides a realistic environment for operators and dispatchers to practice normal, everyday operating tasks and procedures as well as experiencing emergency operating situations. Various training activities can be practiced safely and conveniently with the simulator responding in a manner similar to the actual power system

4.4.3 HVDC and FACTS -Enhancing Power System Performance

High Voltage Direct Current (HVDC), a well-proven technology which used to transmit bulk electricity over long distances by overhead transmission lines or submarine cables. For long distance transmission especially for transmission distances above 600 km, it is economical and provides better security and flexibility. At present HVDC is extensively getting acceptance as a means of transporting bulk energy generated by large renewable energy sources such as offshore wind farms to the load center. It is also used to interconnect separate power systems, where traditional alternating current (AC) connections cannot be used. In an HVDC system, electric power is taken from one point in a three-phase AC network, converted to DC in a converter station, transmitted to the receiving point by an overhead line or cable and then converted back to AC in another converter station and injected into the receiving AC network. Typically, an HVDC transmission has a rated power of more than 100 MW and many are in the 1,000-3,000 MW range. With an HVDC system, the power flow can be controlled rapidly and accurately in terms of both power level and direction. This possibility is often used to improve the performance and efficiency of the connected AC networks. There are three different categories of HVDC transmission and they are,
1. point-to-point transmission,
2. back-to-back stations, and
3. multi-terminal systems.

The reasons for selecting HVDC instead of AC for a specific project are often numerous and complex. The most common arguments in its favor are,
1. lower investment cost,
2. long distance water crossing,
3. lower losses,
4. asynchronous interconnections,
5. controllability,
6. limited short-circuit currents, and
7. environment.

In general, the different reasons for using HVDC falls into two main groups viz. HVDC is necessary or desirable from the technical point of controllability. HVDC results in a lower total investment and/or is environmentally superior. As environmental aspects are increasingly important and HVDC has the advantage of a lower environmental impact than AC since the transmission lines are much smaller and need less space for the same power capacity. One of the most important differences between HVDC and AC is the possibility to accurately control the active power transmitted on a HVDC line. This is in contrast to AC lines, where the power flow cannot be controlled in the same direct way. The controllability of the HVDC power is often used to improve the operating conditions of the AC networks where the converter stations are located.

Another important property of an HVDC transmission is that it allows the interconnection of asynchronous networks. In a transmission system different grids are interconnected but some time it may happen that one or more grid or power system may not be in synchronization with other grid or systems or there exists a weak link between different power systems due to various factors. To overcome these things a HVDC back to back system is used.

In a HVDC back-to-back configuration, two independent neighboring systems with different and incompatible electrical parameters (Frequency / Voltage Level / Short-Circuit Power Level) are connected via a DC link. The basic principle of operation of an HVDC linking system is based on the conversion of AC to DC and vice-versa by means of converter valves. Process involved in this is simple in theory but in practice it is very complex as it involves high voltages and power semiconductor devices. Power from two or more neighboring system is first converted into DC with the help of power rectifier systems and summed up. After summing, the resulted power is again converted into AC with the help of power inverter and supplied to different grids. Thus a HVDC interconnection phase is involved in linking of different power systems to obtain a stable power system with uniform parameters. The main advantage of HVDC back to back system is given below.

- Power can be upgraded to desired frequency,
- Two asynchronous systems can be joined successfully without loss of stability,
- Stability of the system is increased as well as power flow can be maintained within the optimal limits,
- More active power can be added where the AC system already is at the limit of its short-circuit capability, and
- To stabilized weak AC links.

Flexible AC Transmission System (FACTS): FACTS is also based on power electronics and was intended to improve the performance of a weak AC systems so as it is capable of transmitting AC power to long distance. Further as the present Grid is interconnected, FACTS help to solve many of its technical issues. FACTS find application in a parallel connection, in a series connection, and in a combination of both to control load flow and to improve dynamic conditions. In the simplest sense FACTS is a combination of power electronics components with traditional power system components.

FACTS can be series and shunt compensation for the transmission lines using these FACTS devices. In series compensation, line impedance is modified, that means net impedance is decreased and thereby increasing the transmittable active power. For shunt compensation, reactive current is injected into the line so as to regulate the voltage at the point of connection. Thus, the active power transmission is increased. In both types of compensations (series and shunt), more reactive power must be provided.

FACTS can be series and shunt compensation for the transmission lines using these FACTS devices. In series compensation, line impedance is modified, that means net impedance is decreased and thereby increasing the transmittable active power. For shunt compensation, reactive current is injected into the line so as to regulate the voltage at the point of connection. Thus, the active power transmission is increased. In both types of compensations (series and shunt), more reactive power must be provided. The general classification of these FACTS devices is,

1. series controllers,
2. shunt controllers,
3. combined series-series controllers, and
4. combined series-shunt controllers.

Series Controllers: Series controllers are being connected in series with the line as they are meant for injecting voltage in series with the line. These devices could be variable impedances like capacitor, reactor or power electronics based variable source of main frequency, sub synchronous or harmonic frequency, or can be a combination of these, to meet the requirements. If the injected voltage is in phase quadrature with the line current, then only supply or consumption of variable reactive power is possible. In order to handle real power also, any other phase relationship has to be involved. These type of controllers include,

1. SSSC – Static Synchronous Series Compensator,
2. TCSC – Thyristor Controlled Series Capacitor,
3. TCSR – Thyristor controlled series reactor,
4. TSSC – Thyristor Switched Series Capacitor and
5. TSSR – Thyristor Switched Series Reactor.

Shunt controllers: Shunt controllers will be connected in shunt with the line so as to inject current into the system at the point of connection. They can also be variable impedance, variable source, or a combination of these. If the injected line current is in quadrature with the line voltage, variable reactive power supply or consumption could be achieved. But any other phase relationship could involve real power handling as well. This category includes STATCOM (Static Synchronous Compensator) and SVC (Static VAR compensator). The common Static VAR compensators are,

1. TCR – Thyristor Controlled Reactor,
2. TSR – Thyristor Switched Reactor,
3. TSC – Thyristor Switched Capacitor,

Combined Series-Series Controllers: This category comprises of separate series controllers controlled in a coordinated manner in the case of a multiline transmission system. It can also be a unified controller in which the series controllers perform the reactive power compensation in each line independently whereas they facilitates real power exchange between the lines via the common DC link. Because, in unified series-series controllers like Interline Power Flow Controller (IPFC), the DC terminals of the controller converters are all connected together.

Combined Series-Shunt Controllers: It is a combination of separate series and shunt controllers, being operated in a coordinated manner. Hence, they are capable of injecting current into the line using the shunt part and injecting series voltage with the series part of the respective controller. If they are unified, there can be real power exchange between the shunt and series controllers via the common DC power link, as in the case of Unified Power Flow Controllers (UPFC).

The advent of the power electronics, FACTS devices have been proven their speed and flexibility in the power industry. But there exist some cost and complexity issues to be addressed to make it widely acceptable by the power utilities. The main benefits of utilizing FACTS devices are mentioned below.

1. *More Utilization of Existing Transmission System:* In all the countries, the power demand is increasing day by day to transfer the electrical power and controlling the load flow of the transmission system. This can be achieved with more and more load centers. Addition of new transmission lines are very expensive to handle the increased load on the system. Such cases, FACTS devices are much economical to meet the increased load on the same transmission lines.

2. *More Increased Transient and Dynamic Stability of The System*: The Long transmission lines are inter-connected with grids to absorb the changing the loading of the transmission line and it is also seen that there should be no line fault creates in the line / transmission system. By doing this the power flow is reduced and transmission line can be trip. By the use of FACTS devices high power transfer capacity is increased at the same time line tripling faults are also reduces.
3. *Increased More Quality of Supply for Large Industries*: New industries want quality electric supply, constant voltage with less fluctuation and desired frequency as stipulated by CERC. Reduced voltage, variation in frequency or loss of electric power can productivity of the industry and cause high economical loss. FACTS devices can help to provide the required quality of supply.
4. *Beneficial for Environment*: FACTS devices are becoming environmental friendly. FACTS devices does not produce any type of waste hazard material so they are pollution free. These devices help to deliver the electrical power more economically with better use of existing transmission lines while reducing the cost of new transmission line and generating more power.
5. *Increased Transmission System Reliability and Availability*: Transmission system reliability and availability is affected by different factors. FACTS devices have ability to reduce such factors and improve the system reliability and availability.

In fact with the help of HVDC and FACTS technologies, Smart Grid which consists of a number of micro grids will be able to provide an efficient, secure, reliable, and sustainable transmission of bulk renewable energy such as hydro, solar and wind to the load centers.

4.4.4 Hybrid Transmission System

Better stability can be achieved by connecting a VSC based HVDC connected in parallel with an AC line is capable of providing real power flow from the sending to the receiving end. The two converters can also control their reactive power and thereby control the voltage magnitude at the sending and receiving end and can influence the AC power flow.

4.4.5 Phasor Measurement Units (PMU)

Phasor Measurement Unit (PMU) is referred to as a power systems real time health check by monitoring the samples voltage and current multiple times within a second at a given location on the grid. This gives the utility a real time view of the power system performance and disturbances so that appropriate and proactive decisions can be taken to improve the efficiency or prevent the system from failure. Wide Area Monitoring System (WAMS) are designed for optimal capacity of the transmission grid and to prevent the spread of disturbances. It provides real time information on stability, operating safety margins, and gives early warnings of system disturbances for preventing black outs and brown outs. PMUs are the heart of the WAMS and is deployed at critical locations of the grid for sampling the voltage and current phasors. PMUs installed in the power system may be utilized for,

1. real time monitoring and control,
2. state estimation,
3. protection and control for distributed generation,
4. network congestion management,
5. angular and voltage stability monitoring, and
6. postmortem analysis of disturbances and faults.

Most of the utilities are installing PMUs in their high voltage substations at their high voltage bus. Installing the PMUs at the two ends of the transmission line is easy to monitor power and voltage.

Need for PMU in power system: The conventional power grid is designed for dispatchable centralized generation. The loads are largely predictable, allowing essentially open-loop grid control. However environmental concerns, accelerated cost reduction in the renewables have made the integration of renewable energy sources an integral part of the Smart Grid. Many countries, especially those belong to developing countries like India has a huge potential of generating the clean energy. For effective utilization of this potential the renewable sources are required to be integrated with the power grid. However this integration presents great challenges for the system operators as both solar photovoltaic and wind sources whether connected to transmission or distribution are all inverter based. This will alter the dynamics of the system completely.

SCADA systems provide only a steady state slow picture in much longer time intervals. Further modern power system grid monitoring tools use data from Remote Terminal Units (RTUs), protective relays, and transducers to provide information to system operators. This information is vital for the operation of the power system on a daily basis and under system contingencies. However, the mechanism used to retrieve data from the devices is asynchronous and relatively slow. The asynchronous nature of the data does not provide accurate angle difference information from two nodes on the network. Additionally, the low data rate may be too slow to capture many short-duration disturbances on the grid. Alternatively, PMUs sample voltage and current many times a second and accurately time-stamp each sample. This technology can be used to provide high-speed and coherent real-time information of the power system that is not available from legacy SCADA systems.

Components of PMU: PMUs are electronic devices that use digital signal processing components to measure AC waveforms and convert them into phasors, according to the system frequency, and synchronize these measurements under the control of GPS reference sources. The analog signals are sampled and processed by a recursive phasor algorithm to generate voltage and current phasors. Different components of a PMU are shown by a block diagram in Figure 4.3.

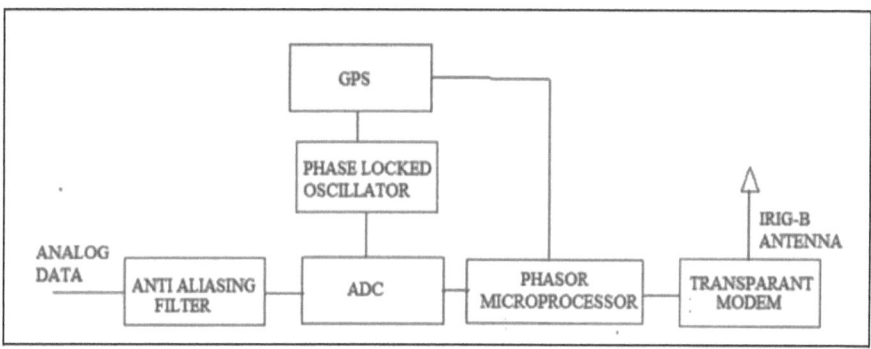

▲ **Figure 4.3:** Components of a PMU.

Indian power grid has evolved into a huge complex infrastructure. The process of making a National grid is almost completed by integrating all the five grids together. The developments in the area of Smart Grid technology around the World have led to a paradigm shift in the way power grids have been continuously monitored and controlled. In order to have a stable secure national Smart Grid, the government of India has also initiated various projects and the installation of Phasor Measurement Units is one such major initiatives.

Implementation of Synchrophasor Technology: Today Synchrophasor technology is used around the world for data visualization and postmortem analysis applications. IEEE C37.118, IEEE Standard for Synchrophasors for Power Systems, defines synchrophasors, provides the requirements for the quality of the measurements, and specifies the protocol for data transfer. The standard defines synchronized phasors as phasors calculated from data samples using a standard time signal as the reference for the measurement. With a universal precise time reference, power system phase angles can be accurately measured throughout a power system. The Global Positioning System (GPS) technology provides an economic option for the same. An important advantage of the GPS technology is that its receiver can automatically detect accurate synchronization. The data frame of the IEEE C37.118 message includes time-quality information from GPS receivers and/or satellite clocks. It is critical to supervise synchrophasor based applications using this information. The standard emphasizes that GPS clocks should support Inter-Range Instrumentation Group-B (IRIG-B) time codes, (a standard format for transferring the timing information), with additional extensions to provide the time quality information to the PMU. Synchrophasors are well suited for steady state and quasi steady state phenomena (not for transient conditions like faults) and for observing low level oscillations. The Figure 4.4 shows the building blocks of a synchrophasor based system.

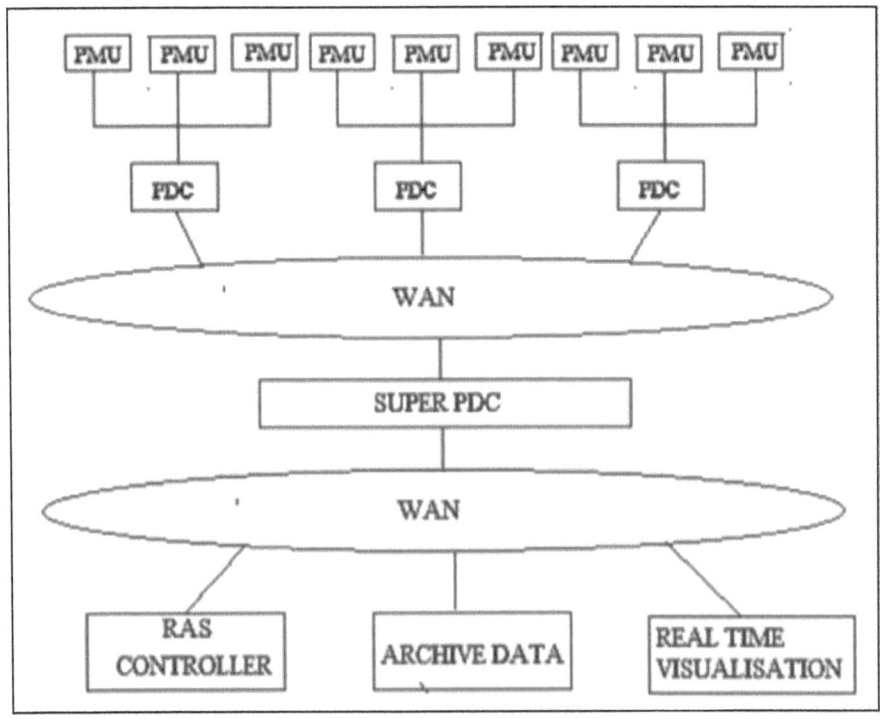

▲ **Figure 4.4:** Building blocks of a synchrophasor based system

PMUs connected to a GPS satellite clock send voltage and current phasor measurements to a phasor data concentrator (PDC) at a rate of 20-60 messages per second. The data from different PMUs distributed in the grid which is transmitted to a PDC located nearby the control center. The PDC collects and sorts the data, time aligns the data until the arrival of the slowest data. The PDC arrange a super data packet consists of phasor measurements from different PMUs with a common time stamp. This superdata packet is then send to the control center for further computation such as State Estimation. Presently PMU applications are not constrained to visualization and State Estimation, rather used for the following applications as well.

1. Automatic generation-shedding application,
2. Islanding detection and control: Today controllers are available which allow utilities to implement Remedial Action Schemes (RASs) and System Integrity Protection Schemes (SIPS) that process coherent data (super packets) and send control commands back to the power system, and
3. Synchrophasors are being used in EMS by many utilities.

Advantages of PMU over SCADA

Phasor Measurement and Control Units (PMCUs), produce accurate time stamped measurements of voltage and current magnitude as well as phasor angles. They also report the status of breakers with timestamps synchronized to those of the measurements. Because PMCUs calculate synchrophasors with respect to a global angle reference, the number of critical measurements is less than when the state estimator uses SCADA measurements.

Another advantage of PMCU measurements for state estimation is that having direct angle measurements diminishes the amount of error introduced by inaccuracies in network parameters.

4.4.6 Wide-Area Monitoring System (WAMS)

WAMS is a real-time monitoring and display of power system components and performance such as GPS synchronized current, voltage and frequency across interconnections over large geographic areas, which help system operators to understand and optimize power system components, behavior and performance. WAMS acquires these data by installing Phasor Measurement Units (PMUs), in selected locations across the power system. Presently the existing power grid is being operated closer to its stability limits, especially with the integration of distributed renewable energy sources without adequate transmission expansion. If the grid is not kept within the operating limits This

may lead to wide area blackouts,. Implementation of WAMS can provide the following benefits.
1. Provides an early warning of worsening system conditions, so that the operators can take remedial actions (Wide Area Situational Awareness (WASA)), and
2. Allows more effective use of automatic controls for self correction, such as controlling the flow of power (Wide Area Adaptive Protection, Control and Automation (WAAPCA).

The data acquired by the WAMS through the PMU can be used to improve the State Estimator (SE). Based on this, a flexible and resilient alternative to the SCADA and EMS, which will function in the most adverse conditions is needed and has been envisaged. WAMS based Power System Stabilizers (PSS) and Special Protection Systems (SPS) are also being developed for,
1. Frequency Stability Control, and Protection,
2. Voltage Stability Control and Protection,
3. Oscillatory Stability Control, and
4. Transient Stability Protection.

4.4.7 Wide Area Monitoring, Protection and Control (WAMPAC)

In WAMPAC, the PMU or synchrophasor measurements collected from the different parts of the network and PMU data based State Estimation (SE) are used for online stability analysis. When an event occurs, its location, time, magnitude and type are first identified. Real-time visualization of the event allows it to be replayed several seconds after it occurs. The future system condition is then analyzed using the information that has been gathered. An online stability assessment algorithm continuously assesses the system to check whether the system is still stable and how quickly the system would collapse if it became unstable. If instability is predicted, then the necessary corrective actions to correct the problem or to avert system collapse are taken.

WAMPAC functionalities helps,
- to initiate actions to correct the system once a voltage, angle or oscillatory instability has been predicted. This may include switching of generators and controlling devices such as the Flexible AC transmission Systems (FACTS), the Power System Stabilizers and HVDC converters, and
- to generate emergency control signals to avoid a large-scale blackout (for example, through selective shedding of load or temporary splitting of the network) in the event of a severe fault.

A configuration of the WAMPAC which is capable of measuring both magnitudes and phase angles, measuring the phases at the same time instantly, capturing the snapshots of the status, observing the steady state and dynamic state, and suitably ascertaining the line load and controlling is shown in Figure 4.5.

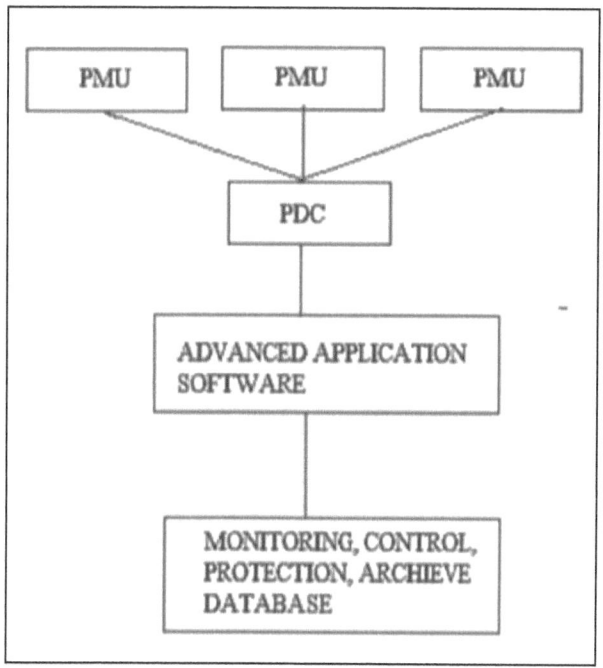

▲ **Figure 4.5:** Configuration of the WAMPAC

4.5 DISTRIBUTION SYSTEMS

In distribution feeder automation, distribution management system, outage management system, etc. are the main components of the distribution automation which are briefly explained below.

4.5.1 Feeder Automation

Distribution Automation (DA) systems today are capable of detecting and locating the fault, isolating the faulty section, restoring power to *healthy* feeder segments is one of the major aspects of smart substation. Here rather than rely on manual switching by field crews, automated feeder devices are used to detect and isolate a faulted feeder section. The data obtained from the breakers/ switches and the level of the fault current flowing in the networks helps one to manage and restore the system in an event of fault. The devices which help in localization and isolation of the fault include Auto Reclosers (AR), Sectionalizes, and Fault Passage Indicators (FPI) etc. A typical example of Fault Location, Isolation, and Service Restoration is demonstrated in Figure 4.6(a) and 4.6(b).

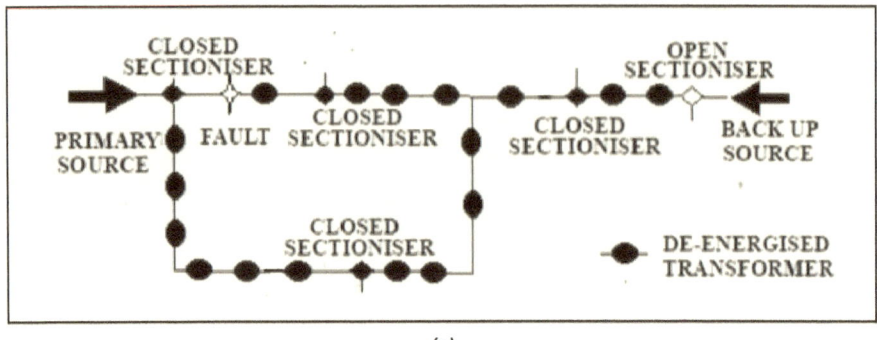

▲ **Figure 4.6:** (a) Initial Outage and Fault Location; (b) Supply Restoration by Feeder Automation

Such feeder automation can be implemented using different strategies and automated devices operating locally without supervision or devices controlled remotely by an automation system.

4.5.1.1 Automated Feeder Switching

Automated feeder switching is the involvement of the automated field devices in such a way that they operate in a coordinated, but unsupervised manner, possibly controlled by a computer system located at a substation or control

center. Remote operation can be manual, semi-automatic, or fully automatic. The following are the various Feeder Automation Architectures employed today.
1. Standalone Automatic Switches using Reclosers and sectionalizes,
2. Centralized System using Switches controlled by central DAS/DMS
3. Substation Centered Approach with substation unit controls switches on associated feeders
4. Peer-to-Peer Arrangement with groups of switches communicates to determine appropriate switching actions.

4.5.2 Distribution Management System (DMS)

DMS is a set of application software and systems which helps to manage distribution assets and plays an important role in ensuring power quality in power distribution making use of real time data acquired from distribution SCADA, Customer Information Systems (CIS), and Geographical Information System (GIS). Advanced DMS operates jointly with Outage Management System (OMS) and Asset Management System (AMS) for optimum reliability and efficiency. Proper distribution and substation sensing, and automation can reduce outage and restoration time, maintain voltage level and improve asset management. Advanced distribution automation processes real-time information from sensors and meters for fault location, automatic reconfiguration of feeders, voltage and reactive power optimization, or to control distributed generation. Sensor technologies can enable condition and performance based maintenance of network components, optimizing equipment performance and hence effective utilization of assets. In the Smart Grid scenario, AMI is an integral part of the DMS for bi-directional information sharing between the utility and the consumer. The following are some of the advantages of implementing the SCADA / DMS in Distribution Automation.
1. Quick isolation of faulty section and fast restoration of healthy section so that only least customers are affected during outage period,
2. All data are available in real time and historical data archive for planning, sharing of data with all stakeholders and MIS
3. Though requires capital investment, but a good SCADA / DMS system implemented in a phased manner brings returns in a shorter period, and
4. All data are available in real time and historical data in archive for planning and other applications of utility.

DMS functions: The following sections briefly describe various Distribution Management System (DMS) functions.

4.5.2.1 Distribution Network Model

The Distribution Automation should represent the various components distribution system of the Utility and include all the primary substation feeders, distribution network and devices with possible islands, which may be formed dynamically. The following devices are generally represented in the Distribution Network Model or Dynamic Mimic Diagram (DMD).

A distribution network is made up of Power Injection Points, Transformers, Feeders, Load (balanced or unbalanced), Circuit Breakers, Sectionalizes, Isolators, Fuses, Capacitor banks, Reactors, Generators, Bus bars, Temporary Jumper, Cut and Ground, Meshed & radial network configuration, Line segments, which can be single-phase, two-phase, or three phase, Conductors, Grounding devices, Fault detectors, IEDs, Operational limits for components such as lines, transformers, and switching devices, etc. The database of the network model of the utility system having an interface with the Geographical Information System (GIS) system of the area which can give a rich visual network presentation for crew management and asset information. The Customer Interface Management can also integrate with the distribution automation system for effective utilization.

4.5.2.2 Network Connectivity Analysis (NCA)

The network connectivity analysis or Topology Analyzer (TA) function provides the connectivity between various network elements. The prevailing network topology will be determined from the status of all the switching devices such as circuit breaker, isolators etc. that affect the topology of the network modeled. NCA runs in real time as well as in study mode. Real-time mode of operation uses data acquired by SCADA. Study mode of operation will use either a snapshot of the real-time data or save cases. NCA can run in real time on event-driven basis. The network topology of the distribution system will be based on
1. tele-metered switching device statuses,
2. manually entered switching device statuses, and
3. modeled element statuses from DA applications.

The NCA will be useful in determining the network topology for the following status of the network.

1. Bus connectivity (Live/ dead status),
2. Feeder connectivity,
3. Network connectivity representing S/S bus as node,
4. Energized /de-energized state of network equipments,
5. Representation of Loops (Possible alternate routes),
6. Representation of parallels, and
7. Abnormal/off-normal state of CB/Isolators.

The NCA also assists the power system operator to know the operating state of the distribution network indicating radial mode, loops and parallels in the network. Distribution networks which are normally operated in radial mode, loops and/or parallel may be intentionally or inadvertently formed.

4.5.2.3 State Estimation (SE)

The Power System State Estimation (SE) is a process used for assessing (estimating) the distribution network state using the data acquired remotely from network measuring points through field devices. As the name implies, the state estimator converts the acquired data and the generated pseudo measurement data into consistent set of network states and state variables viz. voltage phasor of the network data and network topology, known historical and static consumer data. All such measurements are synchronized or time stamped using a common clock and communicated from geographically distant locations to a load dispatch center. In fact state estimator assess loads of all network nodes, and, consequently, assessment of all other state variables such as voltage and current phasors of all buses, sections and transformers, active and reactive power losses in all sections and transformers, etc. in the distribution network. Network state comprises the set of voltage phasors for all buses. State variables comprise all other variables such as voltages drops, section currents, load which can be calculated from the Network state. SE is the basic Distribution Management System (DMS) application, since practically all other DMS applications are based on its results.

A static state estimate is obtained from measurements taken within a time interval of about 0-5 seconds. This is the commonly used state estimator. Obviously, a state estimator of this type essentially gives a steady state snapshot of the system. A dynamic state estimate is obtained from measurements in a relatively shorter time (say 0.01 seconds). These measurements could be used for advanced control schemes which we shall see later. The main concern in state estimation is the reliability of the measured data. Usually to minimize the errors, the data is crosschecked using redundant measurements.

4.5.2.4 VOLT /VAR Control (VVC)

In electrical power system the reactive power can be generated at source generators or can be injected at the substations through Volt-VAR systems. It is more appropriate to inject at substations rather than producing then at generator points and transporting them over long distances. Any power system always tries to optimize on the reactive power flow over their networks. The coordination of voltages and reactive power flows control requires coordination of VOLT and the VAR function. This function shall provide high quality voltage profiles, minimal losses, controlling reactive power flows, minimal reactive power demands from the supply network. The following resources should be taken into account in any voltage and reactive power flow control.
1. TAP changer for voltage control, and
2. VAR control devices such as switchable and fixed type capacitor banks.

4.5.2.5 Load Flow Studies (LFS)

Load Flow Studies detect line loads and bus voltages which are out of range, inappropriately large bus phase angles, system component loads particularly transformer load, proximity to Q-limits at generation buses, and other parameters that can create operating difficulties. Intermediate load and off-peak load studies are also useful, since off-peak loads can result in high voltage conditions that are not identified during peak loads. Load Flow Studies assist system operators in calculating power levels at each generating unit for economic dispatch, analysing outages and other forced operating conditions and coordinating power pools. In most instances, LFS are used to assess system performance and operations under a given condition. The Load Flow Studies provide active and reactive losses on the following network components and physical quantities.
1. Station power transformers,
2. Feeders,
3. Sections,
4. Distribution circuits including feeder regulators and distribution transformers and the total circuit loss,
5. Phase voltage magnitudes and angles at each node,

6. Phase and neutral currents for each feeder and transformers,
7. Total three phases and per phase KW and KVAR losses in each feeder, section, transformer,
8. Active and reactive power flows in all sections, transformers of overloaded feeder, lines, bus bars, transformers loads etc,
9. List of limit violations of voltage magnitudes, overloading, and
10. Voltage drops.

4.5.2.6 Load Flow Application (LFA)

Load Flow Application (LFA) is used for the calculation of steady states of radial and weakly meshed primary MV power grids, as well as state of secondary LV power grids. It is intended for determination of network state variables of unbalanced and balanced networks, on the basis of known root voltage and data regarding consumption (load) of all nodes.

The network state consists of the variables such as complex voltages, currents, flows of real and reactive power, voltage drops, and losses. Usually, among all state variables, the state vector is defined as a set of all network nodes' complex voltages. The state vector is sufficient for calculation of any other network state variable.

Generally, load (power) flow model of power systems represents mathematical description of real and reactive power balance in the system, where power supply needs to be equal to the sum of loads and losses. The model consists of a set of complex equations. Its dimensions are equal to the number of network nodes. It describes the state vector of the considered network particularly the set of complex voltages in all network nodes.

LFA is one of the basic DMS power applications. It is rarely used alone, but as a base component of almost all other DMS power applications, such as Performance Indices (PI), State Estimator (SE), Network Reconfiguration (NR), and Volt VAR Optimization (VVO).

4.5.2.7 Load Shed Application (LSA)

Load Shedding Application is aimed for shedding of load under emergency conditions, as well as restoring load after restoring system conditions especially when demand is higher than supply. The reasons for less supply are several, including the faults, tripping of lines, insufficient generation, etc. In these situations the power system operator tries to distribute available power through Shedding of loads to consumers over small definite periods till the consumer tides over the situation of loss of power. The load-shed application helps to automate and optimize the process of selecting the best combination of switches to be opened and controlling in order to shed the desired amount of load. Given a total amount of load to be shed, the load shed application shall recommend different possible combinations of switches to be opened, in order to meet the requirement. The despatcher is presented with various combinations of switching operations, which shall result in a total amount of load shed, which closely resembles the specified total. The despatcher can then choose any of the recommended actions and execute them. In case of failure of supervisory control for few breakers, the total desired load shed/restore will not be met. Under such conditions, the application will inform the dispatcher the balance amount of load to be shed /restore. The load-shed application runs again to complete the desired load shed /restore process.

4.5.2.8 Fault Management And System Restoration (FMSR)

The availability of real-time data of the field equipments such as breakers, switches, etc. and the level of the fault current flowing in the networks helps the maintenance crew to manage and restore the system in an event of fault. FMSR application also provides assistance to the power system despatcher for detection, localization, isolation and restoration of distribution system, if after a fault in the system has occurred with the help of o SCADA/DMS. The field equipments which are the part of the distribution SCADA which help to localize and isolate the faults are mainly Auto Reclosers(AR), Sectionalisers and Fault Passage Indicators (FPI).

4.5.2.9 Loss Minimization Via Feeder Reconfiguration (LMFR)

Providing power supply by modifying the feeder configuration topology of the distribution network, in the event of a fault occurrence in an appropriate fashion is the functionality of Loss Minimization via Feeder Reconfiguration. The information of network topology and availability of adjacent feeder networks are required for the right selection of feeders with overall aim of reducing the line losses and maximum power delivery to consumers. The LMFR also has the capability of identifying the opportunities to minimize technical losses in the distribution system by reconfiguration of feeders in the network for a given load scenario. It calculates the current losses based on the loading of all elements of the network. In case if the telemetered values are not updated due to telemetry failure, LMFR with the help of LFA, completes its functionality. The LMFR application can be utilized to have the various scenarios for a given scheduled and unscheduled outages, equipment operating limits, tags placed in the SCADA system while recommending the switching operations.

4.5.2.10 Load Balancing Via Feeder Reconfiguration (LBFR)

Load Balancing via Feeder Reconfiguration is used for the optimal balance of the segments of the network that are over and under loaded. This helps to improve utilization of the capacities of distribution facilities such as transformer and feeder ratings. The feeder reconfiguration function can also be used to have a scenario of an overload condition, unequal loadings of the parallel feeders and transformers. The system will help to generate the switching sequence to reconfigure the distribution network for transferring load from some sections to other sections. The advanced LBFR application is capable of managing the scheduled and unscheduled outages, equipment operating limits, tags placed in the SCADA system while recommending the switching operations. The function distribute the total load of the system among the available transformers and the feeders in right proportion to their operating capacities, considering the discreteness of the loads, available switching options between the feeder and permissible intermediate overloads during switching. In the dispatch Training Simulator (DTS), the switching operations can be simulated and visualize the impact on the distribution network by comparing real-time load distribution, voltage profiles, load restored, distribution system losses and the number of affected customer.

4.5.2.11 Distribution Load Forecast (DLF)

The Distribution Automation system keeps logging data periodically of the network. This historical database and weather conditions data collected over a period can be used for prediction and to have forecasting of the requirement of consumer loads. Generally there are two types of forecasting that are resorted too. Short Term Load Forecasting (STLF) will be used for assessment of the sequence of average electrical loads in equal time intervals, from 1 to 7 days ahead. The Long term forecasting is used for forecasting load growths over longer durations. The forecasting techniques are based on different methods such as,

1. autoregressive,
2. least Squares Method,
3. time Series Method,
4. neural Networks,

5. Kalman filter, and
6. weighted Combination of these method.

4.5.2.12 Outage Management System (OMS)

An Outage Management System is network management software that is capable of restoring the network model after an outage. Outage Management Systems are integrated properly with SCADAS/DMS, for timely, correct and precise control actions. OMS is not only capable of performing restoration activities related to service, but also capable of tracking, displaying and grouping outages.

An OMS is mainly used by operators in electric distribution systems. It can help in identifying the portion of the circuit responsible for the interruption. Based on the different criteria present in the network, it can also assist in grouping and prioritizing the resources and indirectly help in minimizing the impacts of outages. An OMS generally uses Field Remote Terminal Units (FRTU) and Fault Passage Indicators (FPI). A typical OMS has the following features,

1. prioritizes restoration efforts and management of resources upon outages,
2. provides supervisors with an estimated timeline of restoration,
3. reports the actual cause of the outage, and
4. provides accurate information about the extent of the outage and its impact on customers and their management.

With the help of fault locating equipments and OMS software, it can carry out the following operations.

1. Locate and identify the faults,
2. Arrange the maintenance crew members to the faulty location,
3. Maintain a proper customer relationship, and
4. Fault restoration.

The benefits of outage management software are,

1. prioritization of resources and planning involved in outage management software results in reduced outages and faster recovery,
2. customer relationship is improved due to better management of outage issues,
3. because of the tracking involved, there is better prediction of outages, allowing them to be handled properly,

4. operational efficiency is increased compared to situations where an outage management system is not in place,
5. operational visibility across the network increases greatly with the use of an outage management system, and
6. decision making is faster for supervisors because of the reports provided by the application, even in cases of complex outages.

Modern OMS operates with the following systems for better results.
1. GIS mapping,
2. Fault Passage Indicators(FPI),
3. Distribution Transformer Monitoring Systems(DTMS),
4. FRTU having redundant communication facility,
5. Advanced Metering Infrastructure (AMI), and
6. CIS and Crew Management System(CMS).

4.5.3 Geographical Information Systems (GIS)

Geographical Information Systems is a technology that captures, stores, manages, analyses and distributes geographic knowledge. It provides tools to interpret that data for understanding the relationships, patterns, or trends intuitively that are not possible to see with traditional charts, graphs and spreadsheets. Power distribution utilities capture and constantly upgrade their consumer and network data. Use of GIS mapping of the consumers and network assets facilitate an easy updatable and accessible database which cater the needs of monitoring and maintaining reliable quality power supply, efficient metering, billing and collections, comprehensive energy audit, theft detection and reduction of Transmission and Distribution losses. All these measures will ultimately improve the overall internal efficiency of the distribution utilities and help accelerate achieving commercial viability.

Components of GIS: GIS is composed of hardware, software and data. GIS software ranges from simple business mapping tools to high-end technology used to manage complex systems. GIS can be divided into four categories.
1. Desktop GIS which helps to analyses, map, manage, share, and publish geographic information on desktop computers.
2. Server GIS, which GIS functionality helps the data to be deployed from a central environment.
3. Embedded GIS, is the technology that lets the GIS functionality and data to be embedded inside other applications.

4. Mobile or field GIS, technologies that run on mobile devices such as PDAs, laptops, and Smart phones. GIS implementation requires careful planning and fairly significant investment for computerization, networking, database connectivity, etc.

Data exchange between DMS and GIS: A typical GIS integration with the DMS is shown in Figure 4.7. Different file formats are used to transfer network topology, spatial data, and asset management data to DMS.

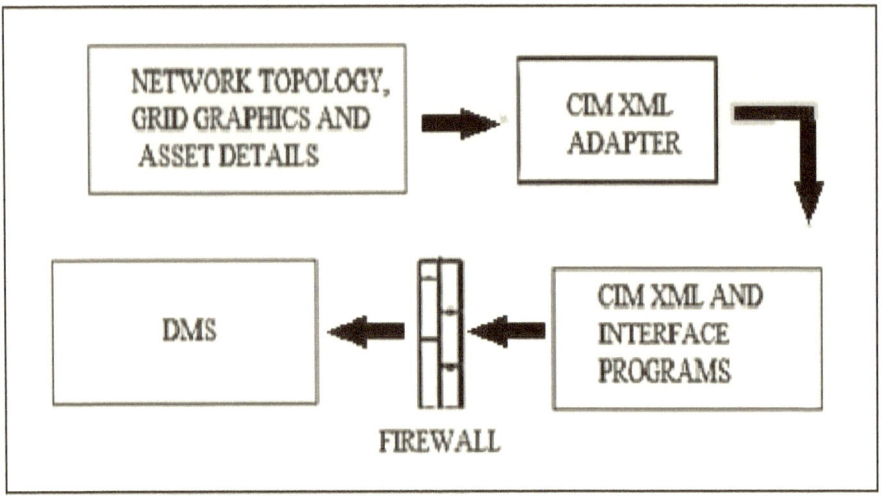

▲ **Figure 4.7:** Data exchange between GIS and DMS

4.5.4 Customer Information Systems (CIS)

The Customer Information System (CIS) is a vital component of the meter-to-cash (M2C) value chain for electric utilities it is the glue that blinds the consumption and metering processing process to payments, collections and other downstream processes that affects a company's top line. Presently CIS systems were designed to handle flat tariffs or volume based tariffs, and monthly, or bi-monthly, or quarterly usage data. Many are not equipped to handle the complexities that accompany the Smart Grid, such as dynamic or time of use tariffs or the surge of usage data gathered in half-hourly or hourly intervals by smart meters.

The Smart Grid technology makes the relationship between customers and utility very complex which requires an advanced and dynamic consumer information system. Though many CIS systems has been developed by various utilities but most of them are not timely proactive in responding to customer

requests, market forces and regulatory changes. A better approach solve this issue is by outsourcing CIS to IT industry experts of niche service providers and access the latest technology while avoiding huge upfront cost. A typical CIS deals with the following data.
1. Customer data such as type, category, connected load, etc.,
2. Meter data, such as type, data, etc.,
3. Billing and payment details, such as method, history, etc.
4. Consumption pattern and history details,
5. Irregularities, and
6. Tariff details.

4.5.5 High-Efficiency Distribution Transformers (HEDT)

Transformer is the most important equipment in the transmission and distribution of electric power. Using better grade materials and optimum design, sudden failures can be reduced along with lower maintenance cost, leading to increased life expectancy. By improving the electrical steel (silicon steel) properties, the losses of a transformer can be made half that of a similar transformer which uses the ordinary steel. With new magnetic materials, and additional copper in the windings, even higher efficiency can be achieved. These benefits add up and balance against the inevitable increase in maintenance and purchase cost.

4.5.6 Phase Shifting Transformers (PST)

Special transformers used to create a phase shift between the primary side voltage and secondary side voltage, are known as Phase Shifting Transformers. Transformers are used to transport electrical power between different voltage levels of the electric grid. Transmission of electric power planning has to be done in such a way that besides conserving Right of Way (ROW) for new lines, optimum utilization of existing lines has to be made. Physical path between generation and load centers for power transmission is generally achieved through a number of alternate paths. There could be uneven loading of parallel transmission lines due to different impedances caused by the tower geometry, conductor sizing, number of sub-conductors and line length. The distribution of the power flow between two parallel lines is dictated by their impedances. A Phase Shifting Transformer (PST) can be employed for power control in transmission lines. The purpose of this phase shift is to control the power flow over transmission lines. Both the magnitude and direction of power flow can be controlled by varying the phase shift.

4.5.7 Solid State Transformer (SST)

Solid State Transformers are one of the exciting and emerging technologies under development. Though the technology is announced in 1980, the availability of the appropriate materials and its cost made it difficult for commercialization. With the advancement in material science and technology, today SST is becoming a reality for commercialization. An SST can take both AC and DC inputs as well as give AC and DC outputs and enable bi-directional power-flows. It can also improve power quality, reactive compensation and harmonic filtering. SST will be only 1% in size and weight of a comparable distribution transformer. SSTs of 11-15kV ratings are expected to be commercialized very soon. Undoubtedly it will bring a radical transformation in the electric grid where AC and DC will merge.

SUMMARY

This chapter begins with elaborating the Substation Automation, by describing the modern components required for realizing digital substations. Then move on to explain the advanced transmission components such as HVDC, FACTS, PMU, and WAMS. This chapter then focuses on the smart distribution systems such as feeder automation, distribution management system in which it explains the State Estimation, Volt/VAR control, Load Flow Application (LFA), Fault Management & System Restoration (FMSR) Application, Load Shed Application (LSA), Fault Management & System Restoration (FMSR) Application, Loss Minimization via Feeder Reconfiguration (LMFR), Load Balancing via Feeder Reconfiguration (LBFR), Distribution Load Forecast (DLF), and Outage Management System(OMS), A brief description of Customer Information Systems(CIS), Geographical Information Systems(GIS), High-Efficiency Distribution Transformers, and Phase Shifting Transformers.

CHAPTER 05
INDUSTRIAL COMMUNICATION BASICS

5.1 INTRODUCTION

The key enabler for the ICS and DCS is the availability of secure and preferably bi-directional data communications and the proper amalgamation of distributed intelligence and communication technologies. Thus, the design and implementation of a modern, reliable communications infrastructure is a fundamental and important requirement for making the ICS or DCS smarter.

The medium by which data is transmitted is known as a communication channel or communication media. The transfer of data takes place in the form of analog signals and the transfer of data is measured in the form of bandwidth, the higher the bandwidth the more the data that will be transferred. Communication media are broadly classified into two categories, namely, guided media (wired) and unguided media (wireless). Both media are used for short distance (LANs, MANs) and long distance (WANs) communication. This chapter elaborates various guided and unguided media used in communication, its merits and demerits, and practical considerations to be taken care while selecting the media for different applications. The chapter also explain the various communication technologies currently available for the deployment in ICS and DCS. Before that, it is better to be familiar with certain terminologies which are briefly explained below.

5.2 TYPES OF TRANSMISSION

There are various transmission classifications depending upon the different technologies employed and they are mainly based on
1. Analog and digital,
2. Synchronous and Asynchronous,
3. Broadcast, Multicast, and Unicast,
4. Simplex, Half Duplex, and Full Duplex, and
5. Baseband and Broadband.

As the communication is the key enabler of the modern SCADA, a basic understanding of these terms become most essential for a automation engineer. Hence brief explanation of these terminologies are given below.

5.2.1 Analog and Digital

There are two types of communication technologies in practice viz analog and digital. Analog communications occur with a continuous signal that varies in frequency, amplitude, phase, voltage, and so on. The variances in the continuous signal produce a wave shape as opposed to the square shape of a digital signal. Digital communications occur through the use of a discontinuous electrical signal and a state change or on-off pulses.In other way the information is encoded digitally as discrete signals and transmitted electronically to the recipients. Digital signals are more reliable than analog signals over long distances or when interference is present. This is because of a digital signal's definitive information storage method employing direct current voltage where voltage on represents a value of 1 and voltage off represents a value of 0. These on-off pulses create a stream of binary data. Analog signals become altered and corrupted because of attenuation over long distances and interference. Since an analog signal can have an infinite number of variations used for signal encoding as opposed to digital's two states, unwanted alterations to the signal make extraction of the data more difficult as the degradation increases. A brief comparison between these two technologies are summarized below.

1. *Bandwidth:* This factor creates the key difference between Analog and digital communication. Analog signal requires less bandwidth for the transmission while digital signal requires more bandwidth for the transmission.
2. *Power Requirement:* Power requirement in case of digital communication is less when compared to analog communication. Since the bandwidth requirement in digital systems is more thus, they consume less power. And Analog communication system requires less bandwidth thus more power.
3. *Fidelity:* Fidelity is a factor which creates a crucial difference between Analog and digital communication. Fidelity is the ability of the receiver which receives the output exactly in coherence with that of transmitted input. Digital communication offers more fidelity as compared to Analog Communication.
4. *Noise Distortion and Error Rate:* Analog systems are affected by Noise while digital systems are immune from Noise and Distortion. Error rate

is another significant difference which separates Analog and Digital Communication. In Analog instruments, there is an error due to parallax or other kinds of observational method.

5. *Synchronization:* Digital communication system offers to synchronize which is not effective in analog communication. Thus, synchronization also creates a key difference between Analog and Digital Communication.
6. *Cost:* Digital communication equipments are costly and digital signal require more bandwidth for transmission.
7. *Hardware Flexibility and Portability:* The hardware of analog communication system is not as flexible as digital communication. Analog systems are less portable as components are heavy while digital systems are more portable as they are compact equipments.

5.2.2 Synchronous And Asynchronous

Communications are either synchronous or asynchronous. Some communications are synchronized with some sort of clock or timing activity and referred to synchronous communications. It rely on a timing or clocking mechanism based on either an independent clock or a time stamp embedded in the data stream. Synchronous communications are typically able to support very high rates of data transfer. On the other hand, asynchronous communications rely on a stop and start delimiter bit to manage the transmission of data. Because of the use of delimiter bits and the stop and start nature of its transmission, asynchronous communication is best suited for smaller amounts of data. Public Switched Telephone Network (PSTN) modems are good examples of asynchronous communication devices.

5.2.3 Broadcast, Multicast, and Unicast

Broadcast, multicast, and unicast technologies determine how many destinations a single transmission can reach.

Unicast technology supports only a single communication to a specific recipient. In this case there is just one sender, and one receiver. Unicast transmission, in which a packet is sent from a single source to a specified destination, is still the predominant form of transmission on LANs and within the Internet. All LANs and IP networks support the unicast transfer mode, and most users are familiar with the standard unicast applications (e.g. http, smtp, ftp and telnet) which employ the TCP transport protocol.

Broadcast technology supports communications to all possible recipients. In this case there is just one sender, but the information is sent to all connected receivers. Broadcast transmission is supported on most LANs (e.g. Ethernet), and may be used to send the same message to all computers on the LAN (e.g. the Address Resolution Protocol (ARP) uses this to send an address resolution query to all computers on a LAN. Network layer protocols (such as IPv4) also support a form of broadcast that allows the same packet to be sent to every system in a logical network (in IPv4 this consists of the IP network ID and an all 1's host number). One example of an application which may use multicast is a video server sending out networked TV channels. Simultaneous delivery of high quality video to a large number of delivery platforms will exhaust the capability of even a high bandwidth network with a powerful video clip server. This poses a major scalability issue for applications which required sustained high bandwidth. One way to significantly ease scaling to larger groups of clients is to employ multicast networking.

Multicast networking technology supports simultaneous communications to multiple specific recipients. In this case there is may be one or more senders, and the information is distributed to a set of receivers. IP multicast provides dynamic many-to-many connectivity between a set of senders (at least 1) and a group of receivers. The format of IP multicast packets is identical to that of unicast packets and is distinguished only by the use of a special class of destination address, which denotes a specific multicast group. Since TCP supports only the unicast mode, multicast applications must use the UDP transport protocol.

Unlike broadcast transmission (used in LAN), multicast clients receive a stream of packets only if they have previously elect to do so. Membership of a group is dynamic and controlled by the receivers. The routers in a multicast network learn which sub-networks have active clients for each multicast group and attempt to minimise the transmission of packets across parts of the network for which there are no active clients.

The multicast mode is useful if a group of clients require a common set of data at the same time, or when the clients are able to receive and store (cache) common data until needed. Where there is a common need for the same data required by a group of clients, multicast transmission may provide significant bandwidth savings (up to 1/N of the bandwidth compared to N separate unicast clients).

The majority of installed LANs are able to support the multicast transmission mode. Shared LANs inherently support multicast, since all packets reach all network interface cards connected to the LAN. The earliest

LAN network interface cards had no specific support for multicast and introduced a big performance penalty by forcing the adaptor to receive all packets (promiscuous mode) and perform software filtering to remove all unwanted packets. Most modern network interface cards implement a set of multicast filters, relieving the host of the burden of performing excessive software filtering.

5.2.4 Simplex, Half Duplex and Full Duplex Communication Channels

In a communication system, there will be a transmitter and a receiver. In between the transmitter and thereceiver, there is a transmission medium of the data/information, usually referred as the communication channel. Although the required information for transmission originates from a single source, there may be more than one destination or receivers. This depends upon how many receiving stations are linked to the channel and how much energy the transmitted signal possesses. If the channel length is more and the transmission power is less, the receiver situated at a long distance cannot receive the data properly. In a digital communications channel, the information can be represented by a stream of bits called bytes. A collection of bytes can be grouped to form a frame or other higher-level message unit. These types of multiple levels of encapsulation facilitate the handling of messages in a complex data communications network. If any communications channel is considered, it has a direction associated with it.

Simplex Channel: It is conventional that the message source is the transmitter, and the destination is thereceiver. A channel whose direction of transmission is unchanging is called as a simplex channel. In other words, a type of data transmission, where message transmission is taken place only in one direction, typical example is the radio station which is a simplex channel because it always transmits the signal to its listeners and never allows them to transmit back. Another example is the television. The advantage of simplex mode of transmission is, since the data can be transmitted only in one direction, the entire band width can be used.

Half Duplex Channel: A half-duplex channel can be considered as a single physical channel in which the direction may be reversed. Messages can flow in two directions in a half-duplex type, but never at the same time. In other words it can be said that at a single time, the transmission of data are done in only one direction. For example, in a telephone call, one party speaks while the other listens. After a pause, the other party speaks and the first party listens. Speaking simultaneously will result in a garbled sound that cannot be understood. The main difficulty of half-duplex mode of transmission is since two channels are used, the band width of the channel should be decreased.

Full Duplex Channel: A full-duplex channel can be used for bi-directional communication. In fact message can be transmitted simultaneously in both directions. It comprises of two simplex channels, a forward channel and a backward (reverse) channel, linking at the same points. The transmission rate of the reverse channel will be very slow if it is used only for flow control of the forward channel. The main problem of the full duplex mode of transmission is, since two channels are required, the band width should be decreased.

5.2.5 Baseband And Broadband

The number of channels thatcan be pushed into a single wire simultaneously over a cable segment depends on whether the customer use baseband technology or broadband technology. Baseband technology can support only a single communication channel. Analog transmission means that data is being moved as waves, and digital transmission means that data is being moved as discrete electric pulses. Baseband uses a direct current applied to the cable. A current that is at a higher level represents the binary signal of 1, and a current that is at a lower level represents the binary signal of 0. Baseband is a form of digital signal. Ethernet is a baseband technology.

Broadband technology divides the communication channel into individual and independent subchannels so that different types of data can be transmitted simultaneously. Broadband uses frequency modulation to support numerous channels, each supporting a distinct communication session. Broadband is suitable for high throughput rates, especially when several channels are multiplexed. Broadband is a form of analog signal. As an example, a coaxial cable TV (CATV) system is a broadband technology that delivers multiple television channels over the same cable. This system can also provide home users with Internet access, but these data are transmitted at a different frequency spectrum than the TV channels. A Digital Subscriber Line (DSL) uses one single phone line and constructs a set of high-frequency channels for Internet data transmissions. A cable modem uses the available frequency spectrum that is provided by a cable TV carrier to move Internet traffic to and from a household. Mobile broadband devices implement individual channels over a cellular connection, and WiFi broadband technology moves data to and from an access point over a specified frequency set. Characteristics of baseband and broadband are summarized in Table 5.1.

▼ **Table 5.1:** Characteristics of baseband and broadband

Baseband	Broadband
Digital signals are used	Analog signals are used
Frequency division multiplexing is not possible	Transmission of data is unidirectional
Baseband is bi-directional transmission	Signal travelling distance is long
Short distance signal travelling	Frequency division multiplexing is possible
Entire bandwidth of the cable is consumed by a single signal in a baseband transmission.	The signals are sent on multiple frequencies but are sent simultaneously

5.3 GUIDED MEDIA

Guided media are more commonly known as wired media or bounded media, or those media in which electrical or optical signals are transmitted through a cables or wires. In fact it needs a physical material medium to propagate and the electrical signals are confined within the cable or wire which transmits them. Typical forms of guided media include copper co-axial cables, fiber-optic cables and twisted-pair copper cables, which can be shielded or unshielded. Transmission of digital data through either guided or unguided communication involves the coding of the data at the sender's end, the modulation of the carrier signal, the demodulation of the signal on the receiving end and the decoding of the binary signal.

5.3.1 Twisted Pair Copper

It is the most widely deployed media type across the world, as the last mile telephone link connecting every home with the local telephone exchange is made of twisted pair copper. It is also used as last mile connectivity to access the internet from home. They are also used in Ethernet LAN cables within homes and offices. They support low to High Data Rates (in order of Giga bits). However, they are effective only upto a maximum distance of a few kilometres/miles, as the signal strength is lost significantly beyond this distance.

They come in two variants, namely UTP (unshielded twisted pair) and STP (shielded twisted pair). Within each variant, there are multiple sub-variants, based on the thickness of the material (like UTP-3, UTP-5, UTP-7 etc.). Twisted-pair cabling has insulated copper wires surrounded by an outer protective jacket. If the cable has an outer foil shielding, it is referred to as *shielded twisted pair (STP)*, which adds protection from radio frequency interference and electromagnetic interference. Twisted-pair cabling, which does not have this extra outer shielding, is called *unshielded twisted pair (UTP)*. Twisted-pair cable is cheaper and easier to work with.

The twisted-pair cable contains copper wires that twist around each other, as shown in Figure 5.1. This twisting of the wires protects the integrity and strength of the signalsthey carry. Each wire forms a balanced circuit, because the voltage in each pair uses the same amplitude, just with opposite phases. The tighter the twisting of the wires, the more resistant the cable is to interference and attenuation. UTP has several categories of cabling, each of which has its own unique characteristics. The twisting of the wires, the type of insulation used, the quality of the conductive material, and the shielding of the wire determine the rate at which data can be transmitted. The UTP ratings indicate which of these components were used when the cables were manufactured. Some types are more suitable and effective for specific uses and environments. Table 5.2 lists the cable ratings.

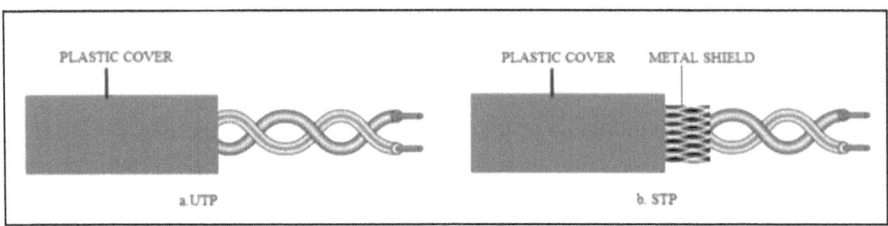

▲ **Figure 5.1:** Twisted Pair Copper

▼ **Table 5.2:** UTP cable ratings

UTP Category	Characteristics	Usage
Category 1	Voice grade telephone cable for up to 1 Mbps transmission rate.	Not recommended for network use, but modems can communicating over it.
Category 2	Data transmission upto 4 Mbps	Used in mainframe and minicomputer terminal connections, but not recommended for high-speed installations.
Category 3	10 Mbps for Ethernet and 4 Mbps for Token Ring	Used in 10 Base-T network installations.
Category 4	16 Mbps	Usually used in Token Ring networks.
Category 5	100 Mbps: has high twisting and thus low crosstalk.	Used in 100 Base-TX, CDDI, Ethernet, and ATM installations: most widely used in network installations.
Category 6	10 Gbps	Used in new network installations requiring high speed transmission. Standard for Gigbit Ethernet.
Category 7	10 Gbps	Used in new network installations requiring high speed transmission.

Copper cable has been around for many years. It is inexpensive and easy to use. A majority of the telephone systems today use copper cabling with the rating of voice grade. Twisted-pair wiring is the preferred network cabling, but it also has its drawbacks. Copper actually resists the flow of electrons, which causes a signal to degrade after it has traveled a certain distance. This is why cable lengths are recommended for copper cables; if these recommendations are not followed, a network could experience signal loss and data corruption. Copper also radiates energy, which means information can be monitored and captured by intruders. UTP is the least secure networking cable compared to coaxial and fiber. If a company requires higher speed, higher security, and cables to have longer runs than what is allowed in copper cabling, fiber-optic cable maybe a better choice.

5.3.2 Copper Co-Axial Cables

Co-axial copper cables have an inner copper conductor and an outer copper shield, separated by a di-electric insulating material, to prevent signal losses as shown in Figure 5.2. This is all encased within a protective outer jacket. The term coaxial comes from the inner conductor and the outer shield sharing a geometric axis. Compared to twisted-pair cable, coaxial cable is more resistant to electromagnetic interference (EMI), provides a higher bandwidth(the bandwidth is 80 times more than twisted pair cable.), and supports the use of longer cable lengths.

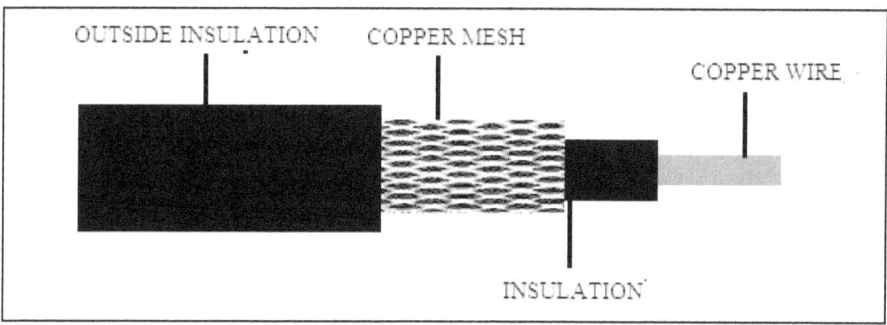

▲ **Figure 5.2:** Co-axial copper cable

It is primarily used in cable TV networks and as trunk lines between telecommunication equipments.
1. It serves as an internet access line from the home.
2. It supports medium to High Data Rates

3. It has much better immunity to noise and hence signal strength is retained for longer distances than in copper twisted pair media.

There are two main types of coaxial cable: thinnet and thicknet. Thinnet, also known as 10Base2, was commonly used to connect systems to backbone trunks of thicknet cabling. Thinnet can span distances of 189 meters and provide throughput up to 10 Mbps. Thicknet, also known as 10Base9, can span 900 meters and provide throughput up to 10 Mbps (megabits per second). The most common problems with coax cable are as follows:
1. Bending the coax cable past its maximum arc radius and thus breaking the center conductor.
2. Deploying the coax cable in a length greater than its maximum recommended length (which is 189 meters for 10Base2 or 900 meters for 10Base9)
3. Not properly terminating the ends of the coax cable with a 90 ohm resistor

5.3.3 Fiber Optic Cables

Here, information is transmitted by propagation of optical signals (light) through fiber optic cables and not through electrical/electromagnetic signals. Due to this, fiber optics communication supports longer distances as there is no electrical interference. As the name indicates, fiber optic cables are made from glass(silica) and is as very thin as the human hair. They are coated with plastic also known as jacket.

As they support very high data rates, fiber optic lines are used as WAN backbone and trunk lines between data exchange equipments. They are also used for accessing internet from home through FTTH (Fiber-To-The-Home) lines. Additionally, they are used even for LAN environment with different LAN technologies like Fast Ethernet, Gigabit Ethernet etc. using optical links at the physical layer.

Fiber-optic cabling has higher transmission speeds that allow signals to travel over longer distances. Fiber cabling is not as affected by attenuation and EMI when compared to cabling that uses copper. It does not radiate signals, as does UTP cabling, and is difficult to eavesdrop on; therefore, fiber-optic cabling is much more secure than UTP, STP, or coaxial.

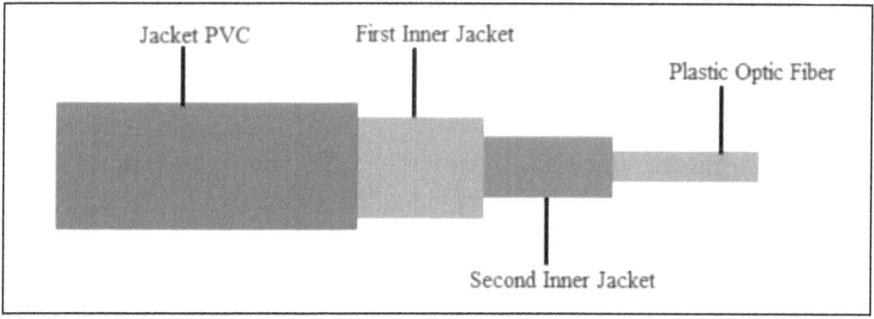

▲ **Figure 5.3:** Fiber Optic Cable

Optical fiber consists of a core and a cladding layer, selected for total internal reflection due to the difference in the refractive index between the two. In practical fibers, the cladding is usually coated with a layer of acrylate polymer or polyimide. This coating protects the fiber from damage but does not contribute to its optical waveguide properties. Individual coated fibers (or fibers formed into ribbons or bundles) then have a tough resin buffer layer and/or core tube(s) extruded around them to form the cable core. Several layers of protective sheathing, depending on the application, are added to form the cable. Rigid fiber assemblies sometimes put light-absorbing (dark) glass between the fibers, to prevent light that leaks out of one fiber from entering another. This reduces cross-talk between the fibers, or reduces flare in fiber bundle imaging applications. For indoor applications, the jacketed fiber is generally enclosed, with a bundle of flexible fibrous polymer *strength members* like aramid (e.g. Twaron or Kevlar), in a lightweight plastic cover to form a simple cable. Each end of the cable may be terminated with a specialized optical fiber connector to allow it to be easily connected and disconnected from transmitting and receiving equipment.

For use in more strenuous environments, a much more robust cable construction is required. In *loose-tube construction* the fiber is laid helically into semi-rigid tubes, allowing the cable to stretch without stretching the fiber itself. This protects the fiber from tension during laying and due to temperature changes. Loose-tube fiber may be *dry block* or gel-filled. Dry block offers less protection to the fibers than gel-filled, but costs considerably less. Instead of a loose tube, the fiber may be embedded in a heavy polymer jacket, commonly called tight buffer construction. Tight buffer cables are offered for a variety of applications, but the two most common are *Breakout* and *Distribution*. Breakout cables normally contain a ripcord, two non-conductive dielectric strengthening members (normally a glass rod epoxy), an aramid yarn, and 3 mm buffer tubing with an additional layer of Kevlar surrounding each fiber. The ripcord is a parallel cord of strong yarn that is situated under the jacket(s) of the cable for jacket removal. Distribution cables have an overall Kevlar wrapping, a ripcord, and a 900 micrometer buffer coating surrounding each fiber. These *fiber units* are commonly bundled with additional steel strength members, again with a helical twist to allow for stretching.

A critical concern in outdoor cabling is to protect the fiber from contamination by water. This is accomplished by use of solid barriers such as copper tubes, and water-repellent jelly or water-absorbing powder surrounding the fiber. Finally, the cable may be armored to protect it from environmental hazards, such as construction work or gnawing animals. Undersea cables are more heavily armored in their near-shore portions to protect them from boat anchors, fishing gear, and even sharks, which may be attracted to the electrical power that is carried to power amplifiers or repeaters in the cable. Modern cables come in a wide variety of sheathings and armor, designed for applications such as direct burial in trenches, dual use as power lines, installation in conduit, lashing to aerial telephone poles, submarine installation, and insertion in paved streets.

5.3.4 Cabling Considerations

Cables are extremely important within networks, and when they experience problems, the whole network can become problematic. This section addresses some of the common cabling issues that many networks experience.

5.3.4.1 Noise

Noise on a line is usually caused from the adjoining devices or from the environment. Noise can be produced by motors, computers, copy machines, fluorescent lighting, and microwave ovens. This background noise can combine with the data being transmitted over the cable and distort the signal.

5.3.4.2 Cabling Connection Types

Cables follow standards to allow for interoperability and connectivity between common devices and environments. The standards are developed and maintained by the Telecommunications Industry Association (TIA) and the Electronic Industries Association (EIA). The TIA/EIA standards enable the

design and implementation of structured cabling systems for commercial buildings. The majority of the standards define cabling types, distances, connections, cable system architectures, cable termination standards and performance characteristics, cable installation requirements, and methods of testing installed cable. The following are commonly used physical interface connection standards.
1. RJ-11 is used for terminating telephone wires,
2. RJ-45 is used for terminating twisted-pair cables in Ethernet LAN, and
3. BNC (British Naval Connector) is used for terminating coaxial cables.

5.3.4.3 Attenuation

Attenuation is the loss of signal strength while it propagates through the medium. The longer a cable, the more attenuation occurs, which causes the signal carrying the data to deteriorate. This is why standards include suggested cable-run lengths. The effects of attenuation increase with higher frequencies. This means that cable used to transmit data at higher frequencies should have shorter cable runs to ensure attenuation does not become an issue.

If a networking cable is too long, attenuation may occur. Basically, the data are in the form of electrons, and these electrons have to swim through a copper wire. However, this is more like swimming upstream, because there is a lot of resistance on the electrons working in this media. After a certain distance, the electrons start to slow down and their encoding format loses form. If the form gets too degraded, the receiving system cannot interpret them any longer. If a network administrator needs to run a cable longer than its recommended segment length, network administrator needs to insert a repeater that will amplify the signal and ensure it gets to its destination in the right encoding format. Attenuation can also be caused by cable breaks and malfunctions. If a cable is suspected of attenuation problems, cable testers can inject signals into the cable and confirm fault.

5.3.4.4 Crosstalk

Crosstalk is a phenomenon caused by the electric or magnetic fields of one communication channel spill over to the signals of adjacent channel. When the different electrical signals mix, the integrity of the signals degrades and data may get corrupted. In a telephone circuit, crosstalk can result in disturbing the hearing part of a voice conversation. It can occur in microcircuits within computers, audio equipment and within network circuits. UTP is much more vulnerable to crosstalk than STP or coaxial, because it does not have extra layers of shielding to protect against crosstalk.

5.3.4.5 Fire Rating of Cables

Fire-resistive or fire-rated cable is a cable that will continue to operate in the presence of a fire. This is commonly known as circuit integrity (CI) cable. Certain cables, when on fire produces hazardous gases that would spread throughout the building quickly. Network cabling that is placed in these types of areas, referred as plenum space, must meet a specific fire rating to ensure it will not produce and release harmful chemicals in case of a fire. Flame-retardant cable is a cable that will not convey or propagate a flame as defined by the flame-retardant or propagation tests. Flame retardant tests measure flame propagation for both horizontal and vertical applications. There are also plenum cable flame tests when it is used in ducts, plenums or other spaces. A flame-retardant cable is not a fire-rated cable. There are certain but specific differences between flame-retardant cables and fire-resistive cables. Flame-retardant cables resist the spread of fire into a new area, whereas fire resistive cables maintain circuit integrity and continue to work for a specific time under defined conditions. These circuit integrity cables continue to operate in the presence of a fire and are sometimes called 1-hour or 2-hour fire-rated cables. CI cables are needed when it is most essential and critical for life safety or to prevent a plant shut down.

Depending upon the situation, while setting up a network, it is important to select the appropriate types of wire. Cables should be installed in unexposed areas so they are not easily tripped over, damaged, or eavesdropped upon. The cables should be strung behind walls and in the protected spaces as in dropped ceiling. In environments that require extensive security, wires are encapsulated within pressurized conduits so if someone attempts to access a wire, the pressure of the conduit will change, causing an alarm to sound and a message to be sent to the security men.

Note: A plenum is the air return path of a central air handling system. It can be either ductwork or open space over a suspended ceiling or raised floor.

5.4 UNGUIDED MEDIA

Unguided media are more commonly known as wireless media or unbounded media, in which data is transferred into electromagnetic wavesand sent through free space without guiding any specific direction. Hence the name unguided media. All unguided media transmission are classified as wireless transmission which is quickly expanding the field of technologies for networking.. The various kinds of unguided media are microwave, cellular radio, radio broadcast and satellite. As wireless technologies continue to proliferate, the organization's security efforts must go beyond locking down its local network. Security should be an end-to-end solution that addresses all forms, methods, and techniques of communication.

Different types of unguided communication are classified based on the frequency spectrum used for communication, the distance between the end stations and the type of encoding used for the communication. Broadband wireless signals occupy frequency bands that may be shared with microwave, satellite, radar, and ham radio use. Unguided communication allows electromagnetic signals to travel between antennas, some of which are on satellites. Antennas can provide point-to-point communication or can send their signals in all directions. These technologies are used for television transmissions, cellular phones, satellite transmissions, spying, surveillance, and garage door openers. The different unguided medias which are employed in distributed SCADA system are briefly explained below.

5.4.1 Microwaves Communication

In this kind the data is transferred via air. The waves travel in a straight line. The data is received and transferred via microwave stations. The speed at which data is transferred is 190 Mbps. The two main microwave wireless transmission technologies are satellite (ground to orbiter to ground) and terrestrial (ground to ground). They are widely used by telephone and cable companies.

5.4.2 Terrestrial Communication

This type of communication is limited to line-of-sight(LOS) transmission. This means that microwaves are transmitted in a straight line and that no obstructions can exist, such as buildings or mountains, between microwave stations as shown in Figure 5.4. To avoid possible obstructions, microwave antennas often are positioned on the tops of buildings, towers, or mountains. It finds applications in long-distance telecommunication service. It requires fewer amplifiers or repeaters than coaxial cable but it is line-of-sight transmission. Short point-to-point links, Data link between local area network, closed-circuit TV, etc.

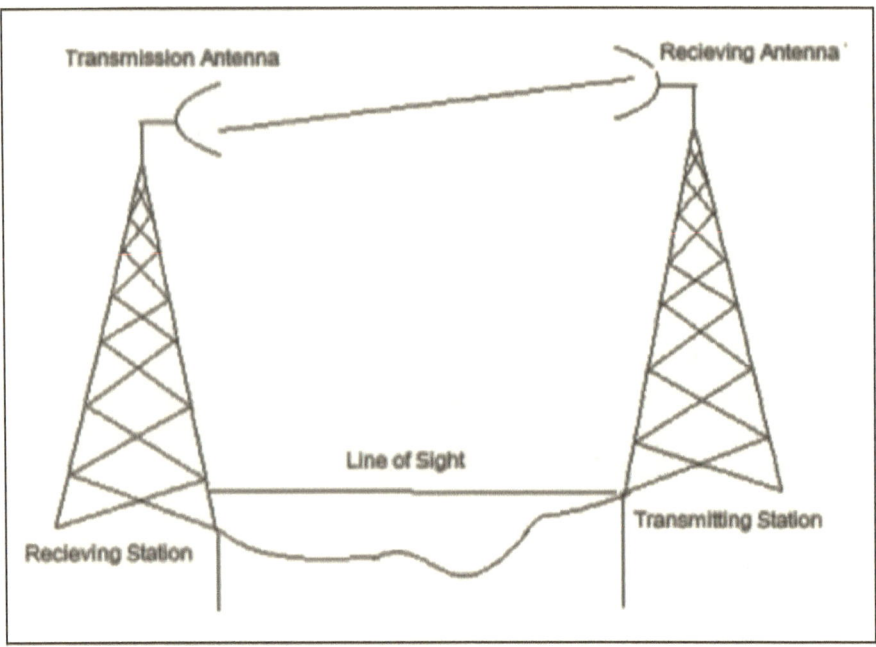

▲ **Figure 5.4:** Terrestrial communication

5.4.3 Satellite Communication

The satellites are located at a distance of 22300 miles above the earth as shown in Figure 5.5. The signals are received from earth stations. Devices like GPS and PDAs also receive signals from these earth based stations. The process of transferring and receiving data takes place within few seconds. The data is transferred at a speed of 1 Gbps. They are used for purposes like weather forecast, military communication, radio transmission, satellite TV, data transmission, etc.

Today the cost of satellite launching has brought down significantly with advanced and reusable launching modules. This made extensive use of satellite communication to provide wireless connectivity between different locations. But for two different locations to communicate via satellite links, they must be bought within the satellite's line-of-sight and area covered by the satellite called footprint. The information or data is appropriately modulated by the ground station and is transmitted to the satellite. A transponder on the satellite receives this signal, amplifies it, and relays it to the receiver. The receiver at the ground station must have an antenna, usually a circular dish like structure normally placed on top of buildings. The antenna contains one or more microwave receivers, depending upon how many satellites it is accepting data from. Satellites provide broadband transmission. If a user is receiving radio or TV data, then the transmission is set up as a one-directional network. If a user is using this connection for internet, then the transmission is set up as a bi-directional network. The available bandwidth depends upon the antenna, terminal type and the service rendered by the service provider.

As the satellite are kilometers above the Earth, time-critical applications may experience delay as the signal has to traverse to and from the satellite. Hence these types of satellites are placed into a low Earth orbit, keeping the distance between the ground stations and the satellites is less when compared to other types of satellites. Further, smaller receivers can be used for reception of signals, which makes low-Earth-orbit satellites ideal for cellular communication, radio and TV stations, Internet, etc use.

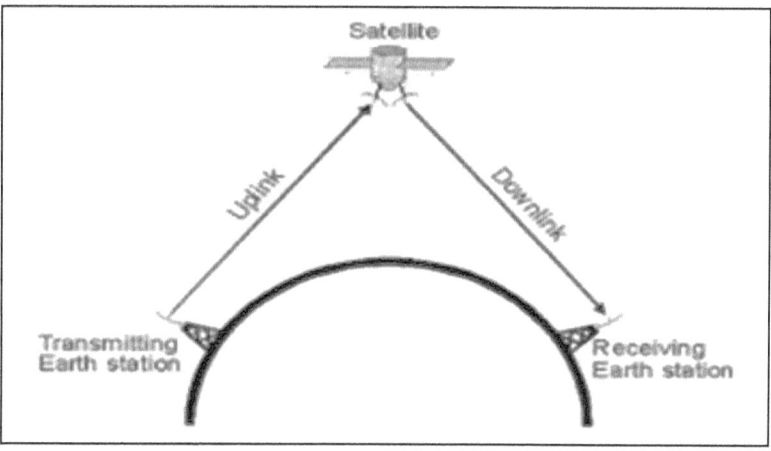

▲ **Figure 5.5:** Satellite communication

5.4.4 Mobile Communication

High frequency radio frequency (RF) waves are used for the transmission of data in mobile communication. Here one can receive and make calls and also access the internet. Earlier, when Gopi wanted to call his girlfriend Jessy, they had to dependthe telephone which had a physical connection. The telephone would only reach as far as the telephone cord that it was attached to, and the handset was also physically connected. So they actually had to sit in one place to carry out a conversation. This model worked for almost 100 years, but once mobile phones were materialized and available to everyone at a low price, there was no going back. Today mobile wireless communication has exploded withits popularity and capabilities.

Mobile phone is a device that can send voice and data over wireless radio links. It connects to a cellular network, which is connected to the Public-Switched Telephone Network (PSTN).So instead of a physical cord and connection that connects the phone and the PSTN, have a device that allows to indirectly connecting to the PSTN. Radio is the transmission of signals via electromagnetic waves within a certain frequency range. A cellular network distributes radio signals over delineated areas, called cells. Each cell has at least one fixed-location transceiver at base station, and is joined toother cells to provide connections over large geographic areas. So as somebody is talking onhis mobile phone and he move out of range of one cell, the base station in theoriginal cell sends his connection information to the next base station so that hiscall is not dropped and he can continue his conversation.

This switching from one cells to another cells, does not arises the requirements of an infinite amount of frequencies to work with. Lots of people around the world may be using their cellphones simultaneously. All of these calls take place with a set of frequencies. An elementary representation of a cellular network is shown in Figure 5.6.

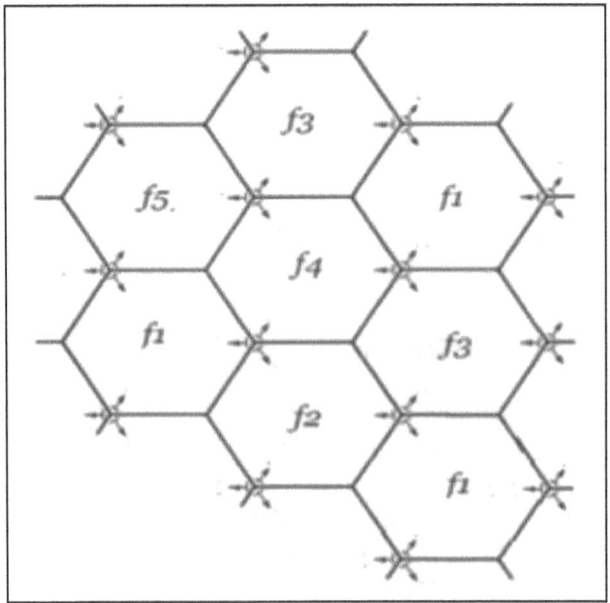

▲ **Figure 5.6:** Nonadjoint cells can use the same frequency ranges

Individual cells can use the same frequency range, as long as they are not right next to each other. So the same frequency range can be used in every other cell, which drastically decreased the amount of ranges required to support simultaneous connections. The industry had to come up with other ways to allow millions of users to be able to use this finite frequency range in a flexible manner.

Presently, mobile wireless communication has been gradually made up of complex and powerful multiple access technologies, which are mentioned below.
1. Frequency division multiple access (FDMA)
2. Time division multiple access (TDMA)
3. Code division multiple access (CDMA)
4. Orthogonal frequency division multiple access (OFDMA)

The characteristics of each of these technologies are briefly described below as they are the foundational constructs of the various cellular network generations.

When the useful bandwidth of the medium exceeds the required bandwidth of a signal, Frequency division multiple access (FDMA) is a better solution. Here a number of signals can be carried simultaneously by modulating the signals into different carrier frequency with sufficient separation between the carrier frequency. In 3G FDMA, the available frequency range is divided into sub-bands called channels, and one channel is assigned to each cell phone. The subscriber has exclusive use of that channel while the call is made, or until the call is terminated, no other calls or conversations can be made on that channel during that call. Using FDMA, multiple users can share the frequency range without the risk of interferece between the simultaneous calls. FMDA was used in the 1G.

When the achievable bit rate of the medium, of the medium exceeds the required data rate of a digital signal, Time Division Multiple Access (TDMA) is a better solution. TDMA increases the speed and efficiency of the cellular network by taking the RF spectrum channels and dividing them into time slots. In TDMA systems, time is divided into frames. Each frame is divided into slots. TDMA requires that each start and end time of each slot which are known to both the source and the destination. Mobile communication systems such as Global System for Mobile Communication (GSM), Digital AMPS (D-AMPS), and Personal Digital Cellular (PDC) use TDMA.

As the term implies, CDMA is a form of multiplexing, which allows numerous signals to occupy a single transmission channel, optimizing theuse of available bandwidth. It assigns a unique code to each voice call ordata transmission to uniquely identify it from all other transmissions sent over the cellular network. In a CDMA spread spectrum network, calls are spread throughout the entire frequency band. CDMA permits all users of the network to simultaneously use every channel in the network. At the same time, a particular cell can simultaneously interact with multiple other cells. These features make CDMA a very powerful technology, for the mobile cellular networks that presently dominate the wireless space. It has improved voice and data communication capability and is more secure.

Orthogonal Frequency Division Multiple Access (OFDMA) is a combination of FDMA and TDMA. In former implementations of FDMA, the different frequencies for each channel were widely spaced to allow analog hardware to separate the different channels. In OFDMA, each of the channels is subdivided into a set of closely spaced orthogonal frequencies with narrow bandwidths (subchannels). Each of the different subchannels can be transmitted and received simultaneously in a Multiple Input and Output (MIMO) manner. The use of orthogonal frequencies and MIMO allows signal processing techniques to reduce the impacts of any interference between different subchannels and to correct for channel impairments, such as noise and selective frequency fading. Mobile wireless technologies have gone through a whirlwind of confusing generations. The first generation (1G) dealt with analog transmissions of voice-only data over circuit-switched networks. This generation provides a throughput of around 19.2 Kbps. The second generation (2G) allows for digitally encoded voice and data to be transmitted between wireless devices, like cell phones.

The third-generation (3G) networks became available by incorporating FDMA, TDMA, and CDMA. In 3G, circuit switching is replaced by packet switching. The flexibility to support a great variety of applicationsand services are the main advantages of 3G. The modular design allows expandability, backward compatibility with 2G networks, interoperability among mobile systems, global roaming, and Internet services. 3G with more enhancements, referred to as 3.9G or mobile broadband, is taking place under the title of Third Generation Partnership Project (3GPP).3GPP has a number of new technologies such as Enhanced Data Ratesfor GSM Evolution (EDGE), High-Speed Downlink Packet Access(HSDPA), CDMA2000, and Worldwide Interoperability for Microwave Access (WiMax). There are two competing technologies that fall under the umbrella of 4G, which are Mobile WiMax and Long-Term Evolution (LTE). A 4G system does not support traditional circuit-switched telephony service as 3G does, but works over a purely packet based network. 4Gdevices are IP-based and are based upon

OFDMA. Communication engineers and scientists are actively involved in developing Fifth-Generation (5G) mobile communication, but standards requirements and implementation are not expected in near feature. Each of the different mobile communication generations has taken advantage of the improvement of hardware technology and processing power. The increase in hardware has allowed for more complicated data transmission between customers as more customers want to use mobile communications. Table 5.3 illustrates some of the main features of the 1G through4G networks. It is important to note that this table does not and cannot easily cover all the aspects of each generation. Earlier generations of mobile communication have considerable variability between countries. The variability was due to country-sponsored efforts before agreed-upon international standards were established. Various efforts between the ITU and countries have attempted to minimize the differences.

The third-generation (3G) networks became available by incorporating FDMA, TDMA, and CDMA. In 3G, circuit switching is replaced bypacket switching. The flexibility to support a great variety of applicationsand services are the main advantages of 3G. The modular design allows expandability, backward compatibility with 2G networks, interoperability among mobile systems, global roaming, and Internet services. 3G with more enhancements, referred to as 3.9G or mobile broadband, is taking place under the title of Third Generation Partnership Project (3GPP).3GPP has a number of new technologies such as Enhanced Data Rates for GSM Evolution (EDGE), High-Speed Downlink Packet Access(HSDPA), CDMA2000, and Worldwide Interoperability for Microwave Access(WiMax). There are two competing technologies that fall under the umbrella of 4G, which are Mobile WiMax and Long-Term Evolution(LTE). A 4G system does not support traditional circuit-switched telephony service as 3G does, but works over a purely packet based network. 4Gdevices are IP-basedand are based upon OFDMA.

Communication engineers and scientists are actively involved in developing fifth-generation (5G) mobile communication, but standards requirements and implementation are not expected until 2020. Each of the different mobile communication generations has taken advantage of the improvement of hardware technology and processing power. The increase in hardware has allowed for more complicated data transmission between users and hence the desire for more users to want to use mobile communications.

Table 5.3 illustrates some of the main features of the 1G through 4G networks. It is important to note that this table does not and cannot easily cover all the aspects of each generation. Earlier generations of mobile communication have considerable variability between countries. The variability was due to country-sponsored efforts before agreed-upon international standards were established. Various efforts between the ITUand countries have attempted to minimize the differences.

▼ **Table 5.3:** Different characteristics of Mobile Technology

	1G	2G	3G	4G
Spectrum	900MHz	1800MHz	2GHz	Various
MultiplxingUsed	FDMA	TDMA	CDMA	OFDMA
Voice Support	Basic Telephony	Caller Id and Voice Mail	Conference Calls and Low Quality Video	High Definition Video
Messaging Features	Not available	Text Only	Graphic and Formatted Text	Full Unified Messaging
Data Support	Not available	Circuit Switched	Packet Switched	Native IPv6
Target Data Rate	Not available	119-128Kbps	2Mbps(10Mbps in 3.9G)	100Mbps1Gbpshaving mobility

5.5 SCADA COMMUNICATION TECHNOLOGIES

SCADA communication technologies broadly classified into two vizwired (guided) or wireless (unguided).

5.5.1 WIRED OR GUIDED MEDIA TECHNOLOGIES

Wired or guided media technologies used in industrial SCADA and power system SCADA are Copper UTP, Coaxial, Optical Fiber, Power Line Carrier Communication (PLCC),Broadband over Power line (BPL) and HomePlug. A brief description of these technologies are given below.

5.5.1.1 Copper Utp

Unshielded Twisted Pair(UTP) is a popular type of cable that consists of two unshielded wires twisted around each other which has been explained earlier in this chapter. Low cost make it attractive, and is used extensively for local-area networks (LANs) and telephone connections. UTP cabling does not offer as high bandwidth or good protection from interference when compared to coaxial or fiber optic cables, but easier to work with.

5.5.1.2 Optical Fiber

Fiber cables offer several advantages over traditional long-distance copper cabling. The bandwidth for an optical fiber which has the same thickness that of a copper cable is much higher. Fiber cables rated at 10 Gbps, 40 Gbps and even 100 Gbps are standard. As light can travel much longer distances in a fiber cable without attenuation, the need for signal boosters are less. Further fiber is less susceptible to interference. Traditional copper network cable requires shielding to protect it from electromagnetic interference. though shielding helps to prevent interference, it is not always sufficient especially,when many cables are strung together in close proximity to each other. But fiber optic cables which are made up of glass or silica avoid most of these problems.

5.5.1.3 Fiber To The Home (FTTH)

Fiber to the Home (FTTH) broadband is one of the fine solutions for providing connectivity. FTTH broadband connections which are referred as fiber optic cable connections to individual houses. It can deliver a large number of digital information including voice, video, and data with more efficiently and economically than coaxial cable. FTTH depend on both active and passive optical networks to provide connectivity. FTTH technology is supposed that it is an apt standard which can provide connectivity without much web traffic congestion.

5.5.1.4 Hybrid Fiber Coax (HFC)

A hybrid fiber coaxial (HFC) network is a guided media technology where optical fiber cable and coaxial cable are used in different portions of a communication network to carry broadband information such as video, data, and voice. Using HFC, service providers install fiber optic cable as the backbone from the cable distribution center to the serving nodes located close to customers. Then using coaxial cables connections are provided to the customers by connecting to the nodes of the fiber optic backbone. The main advantage of using HFC as backbone is that some of the features of fiber optic cable such as high bandwidth, improved noise immunity and interference susceptibility, etc can be brought close to the customer premises without the replacement of the existing coaxial cable if it exists already.

5.5.1.5 Power Line Carrier Communication (PLCC)

Power-line carrier communication (PLCC) is a communication method that uses electrical wiring to simultaneously carry both data and electric power. Although many utilities use this over longer distances to send data, it is very rarely used within the buildings. But recently PLCC is used to achieve load shedding in Advanced Metering Infrastructure (AMI). It uses the existing power lines so that, no additional investment on cables and structural alterations to the building. It is an economical and a reliable technique to achieve bi-directional communications which one of the prime requirement in Smart Grid.

Different PLCC technologies are required for different applications, ranging from home automation to Internet access which is often called broadband over power lines (BPL). Most PLCC technologies limit themselves to one type of wire but some can cross between the distribution network and customer premises wiring. Typically transformers prevent propagating the signal, which requires multiple technologies to form very large networks. Various data rates and frequencies are used in different situations. But the main drawback is low bandwidth and point-to-point communication. Although long distance communication is possible, it poses significant challenges, especially in developing countries where the disturbances on transmission lines cause issues.

In SCADA, PLCC is mainly used in PSS for tele-protection and tele-monitoring between electrical substations through power lines at high voltages, such as 110 kV, 220 kV, 400 kV. But this can also be used by utilities for Energy Management Systems (EMS), fraud detection and network management, Advanced Metering Infrastructure (AMI), Demand Side Management (DSM), load control, and demand response (DR) etc.

5.5.1.6 Broadband Over Power Line (BPL)

PLCC can be broadly grouped as narrow band PLCC and broadband PLCC, also known as low frequency and high frequency respectively. There are four basic forms of PLCC and they are:
1. Narrowband internal applications where home wiring is used for home automation and intercoms but with low bit rate,
2. Narrowband outdoor applications which are mainly used by the power distribution utilities for AMR,
3. Broadband in-house mains power wiring can be used for high speed data transmission for home networking, and
4. Broadband over Power Line which uses outdoor mains power wiring and can be used to provide broadband internet.

Broadband PLCC works at higher frequencies. High data rates (up to 100s of Mbps) is used in shorter-range applications. In fact BPL is a system to transmit two-way data over existing AC medium voltage electrical distribution wiring, between transformers, and AC low voltage wiring between transformer and customer outlets. This makes it suitable for indoor as well as outdoor applications. This avoids the expense of a dedicated network of wires for data communication, and the expense of maintaining a dedicated network of antennas, radios and routers in wireless network.

5.5.1.7 Homeplug

HomePlug is a Power line networking which uses power line communications. The specification of Homeplug is defined by the home networking technology that connects devices to each other through the power lines within a home. HomePlug certified products connect PCs and other devices that use Ethernet, USB and 802.11 WiFi technologies to the power line via a HomePlug bridge or adapter. Some products have HomePlug technology built-in. It is one of the cheapest forms of home networking and has a low start-up cost and minimal IT workload. It also will not have an adverse effect on home electric bills.

As consumers need to connect more devices, the need for high throughput connectivity has tremendously amplified. HomePlug technology enables a home's electrical wires to distribute broadband Internet, Ultra High Definition video streaming, virtual reality, digital music and smart energy applications. HomePlug hybrid networking products are used by consumers and service providers worldwide to provide both wired home networking connectivity and WiFi extension throughout the home in dead spots and areas furthest from the router.. Growing smart home trends will also continue to increase the strain on home networks and the need for a singular network for both entertainment and IoT products.

HomePlug adapters are available advertising physical rates of 200Mbps, 900Mbps and 1Gbps. This claims more bandwidth than an ordinary broadband connection, and perfect for streaming HD videos, downloading large files and online gaming. One benefit of this high capacity is the ability for multiple simultaneous communication streams.

5.5.2 WIRELESS OR UNGUIDED MEDIA TECHNOLOGIES

Presently, the unguided or wireless technology has become the most thrilling area in communications and M2M networking. The advancement of wireless communication has revolutionized Industrial SCADA networks as well. Some of the wireless technologies which find applications in DCS and Smart Grid are Frequency Hopping Spread Spectrum (FHSS), 3G Cellular, WiFi, WiMax, ZigBee, ZWave, and VSAT which are briefly described below.

5.5.2.1 IEEE and Wireless Standards

Standards are developed so that many different vendors can create various products that will work together seamlessly. Standards are usually developed on a consensus basis among the different vendors in a specific industry. The Institute of Electrical and Electronics Engineers (IEEE) develops standards for a wide range of technologies and wireless being one of them.

The 802.11 standard outlines how wireless clients and Access Points communicate, lays out the specifications of their interface, dictates how signal transmission should take place, and describes how authentication, association, and security should be implemented. IEEE created several task groups to work on specific areas within wireless communications. Each group had its own focus and was required to investigate and develop standards for its specific section. The letter suffixes indicate the order in which they were proposed and accepted. As an example, one of the members of the 802.11 family which specifies WiFi is succinctly described below.

802.11b High Rate or WiFi is an extension to 802.11 which applies to wireless LANS and provides 11 Mbps transmission in the 2.4 GHz band. 802.11b uses only Direct Sequence Spread Spectrum (DSSS). 802.11b is ratification to the original 802.11 standard, allowing wireless functionality comparable to Ethernet.

Wireless LAN products are being developed following the stipulations of this 802.11i wireless standard. Customers should review the certification issued by the WiFi Alliance before buying wireless products as it assesses the systems against the 802.11i proposed standard.

5.5.2.2 Frequency Hopping Spread Spectrum (FHSS)

Frequency-hopping spread spectrum (FHSS) transmission is the repeated switching of frequencies during radio transmission to reduce interference and avoid interception. It is one of the better methods to counter eavesdropping, and to block jamming of telecommunications. It also has the capability of minimize the effects of unintentional interference. In FHSS, the transmitter hops between available narrowband frequencies within a specified broad channel in a pseudo-random sequence known to both sender and receiver. A short burst of data is transmitted on the current narrowband channel, and then transmitter and receiver tune to the next frequency in the sequence for the next burst of data. In most systems, the transmitter will hop to a new frequency more than twice per second. Because no channel is used for long, and the odds of any other transmitter being on the same channel at the same time are low, FHSS is often used as a method to allow multiple transmitter and receiver pairs to operate in the same space on the same broad channel at the same time. Direct Sequence Spread Spectrum (DSSS) is a related technique. It also spreads a signal across a wide channel, but it does so all at once instead of in discrete bursts separated by hops. It can achieve higher throughput, but DSSS is more susceptible to interference and less effective as a spectrum-sharing method.

Advantages of Spread-Spectrum Transmission: The frequency hopping spread spectrum over a fixed-frequency transmission method has following primary advantages.

1. The method is very resistant to narrow band interference since the spread signal causes the interfering signal to recede into the background.
2. The signals are very difficult to intercept. FHSS signals make it seem like there has been an increase in background noise when a narrow band receiver detects them. In order to intercept the signal, the pseudorandom transmission hopping sequence has to be known.
3. FHSS transmissions can share frequency bands with a number of other types of conventional transmissions without causing significant interference. Each of these signals causes minimal interference and allows the bandwidth to be used more effectively.

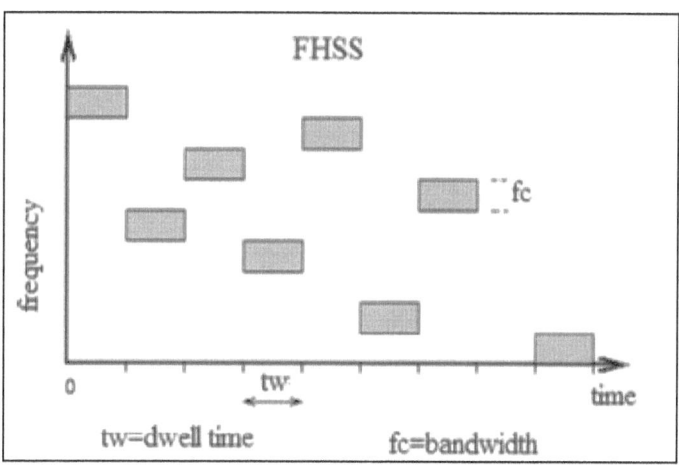

▲ **Figure 5.7:** changing carrier frequency at random

However FHSS faces certain disadvantages such as the requirement greater bandwidth than when a single carrier frequency is used. Also, one of the most problematic aspects of this technology is to initially synchronize the transmitter and receiver. The significant time requirement to master the process of synchronizing the transmitter and receiver.

5.5.2.3 4G Cellular

4G also follows the same pattern of G's which has been introduced in the early 1990's by the ITU. The pattern is actually a wireless initiative called the International Mobile Communications (2000IMT-2000). 3G therefore comes just after 2G and 2.9G, the second generation technologies. 2G technologies include, the Global System for Mobile (GSM). 2.9G brings standards that are midway between 2G and 3G, including the General Packet Radio Service (GPRS), Enhanced Data rates for GSM Evolution (EDGE), Universal Mobile Telecommunications System (UMTS), etc. A characteristic Comparison of the various Gs is given in Table 5.4.

▼ **Table 5.4:** Characteristic Comparison of the various Gs

2G	3G	3.9G(3GPP)	4G
Higher bandwidth than 2G	voice and data are integrated	Higher data rates	IP packet switched network
Always on technology for e-mail	Packet switched technology,	Use of OFDMA technology.	Packet switched technology.

This is a low cost solution to enable long-range communication both within the plant and from field to the MCC. It can be rolled out quickly using the existing cellular infrastructure. Most smart meters in future are expected to use this technology to communicate with the Meter Data Management Systems (MDMS). The constraint is that it can become undependable if natural disaster strikes.

In DCS like Smart Grid, for remote monitoring and control solution, network coverage for a broad range of locations is necessary. In order to achieve coverage, the remote devices deployed are fitted with wireless modems that enabled communication on a range of frequencies and protocols. This allowed selection of either 2G EDGE, 3G HSPDPA, with a flexible architecture to accommodate Long-Term Evolution (LTE) when deployed by carriers.

5.5.2.4 WiFi

WiFi is a trademarked phrase for the open standards, is a popular wireless networking technology that uses radio waves to provide wireless high-speed Internet and network connections. One can attain bandwidth from 9 to 94mbs. The limited distance of 100 to 290m and poor reception in buildings are the main drawbacks. WiFi networks have no physical wired connection between sender and receiver as it uses radio frequency (RF) technology. RT being a member of the electromagnetic spectrum associated with radio wave propagation (2.9GHz for 802.11b or 11g, 9GHz for 802.11a). The RF current supplied to an antenna creates an electromagnetic field which makes radio waves to propagate through space. The cornerstone of any wireless network is an access point (AP). The primary job of an access point is to broadcast a wireless signal that computers can detect and tune into. In order to connect to an access point and join a wireless network, computers and devices must be equipped with wireless network adapters.

WiFi Alliance: WiFi is supported by many applications and devices including video game consoles, home networks, PDAs, mobile phones, major operating systems, and other types of consumer electronics. Any products that are tested and approved as *WiFi Certified* by the WiFi Alliance are certified as interoperable with each other, even if they are from different manufacturers. For example, a user with a WiFi Certified product can use any brand of access point with any other brand of client hardware that also is also WiFi Certified. Products that pass this certification are required to carry an identifying seal on their packaging that states WiFi Certified and indicates the radio frequency band used.

5.5.2.5 WiMax

WiMax (Worldwide Interoperability for Microwave Access) adheres to IEEE 802.16d communication standard. One can get up to 79 MBS bandwidth over 10 to 30 miles. This bandwidth is a savior when an environmental disaster strikes a densely populated area and all smart meters start communicating the outages at same time. WiMax can handle this increased traffic and because of this, WiMax can be an ideal backhaul medium for in-premise WiFi and ZigBee devices. Higher cost and poor market adoption are major constraints at the moment and is not a replacement for WiFi or wireless hotspot technologies. However, all-in-all, it can be cheaper to implement WiMax instead of standard wired hardware like with DSL.

WiMax equipment exists in two basic forms: base stations, installed by service providers to deploy the technology in a coverage area; and receivers, installed in clients. WiMax is developed by an industry consortium, overseen by a group called the WiMax Forum, who certifies WiMax equipment to ensure that it meets technology specifications. Its technology is based on the IEEE 802.16 set of wide-area communications standards. Presently deployments of WiMax are in fixed locations, but a mobile version is under development.

WiMax can be installed faster than other internet technologies because it can use shorter towers and less cabling, supporting even non-line-of-sight (NLoS) coverage across an entire city or country. WiMax isn't just for fixed connections either, like at home. One can also subscribe to a WiMax service for his mobile devices since USB dongles, laptops and phones can have the technology built-in. In addition to internet access, WiMax can provide voice and video transferring capabilities as well as telephone access. Since WiMax transmitters can span a distance of several miles with data rates reaching up to 30-40 Megabits Per Second (Mbps), it's easy to see its advantages, especially in areas where wired internet is impossible or too costly to implement.

However it has certain disadvantages. Because WiMax is wireless by nature, the further away from the source that the client gets, the slower their connection becomes. This means that while a user might pull down 30 Mbps in one location, moving away from the cell site can reduce that speed to 1 Mbps or next to nothing. Frequency hopping. Similar to when several devices suck away at the bandwidth when connected to a single router, multiple users on one WiMax radio sector will reduce performance for the others. WiFi is much more popular than WiMax, so more devices have WiFi capabilities built in than they do WiMax. However, most WiMax implementations include hardware that allow a whole household, for example, to use the service via WiFi, much like how a wireless router provides internet for multiple devices

5.5.2.6 ZigBee

It is a low cost and low power technology and uses unlicensed spectrum which covers only short distance. ZigBee is gaining popular in the home energy market and multiple ZigBee enabled products are available in the market. ZigBee enables the smart meter to communicate with home appliances which helps to shed the load. ZigBee technology enables the coordination of communication among thousands of tiny sensors, which can be scattered throughout the offices, farms, factories, picking up information about the operations and process. They are designed to consume very little energy because will be left in place for five to ten years and their batteries need to last. ZigBee devices communicate efficiently, passing data over radio waves from one to the other like a human chain. At the end of the line the data can be dropped into a computer for analysis or picked up by another wireless technology like WiFi or WiMax.

ZigBee is constituted of mesh technology and this makes it more robust. ZigBee alliance has come up to ensure interoperability among home appliances but the limited distance and inability to penetrate concrete walls are major constraints. Further the low power consumption limits transmission distances to 10–100 meters line-of-sight, depending on power output and environmental characteristics. ZigBee devices can transmit data over long distances by passing data through a mesh network of intermediate devices to reach more distant ones. ZigBee is typically used in low data rate applications that require long battery life and secure networking. ZigBee networks are secured by 128 bit symmetric encryption keys. ZigBee has a defined rate of 290 kbit/s, best suited for intermittent data transmissions from a sensor or input device. ZigBee is an ideal solution for Personnel Area Networks. However, ZigBee is not for situations with high mobility among nodes. Hence, it is not suitable for tactical ad hoc radio networks in the battlefield, where high data rate and high mobility is present and needed.

5.5.2.7 ZWave

ZWave is an RF for signaling and control protocol used for communication among devices preferably in home automation. It is developed by Zensys, Inc. a start-up company based in Denmark and is released in 2004. Its technology is based on the concepts of ZigBee; ZWave strives to build simpler and less expensive devices than ZigBee. Later, Sigma Designs of Milpitas, CA took over Zensys. Today a number of manufacturers make ZWave compatible products, mostly in the building automation and HVAC. ZWave operates at 908.42 MHz in the US and 868.42 MHz in Europe using a mesh networking topology. A ZWave network can contain up to 232 nodes, although reports exist of trouble with networks containing over 30-40 nodes. ZWave operates using a number of profiles, but the manufacturer claims they interoperate. Use care when selecting products as some products from certain manufacturers are not compatible with other manufacturers' products.

ZWave uses GFSK modulation with Manchester channel encoding. A network controller at the center monitors and controls the ZWave network. Each product in the home must be included to the ZWave network before it can be controlled via ZWave. Each ZWave network is identified by a Network

ID and each device is further identified by a Node ID. The Network ID is the common identification of all nodes belonging to one logical ZWave network. Network ID has a length of four bytes and is assigned to each device by the primary controller when the device is added into the network. Nodes with different Network ID's cannot communicate with each other. The Node ID is the address of the device / node existing within network. The Node ID has a length of one byte. ZWave uses a source-routed mesh network topology and has central controller. Devices can communicate to one another by using intermediate nodes to route around and circumvent household obstacles or radio dead spots that might occur though a message called healing. A ZWave network can consist of up to 232 devices with the option of bridging networks if more devices are required.

As a source routed static network, ZWave assumes that all devices in the network remain in their original detected position. Mobile devices, such as remote controls, are therefore excluded from routing. Zwave released later versions with added network discovery mechanisms so that 'explorer frames' could be used to heal broken routes caused by devices that have been moved or removed. A Pruning algorithm is used in explorer frame broadcasts and is therefore supposed to reach the target device, even without further topology knowledge by the transmitter. Explorer frames are used as a last option by the sending device when all other routing attempts have failed.

The interoperability layer of Zwave ensures that devices can share information and allows all ZWave hardware and software to work together. Its wireless mesh networking technology enables any node to talk to adjacent nodes directly or indirectly, controlling any additional nodes. Nodes that are within range communicate directly with one another. If they aren't within range, they can link with another node that is within range of both to access and exchange information. In September 2016, certain parts of the ZWave technology were made publicly available, when Sigma Designs released a public version of ZWave's interoperability layer, with the software added to ZWave's open-source library. The open source availability allows software developers to integrate ZWave into devices with fewer restrictions.

ZWave Alliance: The ZWave Alliance was established in 2009 as a consortium of companies that make connected appliances controlled through apps on smart phones, tablets or computers using ZWave wireless mesh networking technology. The alliance is a formal association focused on both the expansion of ZWave and the continued interoperability of any device that utilizes ZWave.

Zwave Security: Recently, ZWave Alliance announced stronger security standards for devices receiving ZWave Certification. The security standards is known as Security 2 (S2), it provides advanced security for devices to be connected to ZWave gateways and hubs. Encryption standards for transmissions between nodes, and mandates new pairing procedures for each device, with unique PIN on each device are established. The new layer of authentication is intended to prevent hackers from taking control of unsecured or poorly secured devices.

5.5.2.8 VSAT

This is widely used today for remote monitoring and control of transmission and distribution substations; proven and quick implementation. The cost is high and severe weather impacts the reliability. VSAT systems provide high speed, broadband satellite communications for Internet or private network communications. VSAT is ideal for remote monitoring and control such as off- shore wind farms, mining industry, vessels at sea, oil & gas camps or any application that requires a broadband Internet connection at a remote location.

A VSAT is a two-way satellite ground station with a dish antenna that is smaller than 3.8 meters. The majority of VSAT antennas range from 79cm to 1.2m. Data rates, in most cases, range from 4 kbit/s up to 16 Mbit/s. VSATs access satellites in geosynchronous orbit or geostationary orbit to relay data from small remote Earth stations (terminals) to other terminals (in meshtopology) or master Earth station *hubs* (in star topology).VSATs are used to transmit narrowband data, or broadband data. VSATs are also used for transportable, on-the-move (utilising phased array antennas) or mobile maritime communications.

5.6 SECURITY IN WIRELESS COMMUNICATIONS

Wireless communication is a rapidly expanding field of technologies for networking, connectivity, communication, and data exchange. There are literally thousands of protocols, standards, and techniques that can be labeled as wireless. These include cell phones, Bluetooth, cordless phones, and wireless networking. As wireless technologies continue to proliferate, the organization's security efforts must go beyond locking down its local network. Security should be an end-to-end solution that addresses all forms, methods, and techniques of communication. While managing network security with filtering devices such as firewalls and proxies is important, one must not overlook the need for endpoint security. Endpoints are the ends of a network communication link. One end is often at a server where a resource resides, and the other end is often a client making a request to use a network resource. Even with secured communication protocols, it is still possible for abuse, misuse, oversight, or malicious action to occur across the network because it originated at an endpoint. All aspects of security from one end to the other, often called *end-to-end security*, must be addressed. Any unsecured point will be discovered eventually and abused. Endpoint security is the security concept that encourages administrators to install firewalls, malware scanners, and an IDS on every host.

5.6.1 Endpoint Threat Detection and Response (ETDR)

Endpoint security is the concept that each individual device must maintain local security whether or not its network or telecommunications channels also provide or offer security. /in a modest way it can be referred as the protection of an organization's network when accessed via remote devices such as laptops or other wireless or mobile devices. Sometimes this is expressed as the end device is responsible for its own security. However, a clearer perspective is that any weakness in a network, whether on the border, on a server, or on a client, presents a risk to all elements within the organization. Traditional security has depended on the network border sentries, such as appliance firewalls, proxies, centralized virus scanners, and even IDS/IPS/IDP solutions, to provide security for all of the interior nodes of a network. This is no longer considered best business practice because threats exist from within as well as without. A network is only as secure as its weakest element. Lack of internal security is even more problematic when remote access services, including dial-up, wireless, and VPN, might allow an external entity (authorized or not) to gain access to the private network without having to go through the border security gauntlet. Endpoint security should therefore be viewed as an aspect of the effort to provide sufficient security on each individual host. Every system should have an appropriate combination of a local host firewall, antimalware scanners, authentication, authorization, auditing, spam filters, and IDS/IPS services.

Endpoint Threat Detection and Response(ETDR) mainly focuses on the endpoint as opposed to the network, threats as opposed to only malware, and officially declares incident and collection of tools primary usage for both detection and incident response.

5.6.2 Transparency

Just as the name implies, transparency is the characteristic of a service, security control, or access mechanism that ensures that it is unseen by users. Transparency is often a desirable feature for security controls. The more transparent a security mechanism is, the less likely a user will be able to circumvent it or even be aware that it exists. With transparency, there is a lack of direct evidence that a feature, service, or restriction exists, and its impact on performance is minimal. In some cases, transparency may need to function more as a configurable feature than as a permanent aspect of operation, such as when an administrator is troubleshooting, evaluating, or tuning a system's configurations. A security boundary can be the division between one secured area and another secured area, or it can be the division between a secured area and an unsecured area. Both must be addressed in a security policy

5.6.3 Redundancy

In order to make the SCADA most reliable, it is necessary to have redundancy built into the industrial SCADA. Satellite Communications do not only provide a feasible means for distributed SCADA communication but also for redundancy for SCADA as a whole. Given the rate at which the cost for satellite communications is dropping and the available bandwidth is increasing, satellite communications infrastructure can be deployed along with conventional broadband technologies which are used in urban areas. Also, given the fact that satellite communications may not be used most of the time in urban areas, except for times when the main infrastructure goes down, the cost for satellite communications infrastructure for Smart Grid may drop further.

Satellite communications can be used to provide redundancy in SCADA especially when employed in PSS. One of the advantages is that it does not require much terrestrial equipment. The only equipment required to establish communications are the VSATs and modems. The main advantage of employing the satellite communication is that even during the adverse weather conditions or disasters like floods, systems like cable and WiFi fail because of wires or repeaters being knocked down, satellite communications would still continue to function. This is especially important since utilities might need to know the field conditions of the distributed pant/process during such times, to find faults or problems in the industry and fix the solutions.

VSATs can be run on backup batteries, if the main source of its power goes down and still continue to relay information to the utilities. In this respect satellite communications is the only technology that can efficiently deliver the results during times of disasters and rough weather conditions. One of the fears in the past has been satellite communications connectivity during thick cloud cover or during rains and snow.

SUMMARY

This chapter has focused on various communication aspects of the industrial SCADA with an emphasis on DCS and Smart Grid. It begins with discussing various types of transmission technology in very modest way so that it is very apt and most essential for a power engineer who engaged in the design and implementation of SCADA and Smart Grid. The chapter then discusses the guided and unguided media used today for communication in such a manner that it is very useful for a practicing communication professional, which includes the various cabling issues as well. The various but most relevant communication technologies which find space not only in industrial SCADA but also in other smart automation technologies today are discussed comprehensively. Finally the chapter focused on the security issues of the wireless communication technology.

CHAPTER 06

INDUSTRIAL NETWORKING BASICS

6.1 INTRODUCTION

Computer networking and M2M communications are the most complex but exciting topics in the field of computer science and engineering which involves various mechanisms, devices, software, and protocols that are interrelated and integrated. Day by day the functionality of current technology is improving drastically especially the area of network security which records an exponential growth. To catch up with the pace of these new emerging technologies, automation engineers have to learn and understand the relevant technology to implement and secure the system. It is indispensible for an automation engineer to configure networking software, protocols, services, and devices which deal with interoperability issues. They also have to install, configure, and interface with telecommunication software and devices, and troubleshoot them effectively. It will be an added advantage if the automation engineer has the knowledge and capability of understanding these issues and analyze them a few levels deeper to recognize fully where vulnerabilities can arise within each of these components and the mitigation techniques. Obviously it is an overwhelming and challenging task. This chapter elaborates the computer networking and M2M communication concept and structure of protocols such as Open Systems Interconnection (OSI) reference model, TCP/IP Model, Enhanced Performance Architecture (EPA), etc along with a brief introduction of networking devices, network topologies, etc.

6.2 PROTOCOLS-THE RULES THAT GOVERN COMMUNICATIONS

All communication, whether face-to-face or over a network, is governed by predetermined rules called protocols. A network protocol is a standard set of rules that determines how systems will communicate across networks. Two different systems that use the same protocol can communicate and understand each other despite their differences, similar to how two people can communicate and understand each other by using the same language.

6.2.1 Structure Of Communication Protocols

When two network nodes exchange data, the procedures involved in this process can be quite complex. When two nodes transfer a file between them, there must be a data path between them which is either directly or through a communication network. Further the following tasks are also to be performed including the following tasks as described below.

1. The source system must either activate the direct data communication path or inform the communication network of the identity of the desired destination system.
2. The source system must ascertain that the destination system is prepared to receive data.
3. The file transfer application on the source system must ascertain that the file management program on the destination system is prepared to accept and store the file for this particular user.
4. If the file formats used on the two systems are incompatible, one or the other system must perform a format translation function.

Obviously a high level of understanding and cooperation between the two nodes are essential without which an error free data transfer is not possible.
The key features of a protocol are:
1. Set of rules or conventions to exchange blocks of formatted data.
2. Syntax: data format
3. Semantics: control information (coordination, error handling).

6.2.2 Interoperability

Interoperability is the ability of a system or a product to work together (inter-operate) with other systems or products without special effort on the part of the customer. Inter-operability becomes a quality of increasing importance for information technology products as the concept that the network is the computer becomes a reality. For this reason, the term is widely used in product marketing descriptions. Data acquisition protocols need to interface with many vendors equipment. Hence inter-operability is most essential. By having a certification process, the protocol ensures that different manufacturers are able to build equipment to the required interoperable standard. This protects the end user when purchasing a certified interoperable protocol supporting device. As more and more manufacturers produce interoperable certified equipment, the choices and confidence of users will increase.

6.2.3 Open and Closed Architecture

Computer systems can be developed to integrate easily with other systems and products (open systems) or can be developed to be more proprietary in nature and work with only a subset of other systems and products (closed systems). The following sections describe the difference between these approaches.

6.2.3.1 Open Systems

Systems described as *open* are built upon standards, protocols, and interfaces that have published specifications. This type of architecture provides interoperability between products created by different vendors. This interoperability is provided by all the vendors involved who follow specific standards and provide interfaces that enable each system to easily communicate with other systems and allow add-ons to hook into the system easily.

A majority of the systems in use today are open systems. The reason an administrator can have Windows XP, Windows 2008, Macintosh, and UNIX computers communicating easily on the same network is because these platforms are open. If a software vendor creates a closed system, it is restricting its potential sales to proprietary environments.

6.2.3.2 Closed Systems

Systems referred to as *closed* use an architecture that does not follow industry standards. Interoperability and standard interfaces are not employed to enable easy communication between different types of systems and add-on features. *Closed systems* are proprietary, meaning the system can only communicate with like systems.

A closed architecture can potentially provide more security to the system because it may operate in a more secluded environment than open environments. Because a closed system is proprietary, there are not as many predefined tools to thwart the security mechanisms and not as many people who understand its design, language, and security weaknesses and thus exploit them. But just relying upon something being proprietary as its security control is practicing security through obscurity. Attackers can find flaws in proprietary systems or open systems, so each type should be built securely and maintained securely. A majority of the systems today are built with open architecture to enable them to work with other types of systems, easily share.

The open architecture allows different manufactures to have equal input into changes to the protocol. In addition, it means that the cost to develop a system is reduced. The producer does not need to design all parts of the SCADA system. In a proprietary system, the manufacturer usually has to design and produce all parts of the SCADA system, although some of those parts may not be so profitable. One manufacturer is free then to specialize on a few products that are its core business.

6.2.4 Connection Oriented and Connectionless Protocols

The terms connection oriented and connectionless are descriptive words used to describe different kinds of communication.

6.2.4.1 Connection Oriented Communication

When two computers communicate, and if the communication is bidirectional after establishing a handshake by setting up an end-to-end connection the connection is referred as connection oriented. Connection-oriented means that when devices communicate, they perform handshaking to set up an end-to-end connection. The handshaking process may be as simple as synchronization such as in the transport layer protocol TCP or as complex as negotiating communications parameters as with a modem. Connection oriented systems can only work in bi-directional communications environments. To negotiate a connection, both sides must be able to communicate with each other. This will not work in a unidirectional environment. Connection oriented protocol services are more reliable network services, that provide acknowledgment after successful delivery, and automatic repeat request functions in case of missing data or detected bit-errors. ATM, Frame Relay and MPLS are examples of a connection-oriented, unreliable protocol.

6.2.4.2 Connectionless Communication

The alternative to connection oriented transmission is connectionless communication. In the datagram mode communication used by the IP and UDP protocols, where data may be delivered out of order, since different packets are routed independently, and may be delivered over different paths is a typical example for connectionless communications. In fact Connectionless means that no effort is made to set up a dedicated end-to-end connection. It is usually achieved by transmitting information in one direction, from source to destination without checking to see if the destination is still there, or if it is prepared to receive the information. When there is little interference, and plenty of speed available, these systems work fine. In environments where there is difficulty transmitting to the destination, information may have to be re-transmitted several times before the complete message is received. Some protocols allow for error correction by requested retransmission.

A packet transmitted in a connectionless mode is frequently called a datagram. Connectionless protocols are usually described as stateless protocols because the end points have no protocol defined way to remember where they are in a *conversation* of message exchanges. The connectionless communications has the advantage over connection-oriented communications in that it has low overhead. It also allows for multicast and bradcast operations in which the same data are transmitted to several recipients in a single transmission.

6.2.6 ISO Open Systems Interconnection Reference Model

International Standards Organization (ISO) is a worldwide federation that works to provide international standards. In 1984, when the basics of the internet was under developing and implementing, ISO worked to develop a protocol set that would be used by all vendors throughout the world to allow the interconnection of network devices. This movement was fueled with the hope of ensuring that all vendor products and technologies could communicate and interact across international and technical boundaries. This Open Systems Interconnection Reference Model, as described by ISO standard 7498, provides important guidelines used by vendors, engineers, developers, and others. This protocol set did not catch on as a standard, but the model of this protocol set, was adopted and is used as an abstract framework to which most operating systems and protocols adhere. In general it is believed that the OSI Reference Model arrived at the beginning of the computing age. The Transmission Control Protocol/internet Protocol (TCP/IP) suite has its own model that it is often used extensively while examining and understanding.

6.2.6.1 Protocol

A network protocol is a standard set of rules that determines how systems will communicate across networks. Two different systems that use the same protocol can communicate and understand each other despite their physical differences, similar to how two people can communicate and understand each other by using the same language. In fact protocol is a set of rules or standards that defines the syntax, semantics and synchronization of communication and possible error recovery methods and may be implemented by hardware, software, or a combination of both.

6.2.6.2 OSI Reference Model

The main concept of OSI is the process of communication between two endpoints in a telecommunication network by segmenting the networking tasks, protocols, and services which are classified into seven different layers. Each layer has its own responsibilities and functionalities regarding how two computers communicate over a network. So in a given message between users, there will be a flow of data down through the layers in the source computer, across the network and then up through the layers in the receiving computer. The functions of different layers are provided by a combination of applications, operating systems, network card device drivers and networking hardware that enable a system to put a signal on a network cable or out over Wi-Fi or other wireless protocol.

The primary objective of the OSI model is to help others to develop products that will work within an open network architecture, which is not owned by any vendor and is not a proprietary. Hence it can easily integrate various technologies without any changes in the system's internal logic. Vendors have used the OSI model as a stepping stone for developing their own networking frameworks. These vendors use the OSI model as a blueprint and develop their own protocols and services to produce functionality which is entirely a new form, or with overlapping layers. Since these vendors use the OSI model as their starting point, integration of other vendor products is an easier task and the interoperability issues become less cumbersome if the vendors had developed their own networking framework from scratch.

The physical transmission of the data happens in the physical layer as electronic signals passed from one computer to another over a communication media. But the computer can communicate through logical channels. Each protocol operating at the same OSI layer on one computer communicating with a corresponding protocol operates at the same operating layer on another computer. This happens through encapsulation which is a method of designing modular communication protocols in which logically separate functions in the network are abstracted from their underlying structures by, inclusion or information hiding within higher level objects. The concept of the encapsulation is explained below.

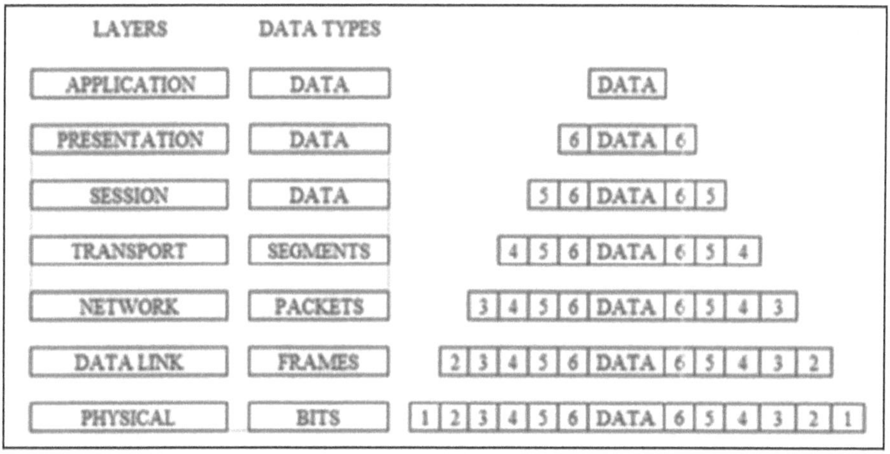

LAYERS	DATA TYPES	
APPLICATION	DATA	DATA
PRESENTATION	DATA	6 DATA 6
SESSION	DATA	5 6 DATA 6 5
TRANSPORT	SEGMENTS	4 5 6 DATA 6 5 4
NETWORK	PACKETS	3 4 5 6 DATA 6 5 4 3
DATA LINK	FRAMES	2 3 4 5 6 DATA 6 5 4 3 2
PHYSICAL	BITS	1 2 3 4 5 6 DATA 6 5 4 3 2 1

▲ **Figure 6.1:** OSI Layer Protocol Data Encapsulation

A message is constructed within a program on one computer and is then passed down through the network protocol's stack. A protocol at each layer adds its own information to the message; hence the message grows in size as it goes down the protocol stack. The message is then sent to the destination computer, and the encapsulation is reversed by taking the packets through the same steps used by the source computer that encapsulated it. At the data link layer, only the information pertaining to the data link layer is extracted, and the message is sent up to the next layer. Then at the network layer, only the network layer data are stripped and processed, and the packet is again passed up to the next layer, and so on. This is how the logical communication between computers takes place. The information stripped off at the destination computer informs to interpret and process the packet properly. Data encapsulation is shown in Figure 6.1.

A protocol at each layer has specific responsibilities and control functions it performs, as well as data format syntaxes it expects. Each layer has a special interface (connection point) that allows it to interact with three other layers viz, (1) communications from the interface of the layer above it, (2) communications to the interface of the layer below it, and (3) communications with the same layer in the interface of the target packet address. The control functions, added by the protocols at each layer, are in the form of headers and trailers of the packet.

The benefit of modularizing these layers, and the functionality within each layer, is that various technologies, protocols, and services can interact with each other and provide the proper interfaces to enable communications.

This means a computer can use an application protocol developed by Novell, a transport protocol developed by Apple, and a data link protocol developed by IBM to construct and send a message over a network. The protocols, technologies, and computers that operate within the OSI model are considered as open systems. They are capable of communicating with other open systems because they implement international standard protocols and interfaces. The specification for each layer's interface is very structured, while the actual code that makes up the internal part of the software layer is not defined. This makes it easy for vendors to write plug-ins in a modularized manner. Systems are able to integrate the plug-ins into the network stack seamlessly, gaining the vendor-specific extensions and functions.

Understanding the functionalities that take place at each OSI layer and the corresponding protocols that work at those layers helps to understand the overall communication process between computers. An in-depth understanding of each protocol is most essential to explore the full range of options each protocol provides and the security weaknesses embedded into each of these options.

Application Layer

The application layer, layer 7, is the top most layer which works closest to the end users. This layer does not include the actual applications, but rather the protocols that support the applications. Application layer provides network access to the user, where services provides file transmissions, message exchanges, terminal sessions, directory services and much more. Network virtual terminal allows a user to log on to the remote host. The application creates software emulation of a terminal at the remote host. When the user tries to log on to the host, the user's computer talks to the software virtual terminal which in turn talks to the host and vice versa. Then the remote host believes it is communicating with one of its own terminals and allows user to log on. This application allows the user to read the remote host files and edit, manage and control host files, and also sometimes fetch the host files for use in the local computer.

To understand the protocol layering and operation, consider that Gopi wants to send a letter to Jessy. The job of Gopi is to write the letter (the application) which creates the content or message and take it to Post Office (application layer protocol). After putting the content into an envelope, write the address of Jessy on the envelope which corresponds to inserting headers and trailers. At the Post Office the envelope is put into the mailbox (pass it on to the next protocol in the network stack). The envelope then passes through various means and finally delivered to Jessy. And she can open the envelop and receive the content. In a similar fashion the different layer protocols perform their assigned responsibilities one by one in a hierarchical fashion and the data reaches the destination, where the data can be stripped off the headers and trailers. More than 15 protocols are presently used in the application layer, including,

1. Simple Mail Transfer Protocol
2. File transfer
3. Web surfing
4. Web chat
5. E-mail clients
6. Network data sharing
7. Virtual terminals
8. Various file and data operations

Mail Services: This layer provides the basis for e-mail forwarding and storage.

Network Virtual Terminal: It allows a user to log on to a remote host. The application creates software emulation of a terminal at the remote host. User's computer talks to the software terminal which in turn talks to the host and vice versa. Then the remote host believes it is communicating with one of its own terminals and allows the user to log on.

Directory Services: This layer provides access for global information about various services.

File Transfer, Access and Management (FTAM): It is a standard mechanism to access and manage the files. Users can access files in a remote computer and manage it. They can also retrieve files from a remote computer.

This layer is not the application itself, it is the set of services an application should be able to make use of directly, although some applications may perform application layer functions. The application layer provides full end-user access to a variety of shared network services for efficient OSI model data flow. This layer has many responsibilities, such as error handling and recovery. It is also used to develop network-based applications. Its major network device or component is the gateway.

Some examples of the protocols working at this layer are the Simple Mail Transfer Protocol (SMTP), Hypertext Transfer Protocol (HTTP), Line Printer Daemon (LPD), File Transfer Protocol (FTP), Telnet, and Trivial File Transfer Protocol (TFTP). Figure 6.2 shows how applications communicate with the underlying protocols through application programming interfaces (APIs).

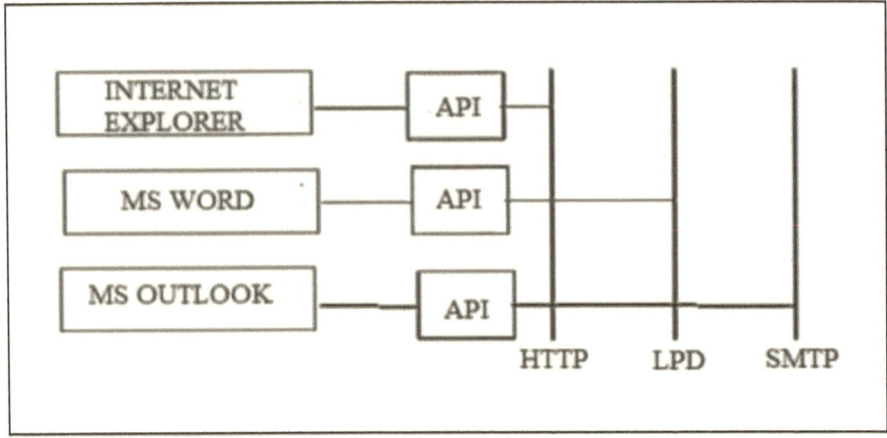

▲ **Figure 6.2:** Application request through API

Supporting interface protocol

If a user makes a request to send an e-mail message through the e-mail client Outlook, the e-mail client sends this information to SMTP which adds its information to the user's message and passes it down to the presentation layer. The protocols that work at this layer include,

1. Hypertext Transfer Protocol (HTTP),
2. File Transfer Protocol (FTP),
3. Line Print Daemon (LPD),
4. Simple Mail Transfer Protocol (SMTP),
5. Trivial File Transfer Protocol (TFTP),
6. Electronic Data Interchange (EDI),
7. Post Office Protocol version 3 (POP3),
8. Internet Message Access Protocol (IMAP),
9. Simple Network Management Protocol (SNMP),
10. Network News Transport Protocol (NNTP),
11. Secure Remote Procedure Call (S-RPC), and
12. Secure Electronic Transaction (SET).

Presentation Layer

The presentation layer, the layer 6, receives information from the application layer protocol and take care of the syntax and semantics of the information exchanged in such a way that all computers following the OSI model can understand. This layer provides a common means of representing data in a structure that can be properly processed by the end system. This means that when a user creates a Word document and sends it out to several people, it does not matter whether the receiving computers have different word processing programs so that each of these computers will be able to receive this file and understand and present it to its user as a document. It is the data representation processing that is done at the presentation layer which enables this to take place.

For example, when a Windows 7 computer receives a file from another computer system, information within the file's header indicates the type of file. The Windows 7 operating system has a list of file types it understands and a table describing the programs to be used to open and manipulate each of these files types. For example, the sender could create a Word file in Word 2010, while the receiver uses Open Office. The receiver can open this file because the presentation layer on the sender's system converted the file to American Standard Code for Information Interchange (ASCII), and the receiver's computer knows and it opens these types of files with its word processor, Open Office.

The presentation layer is not concerned with the meaning of data, but with the syntax and format of those data. It works as a translator, by translating the application format to a standard format which is used for passing message over a network. If a user uses a Corel application to save a graphic, for example, the graphic could be a Tagged Image File Format (TIFF), Graphic Interchange Format (GIF), or Joint Photographic Experts Group (JPEG) format. The presentation layer adds information to inform the destination computer the file type and how to process and present it. In a similar fashion, if the user sends this graphic to another user who does not have the Corel application, the user's operating system can still present the graphic because it has been saved into a standard format. Figure 6.3 illustrates the conversion of a file into different standard file types.

This layer also handles data compression and encryption issues. If a program requests a certain file to be compressed and encrypted before being transferred over the network, the presentation layer provides the necessary information for the destination computer. It provides information regarding the encryption of the file and compression so that the receiving system knows the software and processes which are necessary to decrypt and decompress the file. Suppose a file which is compressed by WinZip is sent to another system which is remotely located. As soon as the system receives this file it looks at the data within the header and comprehends the application which can decompress the file. If the system has WinZip installed, then the file can be decompressed and presented in its original form. If receiving system does not have an application that understands the compression/decompression instructions, the file will be presented with an unassociated icon.

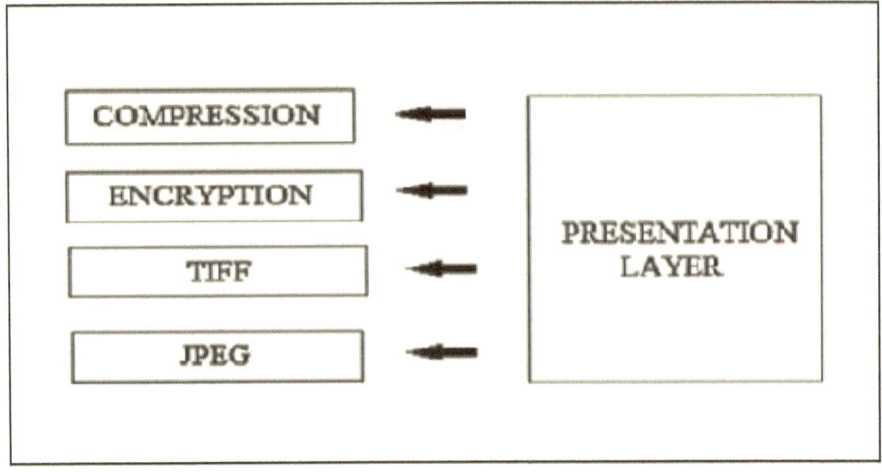

▲ **Figure 6.3:** Data into standard formats by the Presentation Layer

There are no protocols that work at the presentation layer. Network services work at this layer, and when a message is received from a different computer, the service basically informs the application protocol. The following are some of the presentation layer standards.

1. American Standard Code for Information Interchange (ASCII)
2. Extended Binary Coded Decimal Interchange Mode (EBCDIC)
3. Tagged Image File Format (TIFF)
4. Joint Photographic Experts Group (JPEG)
5. Motion Picture Experts Group (MPEG)
6. Musical Instrument Digital Interface (MIDI)

Session Layer

This layer sets up, coordinates and terminates conversations. Services include authentication and reconnection after an interruption. Transfer of data from one destination to another session layer streams of data are marked and are resynchronized properly, so that the ends of the messages are not cut prematurely and data loss is avoided. On the internet, Transmission Control Protocol (TCP) and User Datagram Protocol (UDP) provide these services for most applications.

When two applications need to communicate or transfer data between themselves, a connection may need to be set up between them. The session layer, layer 5, is responsible for establishing a connection between the two applications, maintaining it during the transfer of data, and controlling the release of this connection. A good analogy for the functionality within this layer is a telephone conversation. When Jessy wants to call a friend, she uses the telephone. The telephone network circuitry and protocols set up the connection over the telephone lines and maintain that communication path, and when Jessy hangs up, they release all the resources they were using to keep that connection open.

Similar to how telephone circuitry works, the session layer works in three phases viz. connection establishment, data transfer, and connection release. It provides session restarts and recovery, if necessary and provides the overall maintenance of the session. When the conversation is over, this path is broken down and all parameters are set back to their original settings. This process is known as dialog management. Figure 6.4 depicts the three phases of a session. Some protocols that work at this layer are Structured Query Language (SQL), NetBIOS, and Remote Procedure Call (RPC).

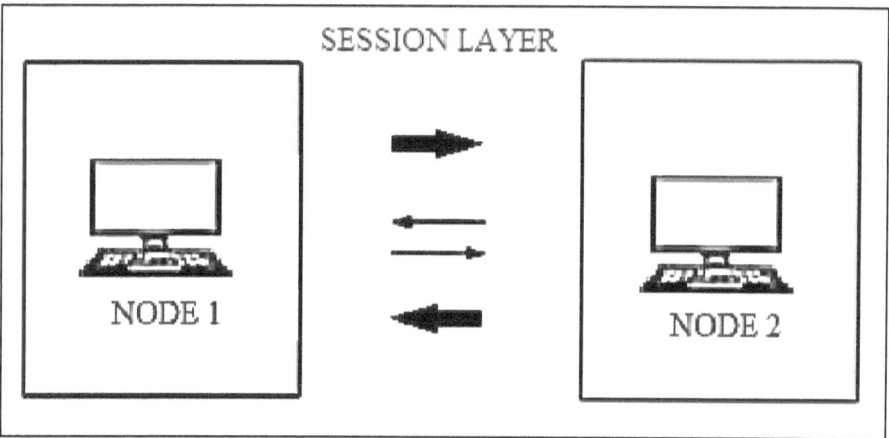

▲ **Figure 6.4:** The three phases of a Session Layer

It can enable communication between two applications to happen in three different modes and they are,
1. Simplex communication which takes place in one direction,
2. Half-duplex communication which takes place in both directions, but only one application can send information at a time, and
3. Full-duplex communication which takes place in both directions, and both applications can send information at the same time.

Many people have difficulty in understanding the difference between what is happening at the session layer versus the transport layer because their similarity in definitions. Session layer protocols control application-to-application communication, whereas the transport layer protocols handle computer-to-computer communication.

It provides inter-process communication channels, which allow a piece of software on one system to call upon a piece of software on another system without the programmer having to know the specifics of the software on the receiving system. The programmer of a piece of software can write a function call that calls upon a subroutine. The subroutine could be local to the system or be on a remote system. If the subroutine is on a remote system, the request is carried over a session layer protocol. The result that the remote system provides is then returned to the requesting system over the same session layer protocol. A piece of software can execute components that reside on another system is the core of distributed computing.

One security issue common to Remote Procedure Call (RPC) is the lack of authentication or the use of weak authentication. Secure RPC can be implemented, which requires authentication to take place before two computers located in different locations can communicate with each other. Authentication can take place using shared secrets, public keys, or Kerberos tickets, session layer protocols need to provide secure authentication capabilities.

Session layer protocols are the least used protocols in a network environment. Thus, many of them should be disabled on systems to decrease the chance of getting exploited. RPC and similar distributed computing need to take place within a network only. Hence, firewalls should be configured so that this type of traffic is not allowed into or out of a network. Firewall filtering rules should be in place to stop this type of unnecessary and dangerous traffic. Some of the protocols that work at this layer include,
1. Network File System (NFS),
2. NetBIOS,
3. Structured Query Language (SQL), and
4. Remote Procedure Calls (RPC).

Transport Layer

When two computers communicate through a connection-oriented protocol, they will first agree on the quantum of information each computer will send at a time, the technique to verify the integrity of the data once received, and the methodology to determine whether a packet was lost along the way. The two computers agree on these parameters through a handshaking process at the transport layer, layer 4. The agreement on these issues before transferring data helps provide more reliable data transfer, error detection, correction, recovery, and flow control, and it optimizes the network services needed to perform these tasks. The transport layer provides end-to-end data transport services and establishes the logical connection between two communicating computers.

The functionality of the session and transport layer is similar so far as both of them set up some type of session or virtual connection for communication. The difference is, protocols that work at the session layer set up connections between applications, whereas protocols that work at the transport layer set up connections between computer systems. For example, one can have three different applications on a computer communicating to three applications on another computer. The session layer protocols keep track of these different sessions. The transport layer protocol can be considered as a bus. It does not know or care the type of applications which are communicating each other. It just provides the mechanism to get the data from one system to another.

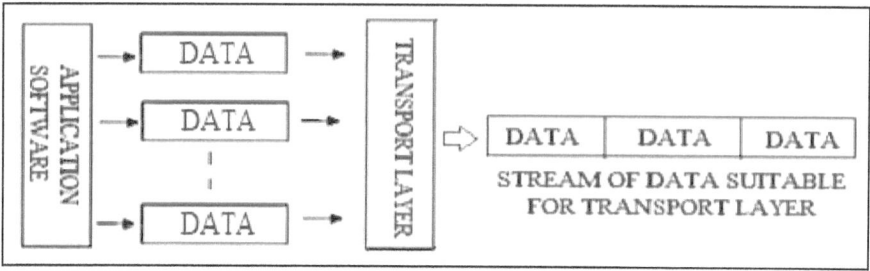

▲ **Figure 6.5:** TCP formats data into a stream suitable for transmission

The transport layer receives data from different applications and assembles the data into a stream to be properly transmitted over the network. The main protocols that work at this layer are TCP, UDP, Secure Socket Layer (SSL), and Sequenced Packet Exchange (SPX). Information is passed down from different entities at higher layers to the transport layer, which assembles the information into a stream, as shown in Figure 6.5. The stream is made up of the various data segments passed to it. Just like a bus can carry variety of people, the transport layer protocol can carry a variety of application data types. In fact functionalities of the transport layer can be summarized as follows,

Segmentation and Reassembling: A message is divided into segments. Each segment contains sequence number, which enables transport layer to reassemble the message. Message is reassembled correctly upon arrival at the destination and replaces packets which were lost in transmission.

Connection Control: It includes Connectionless Transport Layer and Connection Oriented Transport Layer.

Connectionless Transport Layer: Each segment is considered as an independent packet and delivered to the transport layer at the destination machine.

Connection Oriented Transport Layer: Before delivering packets, connection is made with transport layer at the destination machine.

Flow Control: In this layer, flow control is performed end to end.

Error Control: Error Control is performed end to end in this layer to ensure that the complete message arrives at the receiving transport layer without any error. Error Correction is done through re-transmission. The following are some of the protocols that work at this layer.
1. Transmission Control Protocol (TCP)
2. User Datagram Protocol (UDP)
3. Secure Sockets Layer (SSL)/Transport Layer Security (TLS)
4. Sequenced Packet Exchange (SPX)

Network Layer

The main responsibilities of the network layer, layer 3, are to insert information into the packet's header so it can be properly addressed and then route the packets to their proper destination. In a network, many routes can lead to one destination. The protocols at the network layer must determine the best path for the packet to take. Routing protocols build and maintain their routing tables. These tables are maps of the network, and when a packet must be sent from computer A to computer M, the protocols check the routing table, add the necessary information to the packet's header, and send it on its way.

The protocols that work at this layer do not ensure the delivery of the packets. They depend on the protocols at the transport layer to catch any problems and resend packets if necessary. IP is a common protocol working at the network layer, although other routing and routed protocols work there as well. Some of the other protocols are the internet Control Message Protocol (ICMP), Routing Information Protocol (RIP), Open Shortest Path First (OSPF), Border Gateway Protocol (BGP), and internet Group Management Protocol (IGMP). A packet can take many routes but reaches the correct destination as the network layer enters routing information into the header. The Figure 6.6 shows how a network layer takes the most efficient path for each packet to take.

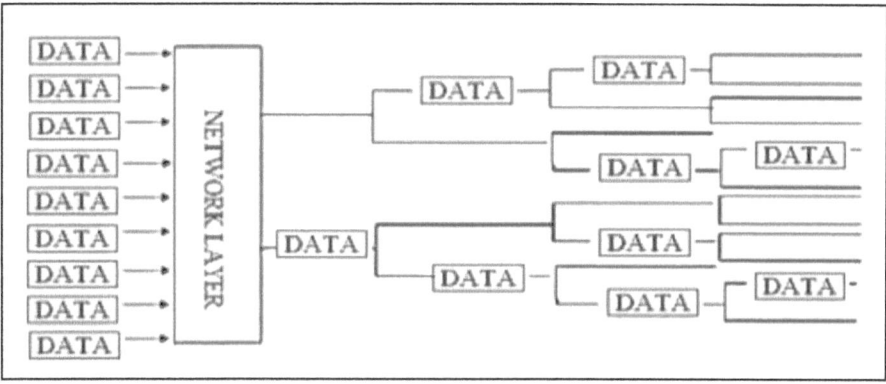

▲ **Figure 6.6:** The most efficient path by the Network Layer

The network layer deliver packets from source to destination across multiple links (networks). If two computers are connected on the same link then there is no need for a network layer. It routes the signal through different channels to the other end and acts as a network controller. It also divides the outgoing messages into packets and to assemble incoming packets into messages for higher levels. Functionalities of network layer are summarized below.

The main function of network layer is, it translates logical network address into physical address. When concerned with the computer networks, message or packet switching, routers and gateways operate in the network layer. Mechanism is provided by Network Layer for routing the packets to final destination. Connection services are provided including network layer flow control, network layer error control and packet sequence control. The following are some of the protocols that work at this layer.

1. Internet Control Message Protocol (ICMP),
2. Routing Information Protocol (RIP),
3. Open Shortest Path First (OSPF),
4. Border Gateway Protocol (BGP),
5. Internet Group Management Protocol (IGMP),
6. Internet Protocol (IP),
7. Internet Protocol Security (IPSec),
8. Internetwork Packet Exchange (IPX),
9. Network Address Translation (NAT), and
10. Simple Key Management for Internet Protocols (SKIP).

Data Link layer

Data link layer which is close to the physical layer where the actual transmission channel is defined is the most reliable node to node delivery of data. It forms frames from the packets suitable to the LAN or wide area network (WAN) technology binary format that are received from network layer and gives it to physical layer for proper line transmission. It also synchronizes the information which is to be transmitted over the data.

LAN and WAN technologies can use different protocols; network interface cards (NICs), cables, and transmission methods. Each of these components has a different header data format structure, and they interpret electric voltages in different ways. The Data link layer determines the format of the data frame to be transmitted properly over Token Ring, Ethernet, ATM, or Fiber Distributed Data Interface (FDDI) networks. If the network is an Ethernet network, for example, all the computers will expect packet headers to be a certain length, the flags to be positioned in certain field locations within the header, and the trailer information to be in a certain place with specific fields. Compared to Ethernet, Token Ring network technology has different frame header lengths, flag values, and header formats. The responsibilities of Data link layer are,

- *Framing*: Frames are the streams of bits received from the network layer into manageable data units. This division of stream of bits is done by Data link layer.
- *Physical Addressing*: The Data link layer adds a header to the frame in order to define physical address of the sender or receiver of the frame, if the frames are to be distributed to different systems on the network.
- *Flow Control*: A mechanism to avoid a fast transmitter from running a slow receiver by buffering the extra bit is provided by flow control. This prevents traffic jam at the receiver side.
- *Error Control*: Error control is achieved by adding a trailer at the end of the frame. Duplication of frames are also prevented by using this mechanism. Data link layers adds mechanism to prevent duplication of frames.
- *Access Control*: Protocols of this layer determines the devices that have control over the link at any given time, when two or more devices are connected to the same link.

The data link is divided into two functional sub-layers viz, the Logical Link Control (LLC) and the Media Access Control (MAC). The LLC, defined in the IEEE 802.2 specification, communicates with the protocol immediately above it, the network layer. The MAC will have the appropriately loaded protocols to interface with the protocol requirements of the physical layer.

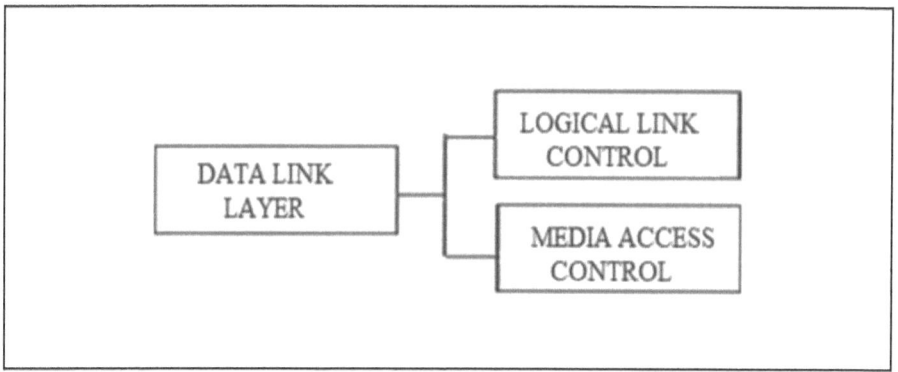

▲ **Figure 6.7:** Two sub-layers of the Data Link Layer

As data is passed down the network stack, it has to go from the network layer to the data link layer. The protocol at the network layer is not aware of the underlying network which may be Ethernet, Token Ring, or ATM. The protocol at the network layer just adds its header and trailer information to the packet and it passes it on to the next layer, which is the LLC sub-layer. The LLC layer takes care of flow control and error checking. Data coming from the network layer passes down through the LLC sub-layer and goes to MAC. The technology at the MAC sub-layer knows if the network is Ethernet, Token Ring or ATM, as it knows how to put the last header and trailer on the packet before it *hits the wire* for transmission.

The IEEE MAC specification for Ethernet is 802.3, Token Ring is 802.5, wireless LAN is 802.11, and so on. A reference to an IEEE standard, such as 802.11, 802.16, or 802.3, refers to the protocol working at the MAC sub-layer of the data link layer of a protocol stack.

Some of the protocols that work at the data link layer are the Point-to-Point Protocol (PPP), ATM, Layer 2 Tunneling Protocol (L2TP), FDDI, Ethernet, and Token Ring. Figure 6.7 shows the two sub-layers that make up the data link layer.

Each network technology such as Ethernet, ATM, FDDI, etc. defines the compatible physical transmission type (coaxial, twisted pair, fiber, wireless) that is required to enable network communication. Each network technology also has defined electronic signaling and encoding patterns. For example, if the MAC sub-layer received a bit with the value of 1 that needed to be transmitted over an Ethernet network, the MAC sub-layer technology would tell the physical layer to create 0.5 volts of electricity. In the *language of Ethernet* this means that 0.5 volts is the encoding value for a bit with the value of 1. If the next bit the MAC sub-layer receives is 0, the MAC layer would tell the physical layer to transmit 0 volts. The different network types will have different encoding schemes. So a bit value of 1 in an ATM network might actually be encoded to the voltage value of 0.85V. It is just a sophisticated Morse code system. The receiving end will know when it receives a voltage value of 0.85V that a bit with the value of 1 has been transmitted.

Network cards bridge the data link and physical layers. Data is passed down through the first six layers and reaches the network card driver at the data link layer. Depending on the network technology being used (Ethernet, Token Ring, FDDI, and so on), the network card driver encodes the bits at the data link layer, which are then turned into electricity states at the physical layer and placed onto the wire for transmission. Some protocols that work at this layer include the following:

1. Address Resolution Protocol (ARP)
2. Reverse Address Resolution Protocol (RARP)
3. Point-to-Point Protocol (PPP)
4. Serial Line Internet Protocol (SLIP)
5. Ethernet
6. Token Ring
7. FDDI
8. ATM

Physical Layer

The electrical and mechanical specifications of the network termination equipment and the transmission medium are defined at this layer which is referred to as layer 1 in the OSI reference model where everything ends up as electrical signals. These include the type of cable and connectors used, the pin assignments for the cable and connectors and the format for the electrical signals. Cable may be coaxial, twisted pair, or fiber optic and the types of connectors depend on the type of cable. Pin assignments depend on the type of cable and also on the network architecture being used, and the encoding scheme used to signal 0 and 1 values in a digital transmission or particular values in an analog transmission depend on the network architecture being used.

Signals can be transmitted as electrical signals, optical signals, or electromagnetic waves. Devices and network components that are associated with the physical layer, are antenna, amplifier, plug and socket for the network cable, the repeater, the transceiver, the T-bar and the terminator. At the physical layer of the communication process, the user data which has been segmented by the transport layer, is placed into packets by the network layer, and further encapsulated as frames by the data link layer. It is then converted to the electrical, optical, or microwave signal that represents the bits in each frame. These signals are then sent on the media one at a time. At the receiving end, the job of the physical layer is to retrieve these individual signals from the media, restore them to their bit representations, and pass the bits up to the data link layer as a complete frame.

Signals and voltage schemes have different meanings for different LAN and WAN technologies. If a user sends data through the dial-up software and transmits out from the modem onto the telephone line, the data format, electrical signals, and control functionality are much different than if that user sends data through the NIC and onto an Unshielded Twisted Pair (UTP) wire for LAN communication. The mechanisms that control this data going onto the telephone line, or the UTP wire, work at the physical layer. This layer controls synchronization, date rates, line noise, and transmission techniques. Specifications for the physical layer include the timing of voltage changes, voltage levels, and the physical connections for electrical, optical, and mechanical transmission. At the destination, the physical layer converts the electrical signals into a series of bit values. These values are grouped into packets and passed up to the data link layer. The following are some of the standard interfaces of Physical layer.

1. EIA-422, EIA-423, RS-449, RS-485,
2. 10BASE-T, 10BASE2, 10BASE5, 100BASE-TX, 100BASE-FX, 100BASE-T,
3. 1000BASE-T, 1000BASE-SX,
4. Integrated Services Digital Network (ISDN),
5. Digital subscriber line (DSL), and
6. Synchronous Optical Networking (SONET)

Attacks on Different Layers

In the different layers of this network model, there can be specific attack types. One concept to understand at this point is that a network can be used as a channel for an attack or the network can be the target of an attack. If the network is a channel for attack, it means the attacker is using the network as a resource. Here, when an attacker sends a virus from one system to another system, the virus travels through the network channel. If an attacker carries out a Denial of Service (DoS) attack, which sends a large amount of bogus traffic over a network link to bog it down, then the network itself is the target. Hence it is important to understand how and where the attacks take place so that appropriate countermeasures can be taken.

6.6.7 TCP/IP MODEL

The most widely used protocol suite is TCP/IP model, but it is not just a single protocol, rather, it is a protocol suite comprising dozens of individual protocols. It is a hierarchical protocol which implies that the lower level protocols support the upper level protocols.

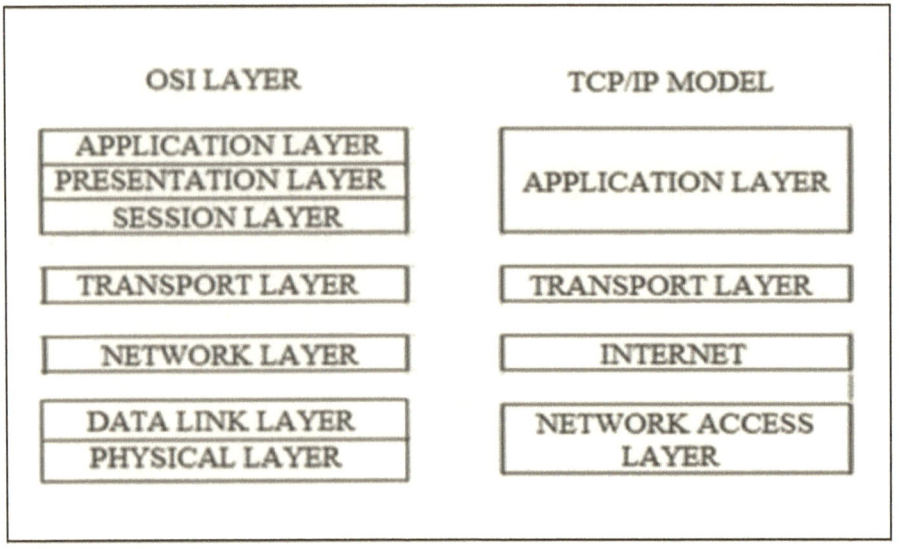

▲ **Figure 6.8:** Comparison of the OSI model with the TCP/IP model

TCP/IP consists of only four layers, as opposed to the OSI Reference Model's seven. The four layers of the TCP/IP model are application, transport, internet, and link or network access layer. The application layer is similar to the combination of session, presentation, and application layers of OSI. Figure 6.8 shows the comparison to the seven layers of the OSI model. TCP/IP is a platform-independent protocol based on open standards. TCP/IP can be found in just about every available operating system, but it consumes a significant amount of resources and is relatively easy to hack into because it was designed for ease of use rather than for security. The TCP/IP protocol suite was developed before the OSI Reference Model was created. The designers of the OSI Reference Model took care to ensure that the TCP/IP protocol suite fit their model because of its established deployment in networking. In fact TCP/IP is a suite of protocols that is the de facto standard for transmitting data across the internet. TCP is a reliable, connection-oriented protocol, while IP is an unreliable, connectionless protocol.

The TCP/IP model's Application layer corresponds to layers 5, 6, and 7 of the OSI model. The TCP/IP model's transport layer corresponds to layer 4 from the OSI model. The TCP/IP model's internet layer corresponds to layer 3 from the OSI model. The TCP/IP model's link layer corresponds to layers 1 and 2 from the OSI model. It has become a common practice to call the TCP/IP model layers by their OSI model layer equivalent names. The TCP/IP model's application layer is already using a name borrowed from the OSI, so that one is a snap. The TCP/IP model's host-to-host layer is sometimes called the transport layer. The TCP/IP model's internet layer is sometimes called the network layer. And the TCP/IP model's Link layer is sometimes called the data link or the network access layer. Since the TCP/IP model layer names and the OSI model layer names can be used interchangeably, it is important to know the model used in various contexts. The four layers of protocol and its components are shown in Figure 6.9.

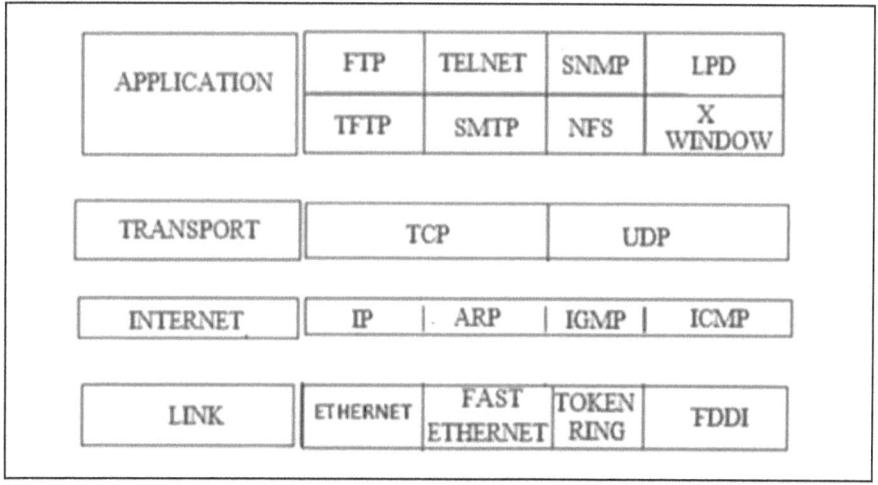

▲ **Figure 6.9:** Four layers of TCP/IP and its component protocols

TCP/IP can be secured using VPN links between systems. VPN links are encrypted to add privacy, confidentiality, and authentication and to maintain data integrity. Protocols used to establish VPNs are Point-to-Point Tunneling Protocol (PPTP), Layer 2 Tunneling Protocol (L2TP), and Internet Protocol Security (IPSec). Another method to provide protocol-level security is to employ TCP wrappers. A *TCP wrapper* is an application that can serve as a basic firewall by restricting access to ports and resources based on user IDs or system IDs. Using TCP wrappers is a form of port-based access control. A brief description of the TCP/IP layers is given below.

Physical and Data Link Layers (Host-to-Network)

The network access layer includes the function of the data link and physical layers of the OSI model. It converts packets into bits for transmission over the physical medium and is responsible for error-free delivery of frames. IEEE 802.2 Logical Link Control (LLC) manages data link communications between devices and performs error checking on frames received. EIA-422-B (RS422) is the Electronic Industry Association standard that defines the electrical characteristic of a balanced interface circuit that is designed to for high common-mode noise rejection and data rates less than 0.5Mbps.

Internet (Network Layer)

Internet layer is the second layer of the four layer TCP/IP model. The position of internet layer is between Network Access Layer and transport layer. Internet protocol assigns IP addresses of the sender and recipient to data packets to be used in routing the message to its intended receiver. IP does not guarantee reliable delivery of data packets. Internet Control Message Protocol (ICMP) is a management protocol used to determine transmission routes from a source to a destination host and to check the availability of a host to receive messages. One of the ICMP utilities is PING, which is used to check the connection of host to the network. ARP is another protocol, which determines the MAC hardware address of a destination host from its IP address. PPP is a full duplex, encapsulation protocol for sending IP messages over Point-to-Point links.

Transport Layer:

Transport layer is the third layer of the four layer TCP/IP model. The position of the Transport layer is between application layer and internet layer. transport layer defines the level of service and status of the connection used when transporting data. The two primary Transport layer protocols of TCP/IP are TCP and UDP. TCP is a full duplex connection-oriented protocol, whereas UDP is a simplex connectionless protocol. When a communication connection is established between two systems, it is done using ports. TCP and UDP each have 65536 ports (ports are explicated in the next section). Since port numbers are 16-digit binary numbers, the total number of ports is 216, or 65536, numbered from 0 through 65535. A port (also called a socket) is little more than an address number that both ends of the communication link agree to use when transferring data. Ports allow a single IP address to support multiple simultaneous communications, each using a different port number. The first 1,024 of these ports (0–1,023) are called the well-known ports or the service ports. This is because they have standardized assignments as to the services they support. For example, port 80 is the standard port for web (HTTP) traffic, port 23 is the standard port for Telnet, and port 25 is the standard port for SMTP.

Application Layer

Application layer is present on the top of the transport layer. It provides applications the capability to access the services of the other layers and defines the protocols that applications use to exchange data. Application layer defines TCP/IP application protocols and helps the host programs for interfacing with transport layer services in using the network. It includes all the higher-level protocols like Domain Naming System, HTTP, Telnet, SSH, File Transfer Protocol, Trivial File Transfer Protocol, SNMP, SMTP, Dynamic Host Configuration Protocol, X Windows, Remote Desktop Protocol etc.

The TCP layer requires a *port number* to be assigned to each message. Similar way it can determine the type of service being provided. It is important to note that whenever *ports* are referred in computer networking, they are not the ports that are used in serial and parallel devices, or ports used for computer hardware control. These ports are merely reference numbers used to define a service. For instance, port 23 is used for telnet services, and HTTP uses port 80 for providing web browsing service. There is a group called the Internet Assigned Numbers Authority (IANA) that controls the assigning of ports for specific services. There are some ports that are assigned, some reserved and many unassigned which may be utilized by application programs. Port numbers are straight unsigned integer values which range up to a value of 65535.

6.6.8 PORTS AND PORT NUMBERS

Ports 1024 to 49151 are known as the registered software ports. These are ports that have one or more networking software products specifically registered with the Internet Assigned Numbers Authority (IANA) in order to provide a standardized port-numbering system for clients attempting to connect to their products. Ports 49152 to 65535 are known as the random, dynamic, or ephemeral ports because they are often used randomly and temporarily by clients as a source port. These random ports are also used by several networking services when negotiating a data transfer pipeline between client and server outside the initial service or registered ports, such as performed by common FTP.

The IANA recommends that ports 49152 to 65535 be used as dynamic*c and/or private ports. However, not all Operating Systems abide by this, especially, the Berkeley Software Distribution (BSD) which uses ports 1024 through 4999. Many Linux kernels use 32768 to 61000. Microsoft, uses the range 1025 to 5000 in Windows Server 2003. Windows Vista, Windows 7, and Windows Server 2008 use the IANA range. Transmission Control Protocol (TCP) operates at layer 4 (the Transport layer) of the OSI model. It supports full-duplex communications, is connection oriented, and employs reliable sessions. TCP is connection oriented because it employs a handshake process between two systems to establish a communication session. Upon completion of this handshake process, a communication session that can support data transmission between the client and server is established.

6.6.9 TCP/IP VULNERABILITIES

TCP/IP's vulnerabilities are numerous. Improperly implemented TCP/IP stacks in various operating systems are vulnerable to buffer overflows, SYN flood attacks, various DoS attacks, fragment attacks, oversized packet attacks, spoofing attacks, man-in-the middle attacks, hijack attacks, coding error attacks, etc. TCP/IP (as well as most protocols) is also subject to passive attacks via monitoring or sniffing. Network monitoring is the act of monitoring traffic patterns to obtain information about a network. Packet sniffing is the act of capturing packets from the network in hopes of extracting useful information from the packet contents. Effective packet sniffers can extract usernames, passwords, email addresses, encryption keys, credit card numbers, IP addresses, system names, and so on.

6.6.10 ENHANCED PERFORMANCE ARCHITECTURE (EPA)

Since all the seven layers of the OSI model are not always necessary, IEC has introduced a reduced form of the OSI seven layer model which is a three layer model called the Enhanced Performance Architecture(EPA). The three layers are,
1. Physical layer
2. Data Link layer, and
3. Application layer

Obviously the physical and data link layer are the hardware layers and the Application layer is the software layer. EPA when used over a network adds a pseudo transport layer to assist network communication.

Physical layer: It is the bottom layer of the EPA model. This converts each frame into a bit stream to be sent over the physical media and keeps track of transmission of one bit at a time. The physical layer provides a physical medium between sender and receiver for sending and receiving information. The connection topology could be point to point, multi-drop, hierarchical, or with multiple masters. The communication can be half duplex or full duplex.

Data Link layer: It deals with the message frame which is the group of bit stream appropriately formed from the bits of data through the physical layer. An acknowledgement is sent for the receipt of data, for reliable and secure transmission.

Pseudo Transport layer: This does the combined function of network and transport layer of the OSI model. Network function is concerned with the routing and data flow over the network from sender to receiver. Transport function includes proper delivery of the message from sender to receiver, message sequencing, and error correction. This function of the transport layer is limited when compared to the OSI layer.

Application Layer: This layer serves the end user directly and will add meaning to the data received from the process. It helps the user to do functions like file transfer and network access. The Figure 6.10 gives a comparison of the EPA model with the OSI model.

OSI MODEL	EPA MODEL
APPLICATION LAYER	APPLICATION LAYER
PRESENTATION LAYER	
SESSION LAYER	
TRANSPORT LAYER	PSEUDO TRANSPORT LAYER
NETWORK LAYER	
DATA LINK LAYER	DATA LINK LAYER
PHYSICAL LAYER	PHYSICAL LAYER

▲ **Figure 6.10:** Comparison of the EPA model with the OSI model

6.3 NETWORK DEVICES

HUB

Hub is an inexpensive way to connect devices on a network which works at physical layer and hence connect networking devices physically together. Hubs are fundamentally used in networks that use twisted pair cabling to connect devices. They are designed to transmit the packets to the other appended devices without altering any of the transmitted packets received. They act as pathways to direct electrical signals to travel along. They transmit the information regardless of the fact if data packet is destined for the device connected or not. Hub falls in two categories, and they are Active Hub and Passive Hub.

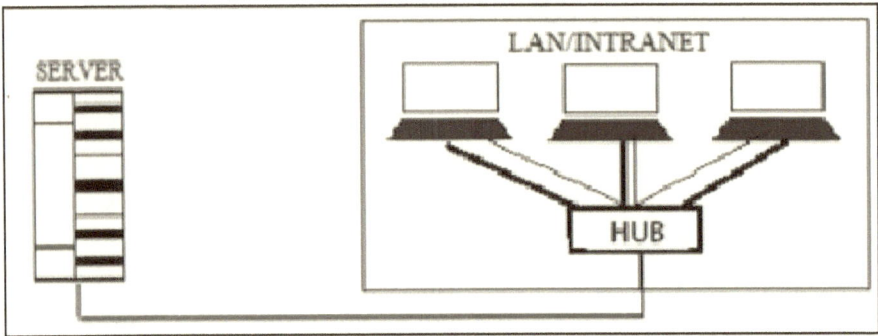

▲ **Figure 6.11:** Data packets go to all workstations reaching the HUB

Active Hubs are smarter than the passive hubs. They not only provide the path for the data signals in fact they regenerate, concentrate and strengthen the signals before sending them to their destinations. Active hubs are also termed as 'repeaters.

Passive Hub are more like point contact for the wires to build in the physical network. They have nothing to do with modifying the signals. Figure 6.11 depicts how data packets go to all workstations reaching the HUB.

Ethernet Hub

It is a device connecting multiple Ethernet devices together and makes them perform the functions as a single unit. They vary in speed in terms of data transfer rate. Ethernet switch do not share the transmission media, do not experience collisions or have to listen for them, can operate in a full-duplex mode, and have bandwidth as high as 200 Mbps, 100 Mbps each way. Either utilizes Carrier Sense Multiple Access with Collision Detect (CSMA/CD) to control Media access. Ethernet hub communicates in half-duplex mode where the chances of data collision are inevitable at most of the times.

An Ethernet switch automatically divides the network into multiple segments, acts as a high-speed, selective bridge between the segments, and supports simultaneous connections of multiple pairs of computers which don't compete with other pairs of computers for network bandwidth. It accomplishes this by maintaining a table of each destination address and its port.

Switches

A switch is more sophisticated than a hub, giving more options for network management, as well as greater potential to expand. They are the connection points of an Ethernet network. Just as in hub, devices in switches are connected to them through twisted pair cabling. But the difference shows up in the manner both the devices; hub and a switch treat the data they receive. Hub works by sending the data to all the ports on the device whereas a switch transfers it only to that port which is connected to the destination device. A switch does so by having an in-built learning of the MAC address of the devices connected to it. Since the transmission of data signals are well defined in a switch hence the network performance is consequently enhanced. Switches operate in full-duplex mode where devices can send and receive data from the switch at the simultaneously unlike in half-duplex mode.

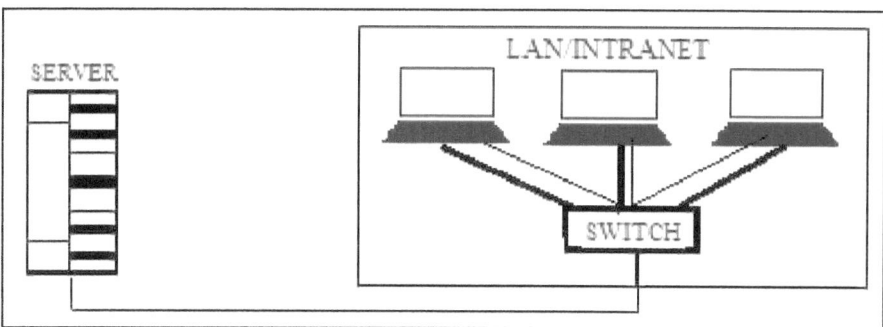

▲ **Figure 6.12:** Data packets sent to only destination workstations through Switch

The transmission speed in switches is much better than in Ethernet hub. The following methods clarify how data transmission takes place in switches. Figure 6.12 how data packets sent to only destination workstations through SWITCH.

Cut-through transmission: It allows the packets to be forwarded as soon as they are received. The method is prompt and quick but the possibility of error checking gets overlooked in such kind of packet data transmission.

Store and forward: In this switching environment the entire packet are received and 'checked' before being forwarded ahead. The errors are thus eliminated before being propagated further. The downside of this process is that error checking takes relatively longer time consequently making it a bit slower in processing and delivering.

Fragment Free: In a fragment free switching environment, a greater part of the packet is examined so that the switch can determine whether the packet has been caught up in a collision. After the collision status is determined, the packet is forwarded.

Layer 3 and 4 Switches

Layer 2 switches only have the intelligence to forward a frame based on its MAC address and do not have a higher understanding of the network as a whole. A layer 3 switch has the intelligence of a muter. It not only can route packets based on their IP addresses, but also can choose routes based on availability and performance. A layer 3 switch is basically a router with certain added capabilities because it moves the route lookup functionality to the more efficient switching hardware level.

The basic distinction between layer 2, 3, and 4 switches is the header information the device looks at to make forwarding or routing decisions (data link, network, or transport OSI layers). But layer 3 and 4 switches can use tags, which are assigned to each destination network or subnet. When a packet reaches the switch, the switch compares the destination address with its tag information base, which is a list of all the subnets and their corresponding tag numbers. The switch appends the tag to the packet and sends it to the next switch. All the switches in between this first switch and the destination host just review this tag information to determine which route it needs to take, instead of analyzing the full header. Once the packet reaches the last switch, this tag is removed and the packet is sent to the destination. This process increases the speed of routing of packets from one location to another. The use of these types of tags, referred to as Multiprotocol Label Switching (MPLS) which will be explained in detail in later sections, not only allows for faster routing, but also addresses service requirements for the different packet types. Some time-sensitive traffic (such as video conferencing) requires a certain level of service (QoS) that guarantees a minimum rate of data delivery to meet the requirements of a user or application When MPLS is used, different priority information is placed into the tags to help ensure that time-sensitive traffic has a higher priority than less sensitive traffic. Many enterprises today use a switched network in which computers are connected to dedicated ports on Ethernet switches, Gigabit Ethernet switches, ATM switches, and more. This evolution of switches, added services, and the capability to incorporate repeater, bridge, and router functionality have made switches an important part of today's networking world.

Because security requires control over whom can access specific resources, more intelligent devices can provide a higher level of protection because they can make more detail-oriented decisions regarding who can access resources. When devices can look deeper into the packets, they have access to more information to make access decisions, which provides more granular access control. Switching makes it more difficult for intruders to sniff and monitor network traffic because no broadcast and collision information is continually traveling throughout the network. Switches provide a security service that other devices cannot provide.

Virtual Local Area Networks (VLANs) are a great way to segment a LAN based on users, device types and functions on the network and are an important part of switching networks. It enables the administrators to have more control over their environment by segmenting into logical groups and manageable entities. VLANs are described later in this chapter. Layer 3 switches help the VLAN routing easier and faster. They make VLANs easier to configure, because a separate router isn't required between each VLAN; all the routing can be done right on the switch. Layer 3 switches also improve VLAN performance, because they eliminate the bottleneck that results from a router forming a single link between VLANs.

Selecting criteria for Switches

As switches improve the performance and efficiency of a network, they should be used when someone
1. Need to make best use of the available bandwidth
2. Have multiple file servers
3. Require improved performance from file servers, web servers or workstations
4. Use high speed multi-media applications
5. Are adding a high speed workgroup to a 10Mbit/sec LAN
6. Plan to upgrade from 10 to 100Mbit/sec or Gigabit network

Bridges

A bridge is a computer networking device used to connect LAN segments which use the same protocol. It works at the data link layer and therefore works with MAC addresses and connects the different networks together and develops communication between them. It connects two local-area networks; two physical LANs into larger logical LAN or two segments of the same LAN that use the same protocol.. When a frame arrives at a bridge, the bridge determines whether or not the MAC address is on the local network segment. If the MAC address is not on the local network segment, the bridge forwards the frame to the necessary network segment.

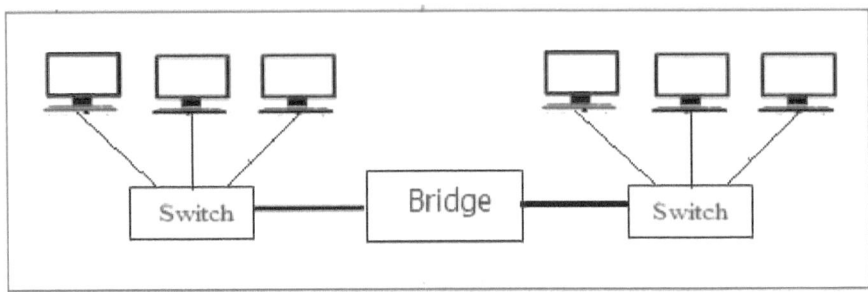

▲ **Figure 6.13:** A Bridge connects two LAN segments

A bridge is used to divide overburdened networks into smaller segments to ensure better use of bandwidth and traffic control. A bridge amplifies the electrical signal, as does a repeater, but it has more intelligence than a repeater and is used to extend a LAN' and enable the administrator to filter frames so he can control which frames go where. The bridge does so by placing itself between the two portions of two physical networks and controlling the flow of the data between them. Bridges nominate to forward the data after inspecting into the MAC address of the devices connected to every segment. The forwarding of the data is dependent on the acknowledgement of the fact that the destination address resides on some other interface. It has the capacity to block the incoming flow of data as well. Today Learning bridges have been introduced that build a list of the MAC addresses on the interface by observing the traffic on the network. This is a leap in the development field of manually recording of MAC addresses. Figure 6.13 demonstrate how a Bridge connects two LAN segments

When using bridges, one has to watch carefully for broadcast storms. Because bridges can forward all traffic, they forward all broadcast packets as well. This can overwhelm the network and result in a broadcast storm, which degrades the network bandwidth and performance.

Types of Bridges

There are mainly three types in which bridges and they are Transparent Bridge, Source Route Bridge, and Translational Bridge. They explained below.

Transparent Bridge: As the name signifies, it appears to be transparent for the other devices on the network. The other devices are ignorant of its existence. It only blocks or forwards the data as per the MAC address.

Source Route Bridge: It derives its name from the fact that the path which packet takes through the network is implanted within the packet. It is mainly used in Token ring networks.

Translational Bridge: A translation bridge is needed if the two LANs being connected are different types and use different standards and protocols. For example, consider a connection between a Token Ring network and an Ethernet network. The frames on each network type are different sizes, the fields contain different protocol information, and the two networks transmit at different speeds. If a regular bridge were put into place, Ethernet frames would go to the Token Ring network, and vice versa, and neither would be able to understand messages that came from the other network segment. A translation bridge does what its name implies—it translates between the two network types.

Another way of categorizing is as local, remote, and translation. A local bridge connects two or more LAN segments within a local area, which is usually a building. A remote bridge can connect two or more LAN segments over a MAN by using telecommunications links. A remote bridge is equipped with telecommunications ports, which enable it to connect two or more LANs, separated by a long distance and can be brought together via telephone or other types of transmission lines.

Routers

Routers are network layer devices and are particularly identified as Layer- 3 devices of the OSI Model. When a router receives a packet, it processes the logical addressing information at the Layer 3 source and destination to determine the path the packet should take. Router is used to create larger composite networks by complex traffic routing. It has the ability to connect dissimilar LANs on the same protocol and also has the ability to limit the flow of broadcasts. A router primarily comprises of a hardware device or a system of the computer which has more than one network interface and routing software.

As it has been mentioned when a router receives the data, it determines the destination address by reading the header of the packet. Once the address is determined, it searches in its routing table to get know how to reach the destination and then forwards the packet to the higher hop on the route. The hop could be the final destination or another router.

Routing tables play a very crucial role for the router to makes a decision. Hence a routing table is ought to be updated and complete. The two ways through which a router can receive information are Static routing and Dynamic routing.

In static routing the routing information is fed into the routing tables manually. It does not only a time consuming and tedious task but can be prone to errors as well. The manual updating is also required in case of statically configured routers when change in the topology of the network or in the layout takes place. Thus static routing is not feasible for complex and large environments but used when number of routers are limited to a minimum of one or two.

For larger environment dynamic routing proves to be the useful solution. The process involves use of peculiar routing protocols to hold communication. The purpose of these protocols is to enable the other routers to transfer information about to other routers, so that the other routers can build their own routing tables.

Brouters

Brouters are the networking devices which combines the functionality of both the bridge and the router. It takes up the functionality of the bridge when forwarding data between networks, and serving as a router when routing data to individual systems. Brouter functions as a filter that allows some data into the local network and redirects unknown data to the other network. Brouters are rare and their functionality is embedded into the routers functioned to act as bridge as well.

Gateways

Gateway is a device which is used to connect multiple networks and passes packets from one packet to the other network. Acting as the 'gateway' between different networking systems or computer programs, a gateway is a device which forms a link between them. It allows the computer programs, either on the same computer or on different computers to share information across the network through protocols. A router is also a gateway, since it interprets data from one network protocol to another. Depending on the types of protocols that the networks support, network gateways can operate at any level of the OSI model and act as a protocol converter.

The different network devices and their functionality are tabulated in Table 6.1.

▼ **Table 6.1:** Network Devices and Functionalities

Device	OSI Layer	Functionality
Repeater	Physical	Amplifies the signal and extends networks.
Hub	Physical	Serve as a central connection for the network equipment and handles a data type known as frames.
Bridge	Data Link	Forwards packets and filters based on MAC addresses; forwards broadcast traffic, but not collision traffic.
Router	Network	Separates and connects LANs creating internetworks; routers filter based on IP addresses.
Switch	Data Link	Provides a private virtual link between communicating devices; allows for VLANs; reduces collisions; impedes network sniffing
Gateway	Application	Connects different types of networks; performs protocol and format translations.

Network Interface Card (NIC)

The hardware devices through which the computers are allowed to communicate or exchange data over a computer network is the Network Interface Card or NIC. It is both an OSI layer 1 (physical layer) and layer 2 (data link layer) device, as it provides physical access to a networking medium and provides a low-level addressing system through the use of MAC addresses. It allows users to connect to each other either by using cables or wirelessly.

Most new computers have either Ethernet capabilities integrated into the motherboard chipset, or use an inexpensive dedicated Ethernet chip connected through the PCI or PCI Express bus, eliminating the need for a standalone card. Computer data is translated into electrical signals send to the network via Network Interface Cards. If the card is not integrated into the motherboard, it may be an integrated component in a router, printer interface or USB device. Typically, there is an LED next to the connector informing the user if the network is active or whether or not data is being transferred on it. Depending on the card or motherboard, transfer rates may be 10, 100, or 1000 Megabits per second.

Today's Network Interface Cards are capable to manage some important data-conversion functions and are mostly software configured unlike in olden days when drivers were needed to configure them. Even if the NIC doesn't come up with the software then the latest drivers or the associated software can be downloaded from the internet.

Modems

Modem is a device which converts the computer-generated digital signals of a computer into analog signals to enable their travelling via a suitable communication medium. The 'modulator-demodulator' or modem can be used as a dial up for LAN or to connect to an ISP. Modems can be both external, as in the device which connects to the USB or the serial port of a computer, or proprietary devices for handheld gadgets and other devices, as well as internal; in the form of add-in expansion cards for computers and Personal Computer Memory Card International Association (PCMCIA) cards for laptops.

Configuration of a modem differs for both the external and internal modem. For internal modems, IRQ – Interrupt request is used to configure the modem along with I/O, which is a memory address. Typically before the installation of built-in modem, integrated serial interfaces are disabled, simultaneously assigning them the COM2 resources. For external connection of a modem, the modem assigns and uses the resources itself. This is especially useful for the USB port and laptop users as the non-complex and simpler nature of the process renders it far much more beneficial for daily usage. Upon installation, the second step to ensure the proper working of a modem is the installation of drivers. The modem working speed and processing is dependent on two factors:
- Speed of UART–Universal Asynchronous Receiver or Transmitter chip which installed in the computer to which the modem connection is made.
- Speed of the modem itself.

TRANSPARENCY of the MODEM is the most critical parameter while selecting for power system SCADA and smart grids from the perspective of security. Transparency of a device ensures that the device achieves the specific task for which it is designed, and must not perform any other task other than predestined especially if the payload is encrypted. Non-transparent MODEM has many drawbacks as severe vulnerability points can be introduced for War dialers, War divers, etc. If a war dialing hacker plant fake historical data and send it to the alarm system, it will affect the way engineers do troubleshooting; they will do something wrong and, in a worst case scenario, cause a horrific incident. Further non-transparent MODEM may not be real time and having a memory element and chances of espionage is very high. Hence MODEMs. when selected for smart grid or power system SCADA must be transparent without any compromise.

Today the presence of a modem on a user system is often one of the greatest woes of a security administrator. Modems allow users to create uncontrolled access points into utility's network. In the worst case, if improperly configured, they can create extremely serious security vulnerabilities that allow an outsider to bypass all perimeter protection mechanisms and directly access the network resources. At best, they create an alternate egress channel that insiders can use to funnel data outside the organization. But these vulnerabilities can only be exploited if the modem is connected to an operational communication channel. One should seriously consider an outright ban on modems in organization's security policy unless utility truly need them for business reasons. In those cases, security officials should know the physical and logical locations of all modems on the network, ensure that they are correctly configured, and make certain that appropriate protective measures are in place to prevent their illegitimate use.

Repeaters

A repeater provides the simplest type of connectivity, because it only repeats electrical signals between cable segments, which enable it to extend a network. Repeaters work at the physical layer and are add-on devices for extending a network connection over a greater distance. The device amplifies signals because signals attenuate the farther they have to travel. A typical signal regeneration from an attenuated signal is shown in Figure 6.4.

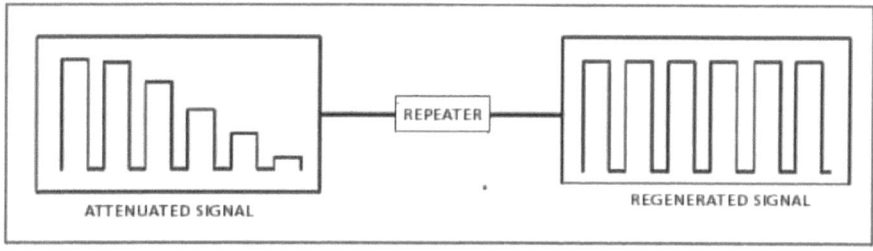

▲ **Figure 6.14:** Repeater showing its functionality

Repeaters can also work as line conditioners by actually cleaning up the signals. This works much better when amplifying digital signals than when amplifying analog signals, because digital signals are discrete units, which makes extraction of background noise from them much easier for the amplifier. If the device is amplifying analog signals, any accompanying noise often is amplified as well, which may further distort the signal. Figure 6.14 shows the functionality of a Repeater.

A hub is a multiport repeater. A hub is often referred to as a concentrator because it is the physical communication device that allows several computers and devices to communicate with each other. A hub does not understand or work with IP or MAC addresses. When one system sends a signal to go to another system connected to it, the signal is broadcast to all the ports, and thus to all the systems connected to the concentrator.

Wireless repeater
A repeater is an electronic device that receives a signal and retransmits it at a higher level and/or higher power, or onto the other side of an obstruction, so that the signal can cover longer distances without degradation. Because repeaters work with the actual physical signal, and do not attempt to interpret the data being transmitted, they operate on the physical layer, the first layer of the OSI model. Repeaters are majorly employed in long distance transmission to reduce the effect of attenuation. It is important to note that repeaters do not amplify the original signal but simply regenerate it.

Wireless access point:
A wireless access point (WAP or AP) is a device that allows wireless communication devices to connect to a wireless network using Wi-Fi, Bluetooth or related standards. The WAP usually connects to a wired network, and can relay data between the wireless devices (such as computers or printers) and wired devices on the network.

A typical corporate use involves attaching several WAPs to a wired network and then providing wireless access to the office Local Area Network. Within the range of the WAPs, the wireless end user has a full network connection with the benefit of mobility. In this instance, the WAP functions as a gateway for clients to access the wired network.

A Hot Spot is a common public application of WAPs, where wireless clients can connect to the Internet without regard for the particular networks to which they have attached for the moment. The concept has become common in large cities, where a combination of coffeehouses, libraries, as well as privately owned open access points, allow clients to stay more or less continuously connected to the Internet, while moving around. A collection of connected Hot Spots can be referred to as a lily-pad network.

Home networks generally have only one WAP to connect all the computers in a home. Most are wireless routers, meaning converged devices that include a WAP, router, and often an Ethernet switch in the same device. In places where most homes have their own WAP within range of the neighbors' WAP, it's possible for technically savvy people to turn off their encryption and set up a wireless community network, creating an intra-city communication network without the need of wired networks.

VLANs

The technology within switches has introduced the capability to use VLANs. VLANs enable administrators to separate and group computers logically based on resource requirements, security, or business needs instead of the standard physical location of the systems. When repeaters, bridges, and routers are used, systems and resources are grouped in a manner dictated by their physical location. Figure 6.15 shows how computers that are physically located next to each other can be grouped logically into different VLANs. Administrators can form these groups based on the users' and company's needs instead of the physical location of systems and resources.

An administrator may want to place the computers of all users in the marketing department in the same VLAN network, for example, so all users receive the same broadcast messages and can access the same types of resources. This arrangement could get tricky if a few of the users are located in another building or on another floor, but VLANs provide the administrator with this type of flexibility. VLANs also enable an administrator to apply particular security policies to respective logical groups. This way, if tighter security is required for the engineering department, the administrator can develop a policy, add all engineering systems to a specific VLAN, and apply the security policy only to the engineering VLAN. A VLAN exists on top of the physical network, as shown in Figure 6.15. If workstation PI wants to communicate with workstation El, the message has to be routed—even though the workstations are physically next to each other—because they are on different logical networks.

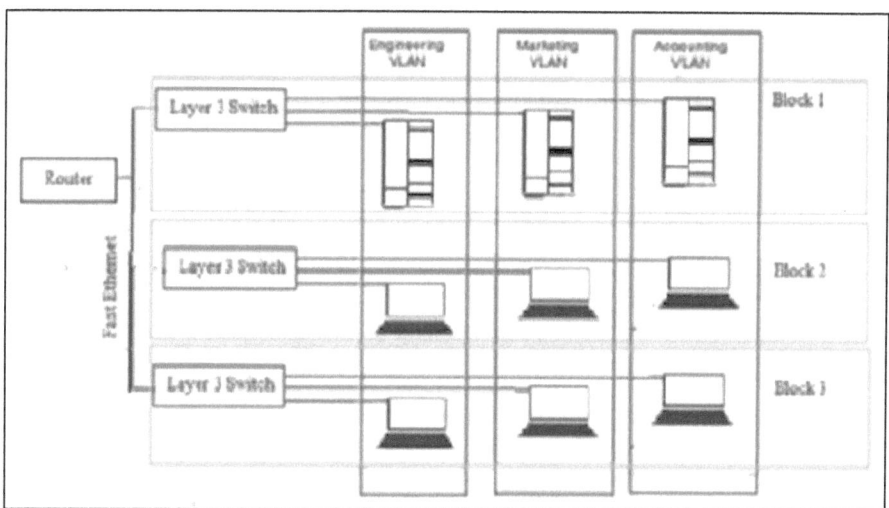

▲ **Figure 6.15:** VLANs enable administrators to manage logical networks

IEEE 802.1Q is the IEEE standard that defines how VLANs are to be constructed and how tagging should take place to allow for interoperability. While VLANs are used to segment traffic, attackers can still gain access to traffic that is supposed to be walled off in another VLAN segment. VLAN hopping attacks allow attackers to gain access to traffic in various VLAN segments. An attacker can have a system act as though it is a switch. The system understands the tagging values being used in the network and the trunking protocols and can insert itself between other VLAN devices and gain access to the traffic going back and forth. Attackers can also insert tagging values to manipulate the control of traffic at the data link layer.

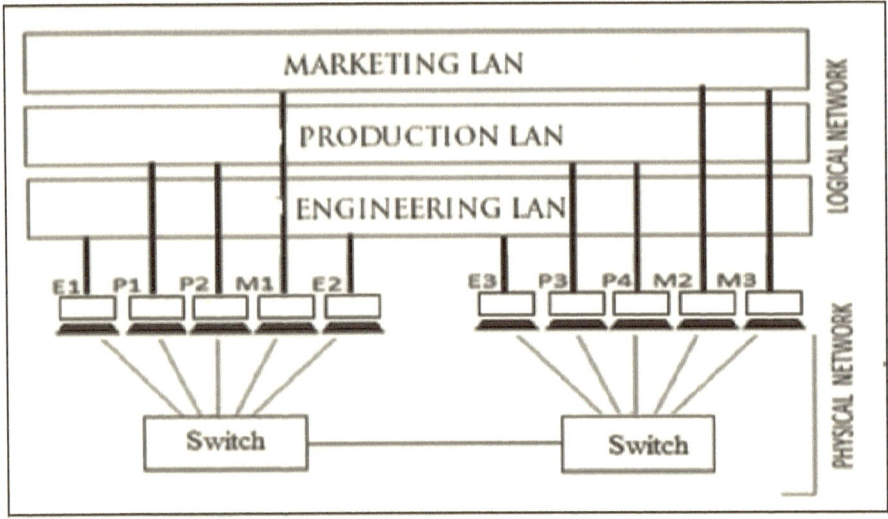

▲ **Figure 6.16:** VLANs showing the existence in a higher level than the physical network

VLAN and SCADA Security

VLANS are good for traffic management but not good for security. It work by having Ethernet switches insert a tag (basically a 4-byte field) in to the header of each Ethernet message. Other switches on the network can read this tag and make decisions on whether a message should be forwarded. In fact this allows the switches to provide limited traffic filtering, primarily for managing broadcast traffic. And managing broadcast traffic is important. VLANs are great traffic management tools. But switches with VLANs are not firewalls. They operate at layer 2 (the Ethernet layer) and don't understand the state of the messages flowing through them. This makes the spoofing of VLAN tags trivial – there is no check to detect if a tag has been adjusted by a *hacker.*

Thus the hacking community has lots of tools designed to bypass switch-based security. It is always recommended to use VLANs as a mechanism for enforcing security policy. They are great for segmenting networks, reducing broadcasts and collisions and so forth, but not as a security tool.

Usually the interest in VLANs is because of IT teams wanting to use IT network technology to solve a plant floor security issue. VLANs are good tools, but deploying them only for security is categorically not a solution.

Media Converters

Media converters are simple networking devices that make it possible to connect two dissimilar media types such as twisted pair with fiber optic cabling. They were introduced to the industry nearly two decades ago, and are important in interconnecting fiber optic cabling-based systems with existing copper-based, structured cabling systems. Media converters support many different data communication protocols including Ethernet, T1/E1, T3/E3, as well as multiple cabling types such as coaxial, twisted pair, multimode and single-mode fiber optics. When expanding the reach of a Local Area Network to span multiple locations, media converters are useful in connecting multiple LANs to form one large *campus area network* that spans over a limited geographic area. As local networks are primarily copper-based, media converters can extend the reach of the LAN over single-mode fiber up to 130 kilometers with 1550 nm optics.

Data Diodes

A data diode is a computer security device that restricts the communication along a network connection between two computers so that data can only be transmitted in one direction. This enables a more sensitive or highly classified computer network to receive data directly from a less secure source while prohibiting the transmission of data in the opposite direction.

Data diodes provide a physical mechanism for enforcing strict unidirectional communication between two networks. They are often implemented by removing transmitting component from one side and receiving component from another side of a bidirectional communication system. Data diodes can only send information from one network (the *low* network) to another network (the *high* network.) The high network often contains data with higher classification level than the low network.

A major limitation of the data diode is that it does not work with the standard TCP/IP protocols. It needs proprietary unidirectional protocols that do not require acknowledgments. On both sides of a data diode, gateways translate unidirectional protocols to standard bidirectional protocols to connect the diode to the rest of the network. However, more high-end products also accept TCP or UDP packets as input. Data diodes can be used to enhance security if placed carefully in combination with other defensive mechanisms.

Firewalls

Today a firewall is an indispensable component of a secure M2M communication which is designed to block unauthorized access while permitting authorized communication. It comprises a device or set of devices configured to permit, deny, encrypt, decrypt, or proxy all computer traffic between different security domains based upon a set of rules called ruleset. Firewalls can be implemented in both hardware and software, or a combination of both. Firewalls are frequently used to prevent unauthorized Internet users from accessing private networks connected to the Internet. All messages entering or leaving the computer network pass through the firewall, which examines each message and blocks those that do not meet the specified security criteria. Without proper configuration, a firewall can often become insignificant. Standard security practices dictate a *default-deny* firewall ruleset, in which the only network connections which are allowed are the ones that have been explicitly allowed.

A variety of firewalls are available in the market, depending upon the requirements and security goals to be achieved. Firewalls have gone through an evolution of their own and have grown in complexity and functionality. The generally deployed firewalls today are:
1. Packet Filtering Firewalls
2. Stateful Firewalls
3. Proxy Firewalls
4. Dynamic Packet Filtering Firewalls,
5. Kernel Proxy Firewalls, and
6. Deep Packet Inspection Firewalls

The following sections give brief descriptions of these firewalls.

Packet Filtering Firewalls

Packet filtering was the first generation of firewalls and it is the most elementary type of all of the firewall technologies. It is a firewall technology that makes access decisions by examining the network-level protocol header values. The device that is carrying out packet filtering processes is configured with Access Control Lists (ACL), which dictates the type of traffic that is allowed into and out of specific networks. The filters review the protocol header information at

the network and transport levels and carrying out PERMIT or DENY actions on individual packets. This means the filters can make access decisions based upon the following basic criteria,
- Source and destination IP addresses
- Source and destination port numbers
- Protocol types
- Inbound and outbound traffic direction

Packet filtering is built into a majority of the firewall products and routers. Packet filtering is also known as stateless inspection because the device does not understand the context that the packets are working within. This means that the device does not have the capability to understand the *full picture* of the communication that is taking place between two systems, but can only focus on individual packet characteristics. Some of the disadvantages of packet filtering firewalls are as follows.
- They cannot prevent attacks that employ application-specific vulnerabilities or functions.
- The logging functionality present in packet filtering firewalls is limited.
- Most packet filtering firewalls do not support advanced user authentication schemes.
- Many packet filtering firewalls cannot detect spoofed addresses.
- They may not be able to detect packet fragmentation attacks.

The advantages to using packet filtering firewalls are that they are scalable, they are not application dependent, and they have high performance because they do not carry out extensive processing on the packets. They are commonly used as the preliminary defense to strip out all the network traffic that is obviously malicious or unintended for a specific network.

Stateful Firewalls
The following are some of the important characteristics of a stateful-inspection firewall.
- Maintains a state table that tracks each and every communication session,
- Provides a high degree of security,
- Is scalable and transparent to users,
- Provides data for tracking connectionless protocols like UDP& ICMP, and
- Stores and updates the state and context of the data within the packets.

When packet filtering is used, a packet arrives at the firewall, and it runs through its ACLs to determine whether this packet should be allowed or denied. If the packet is allowed, it is passed on to the destination host, or to another network device, and the packet filtering device forgets about the packet.

When a connection begins between two systems, the firewall investigates all layers of the packet such as headers, payload, and trailers. All of the necessary information about the specific connection such as source and destination IP addresses, source and destination ports, protocol type, header flags, sequence numbers, timestamps, etc. are stored in the state table. Once the initial packets undergo this in-depth inspection the firewall then reviews the network and transport header portions for the rest of the session. The values of each header for each packet are compared to what is in the current state table, and the table is updated to reflect the progression of the communication process. Scaling down the inspection of the full packet to just the headers for each packet is done to increase performance. TCP is considered a connection-oriented protocol, and the various steps and states this protocol operates within are very well defined. A connection progresses through a series of states during its lifetime.

If the acknowledgment and/or sequence numbers are out of order, this could imply that a replay attack is underway and the firewall will protect the internal systems from this activity. Stateful inspection firewalls unfortunately have been the victims of many types of DoS attacks. Several types of attacks are aimed at flooding the state table with bogus information. The state table is a resource, similar to a system's hard drive space, memory, and CPU. When the state table is stuffed full of bogus information, the device may either freeze or reboot. In addition, if this firewall must be rebooted for some reason, it will lose its information on all recent connections; thus, it may deny legitimate packets.

Proxy Firewalls

A proxy firewall stands between a trusted and untrusted network and makes the connection, each way, on behalf of the source. What is important is that a proxy firewall breaks the communication channel; there is no direct connection between the two communicating devices. Where a packet filtering device just monitors traffic as it is traversing a network connection, a proxy ends the communication session and restarts it on behalf of the sending system. Figure 6.17 illustrates the steps of a proxy-based firewall. The firewall is not just applying ACL rules to the traffic, but stops the user connection at the internal interface of the firewall itself and then starts a new session on behalf of this user on the external interface. When the external web server replies to the request, this reply goes to the external interface of the proxy firewall and ends. The proxy firewall examines the reply information and if it is deemed safe, the firewall starts a new session from itself to the internal system. This is just like our analogy of what the delivery man does between the customer and the president.

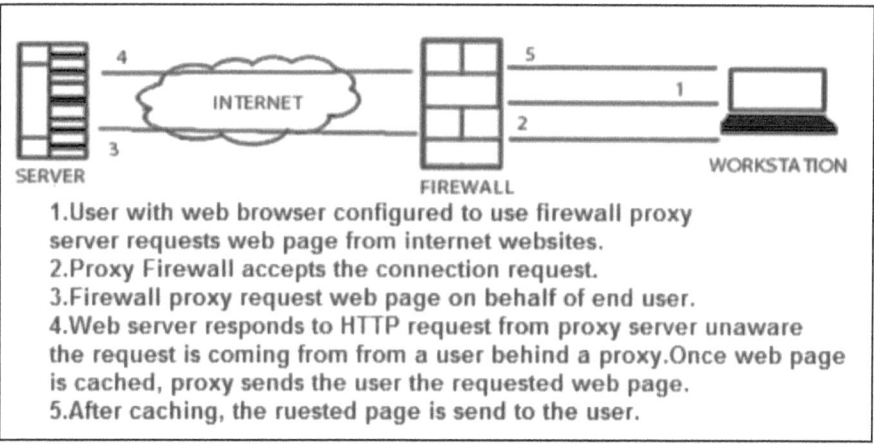

▲ **Figure 6.17:** Functional steps of a Proxy Firewall

A circuit-level proxy creates a connection (circuit) between the two communicating systems. It works at the session layer of the OSI model and monitors traffic from a network based view. This type of proxy cannot *look into* the contents of a packet; thus, it does not carry out deep-packet inspection. It can only make access decisions based upon protocol header and session information that is available to it. While this means that it cannot provide as much protection as an application-level proxy, because it does not have to understand application layer protocols, it is considered application independent. So it cannot provide the detail-oriented protection that a proxy that works at a higher level can, but this allows it to provide a broader range of protection where application layer proxies may not be appropriate or available.

Dynamic Packet Filtering

When an internal system needs to communicate to an entity outside its trusted network, it must choose a source port so the receiving system knows how to respond properly. Ports up to 1023 are called *well-known ports* and are reserved for server-side services. The sending system must choose a dynamic port higher than 1023 when it sets up a connection with another entity. The dynamic packet-filtering firewall then creates an ACL that allows the external entity to communicate with the internal system via this high port. If this were not an available option for the customers dynamic packet-filtering firewall, customer would have to allow punch holes in customers firewalls for all ports above 1023, because the client side chooses these ports dynamically and the firewall would never know exactly on which port to allow or disallow traffic.

Kernel Proxy Firewalls

A *kernel proxy firewall* is considered a fifth-generation firewall. It differs from all the previously discussed firewall technologies because it creates dynamic, customized network stacks when a packet needs to be evaluated. When a packet arrives at a kernel proxy firewall, a new virtual network stack is created, which is made up of only the protocol proxies necessary to examine this specific packet properly. If it is an FTP packet, then the FTP proxy is loaded in the stack. The packet is scrutinized at every layer of the stack. This means the data link header will be evaluated along with the network header, transport header, session layer information, and the application layer data. If anything is deemed unsafe at any of these layers, the packet is discarded. Kernel proxy firewalls are faster than application layer proxy firewalls because all of the inspection and processing takes place in the kernel and does not need to be passed up to a higher software layer in the operating system. It is still a proxy-based system, so the connection between the internal and external entity is broken by the proxy acting as a middleman, and it can perform NAT by changing the source address, as do the preceding proxy-based firewalls.

Deep Packet Inspection Firewall

Deep Packet Inspection (DPI) is an advanced method of examining and managing network traffic. It is a form of packet filtering that locates, identifies, classifies, reroutes or blocks packets with specific data or code payloads that conventional packet filtering, which examines only packet headers, cannot detect. DPI usually examines the data part (and possibly also the header) of a packet as it passes an inspection point, searching for protocol non-compliance, viruses, spam, intrusions, or defined criteria to decide whether the packet may pass. DPI enables advanced network management, user service, and security functions as well as internet data mining, eavesdropping, and internet censorship. It is used in a wide range of applications, at the enterprise level, in telecommunications service providers, and in governments. DPI combines the functionality of an Intrusion Detection System (IDS) and an Intrusion Prevention System (IPS) with a traditional stateful firewall. This combination makes it possible to detect certain attacks that neither the IDS/IPS nor the stateful firewall can catch on their own.

▼ **Table 6.2:** Comparison of Different Types of Firewalls

Firewall Type	OSI Layer	Characteristics
Packet filtering	Network layer	Verifies the destination and source addresses level, port numbers, services requested, etc at the routers using ACLs to monitor the network traffic
		services requested, etc at the routers using ACLs to monitor the network traffic.
Application-level	Application layer	Verifies deep into packets and makes granular
		access control decisions. It requires one proxy
		per protocol. They filter packets not only according to the service for which they are intended (as specified by the destination port), but also by certain other characteristics such as HTTP request string. While application-level gateways provide considerable data security, they can dramatically impact network performance.
Circuit-level proxy	Session layer	Verifies only at the header packet information. It protects a wider range of protocols and services than an application-level proxy, but does not provide the detailed level of control available to an application-level proxy. However monitor the TCP handshaking going on between the local and remote hosts to determine whether the session being initiated is legitimate — whether the remote system is considered trusted.
		is considered trusted.
Stateful	Network layer	It not only examines each packet, but also keep track of whether or not that packet is part of an established TCP session. This offers more security than either packet filtering or circuit monitoring alone, but exacts a greater toll on network performance.
Kernel proxy	Application layer	Faster because processing is done in the kernel. One network stack is created for each packet

6.4 DE MILITARIZED ZONE (DMZ)

The word zone is used to mean any area that is separated from another, or is distinguished from another based on distinctive circumstances. In computer networking, levels of administration and access are referred as zones. A De Militarized Zone (DMZ) is which is also referred to as a screened subnet, or a perimeter network is a special local network configuration designed to improve security by segregating computers on each side of a firewall. It is a computer or small sub network that sits between a trusted internal network, such as a corporate private LAN, and an untrusted external network, such as the public Internet. Typically, the DMZ contains devices accessible to Internet traffic, such as Web (HTTP) servers, FTP servers, SMTP (e-mail) servers and DNS servers. The phrase demilitarized zone comes from military use, meaning a buffer area between two enemies. The most common services which are provided to the users typically placed in the DMZ are,

1. Web servers
2. Mail servers
3. FTP servers
4. VoIP servers

Web servers that communicate with an internal database require access to a database server, which may not be publicly accessible and may contain sensitive information. The web servers can communicate with database servers either directly or through an application firewall for security reasons.

E-mail messages and particularly the user database are confidential, so they are typically stored on servers that cannot be accessed from the internet, but can be accessed from email servers that are exposed to the Internet. The mail server inside the DMZ passes incoming mail to the secured/internal mail servers. It also handles outgoing mail.

For security compliance and monitoring reasons, in a business environment, some enterprises install a proxy server within the DMZ. This has the following benefits.

1. Obliges internal users (usually employees) to use the proxy server for Internet access.
2. Reduced Internet access bandwidth requirements since some web content may be cached by the proxy server.
3. Simplifies recording and monitoring of user activities.
4. Centralized web content filtering.

6.5 NETWORK TOPOLOGIES

The physical layout and arrangement of computers and networking devices is known as the network topology. Topology refers to the manner in which a network is physically connected and shows the layout of resources and systems. The logical topology is the grouping of networked systems into trusted collectives. The physical topology is not always the same as the logical topology. A network can be configured as a physical star but work logically as a ring, as in the Token Ring technology. The four basic network topologies are ring, bus, star, and mesh. The best topology for a particular network depends on how nodes are supposed to interact, the protocols used, the types of applications, the reliability, and expandability, physical layout of a facility, existing wiring, and the technologies implemented. The wrong topology or combination of topologies can negatively affect the network's performance, productivity, and growth possibilities. Most networks are very complex and are usually implemented using a combination of topologies.

Ring Topology

A ring topology connects each system as points on a circle as shown in Figure 6.17. The connection medium acts as a unidirectional transmission loop. These links form a closed loop and do not connect to a central system, as in a star topology which is discussed later in this chapter. In a physical ring formation, each node is dependent upon the preceding nodes. In simple networks, if one system fails, all other systems could be negatively affected because of this interdependence.

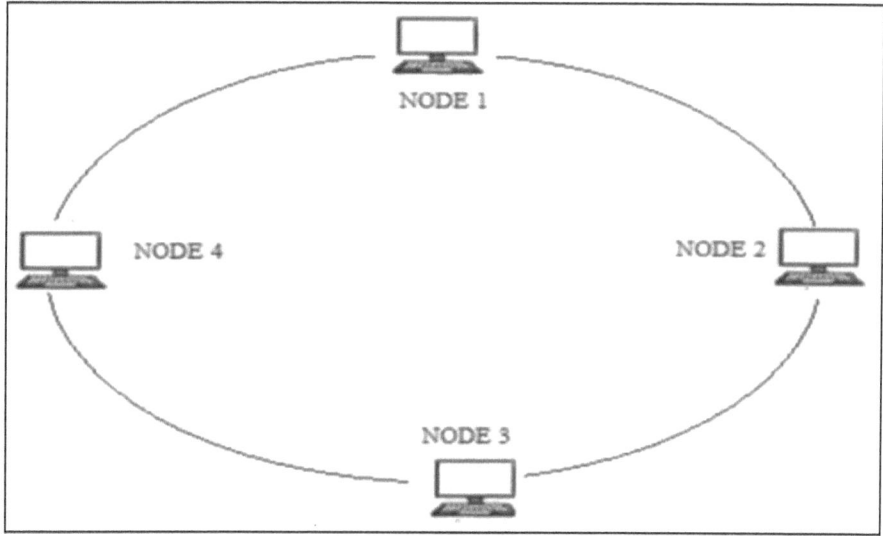

▲ **Figure 6.18:** Ring Topology

In ring topology only one system can transmit data at a time. Traffic management is performed by a token. The *token,* which is a special bit pattern, travels around the ring until a system catches it. The system which possesses the token can transmit the data. Data and the token are transmitted to a specific destination. As the data travels around the loop, each system checks whether it is the intended recipient of the data. If not, it passes the token else, it reads the data. Once the data is received, the token is released and returns to travels around the loop until another system grabs it. If any one segment of the loop is broken, all communication around the loop ceases.

Today, most networks have redundancy or other mechanisms that protects the whole network from being affected by malfunctioning of one workstation. Network engineers generally designs ring topologies incorporating a fault tolerance mechanism, such as dual loops running in opposite directions, to prevent single points of failure.

Bus Topology

A bus topology connects each system to a trunk or backbone cable. All systems on the bus can transmit data simultaneously, which can result in collisions. To avoid this, the systems employ a collision avoidance mechanism that basically *listens* for any other currently occurring traffic. If traffic is observed, the system waits a few moments and listens again. If no traffic is observed, the system transmits its data. On a bus topology, all systems on the network hear the data which is transmitted. If the data is not addressed to a specific system, that system just ignores the data. The main benefit of a bus topology is that if a single segment fails, communications on all other segments continue uninterrupted. However, the central trunk line remains a single point of failure.

There are two types of bus topologies: linear and distributed (tree). A linear bus topology employs a single trunk line which has explicitly two end points. In distributed(tree) topology all the nodes of the network are connected to a common transmission medium which has more than one end points as shown in Figure 6.19(a) and 6.19(b). The primary reason a bus is rarely if ever used today is that it must be terminated at both ends and any disconnection can take down the entire network. A fully connected bus topology is having only one linking channel.

In a simple bus topology, a single cable runs the entire length of the network. Nodes are attached to the network through drop points on this cable. Data communications transmit the length of the medium, and each packet transmitted has the capability of being looked at by all nodes. Depending upon the packet's destination address, each node decides to accept or ignore the packet.

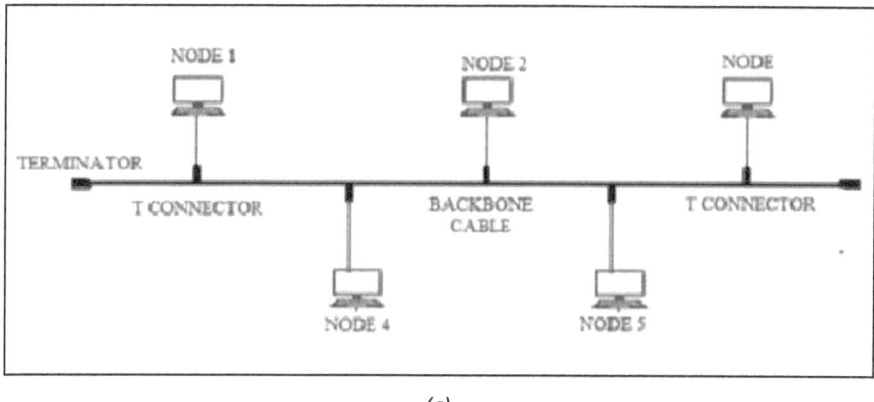

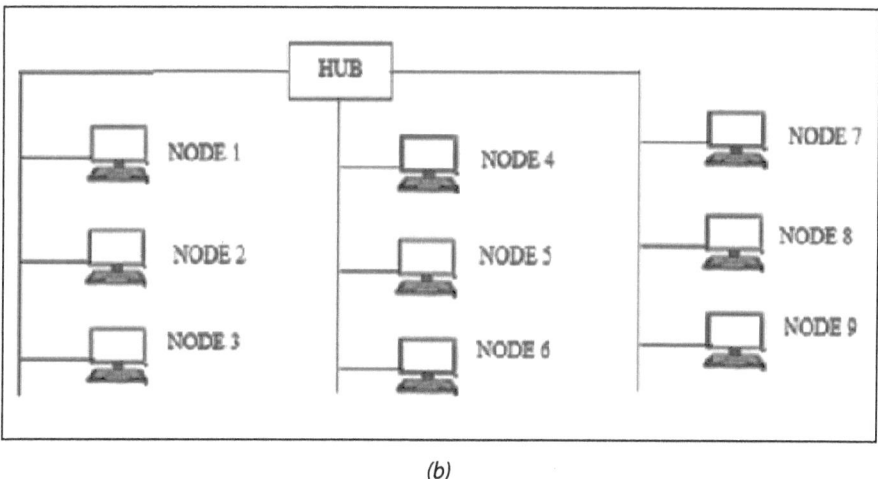

▲ **Figure 6.19:** (a) Linear bus topology (b) Distributed (tree) bus topology

Star Topology

A star topology employs a centralized connection device. This device can be a simple hub or switch. Each node has a dedicated link to the central device as shown in Figure 6.20. The central device needs to provide enough throughput so that it does not turn out to be a detrimental bottleneck for the network as a whole. As a central device is required in star topology, it is a potential single point of failure; redundancy may need to be implemented.

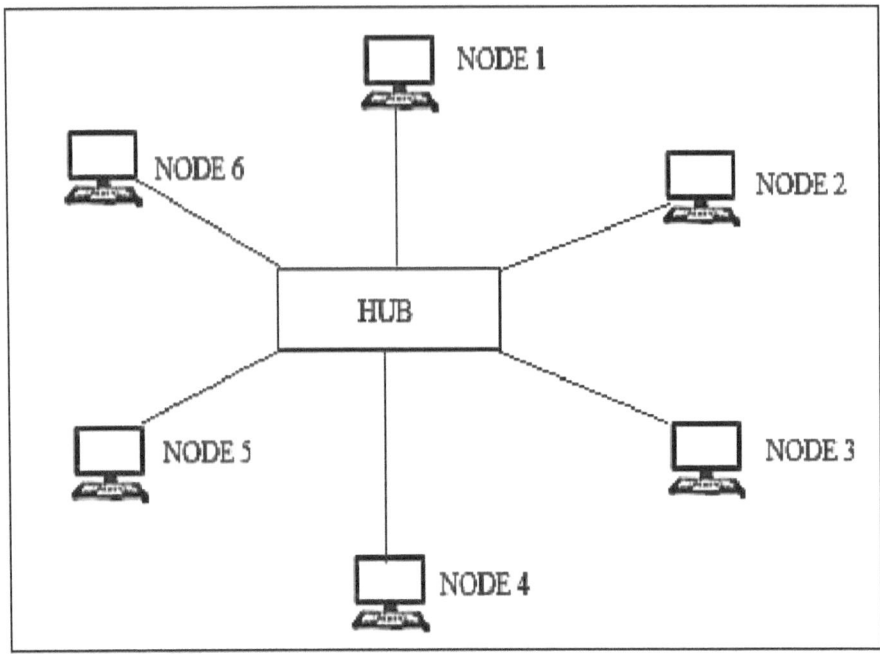

▲ **Figure 6.20:** Star Topology

If any one segment fails, the other segments can continue to function. However, the central hub is a single point of failure. Generally, the star topology uses less cabling than other topologies and helps the identification of damaged cables easier.

A logical bus and a logical ring can be implemented as a physical star. Ethernet is a bus based technology. It can be deployed as a physical star, but the hub or switch device is actually a logical bus connection device. Likewise, Token Ring is a ring-based technology. It can be deployed as a physical star using a Multistation Access Unit (MAU). MAU allows the cable segments to be deployed as a star while the device makes logical ring connections internally. Not many networks use true linear bus and ring topologies anymore. A ring topology can be used for a backbone network, but most networks are constructed in a star topology because it enables the network to be more resilient and not as affected if an individual node experiences a problem.

Mesh Topology

A mesh topology connects systems to other systems using numerous paths as shown in Figure 6.21. This arrangement is usually a network of interconnected routers and switches that provides multiple paths to all the nodes on the network. In a full mesh topology, every node is directly connected to every other node, which provides a great degree of redundancy. A partial mesh topology connects many systems to many other systems. A full mesh topology provides redundant connections to systems, allowing multiple segment failures without seriously affecting connectivity. The Internet is an example of a partial mesh topology.

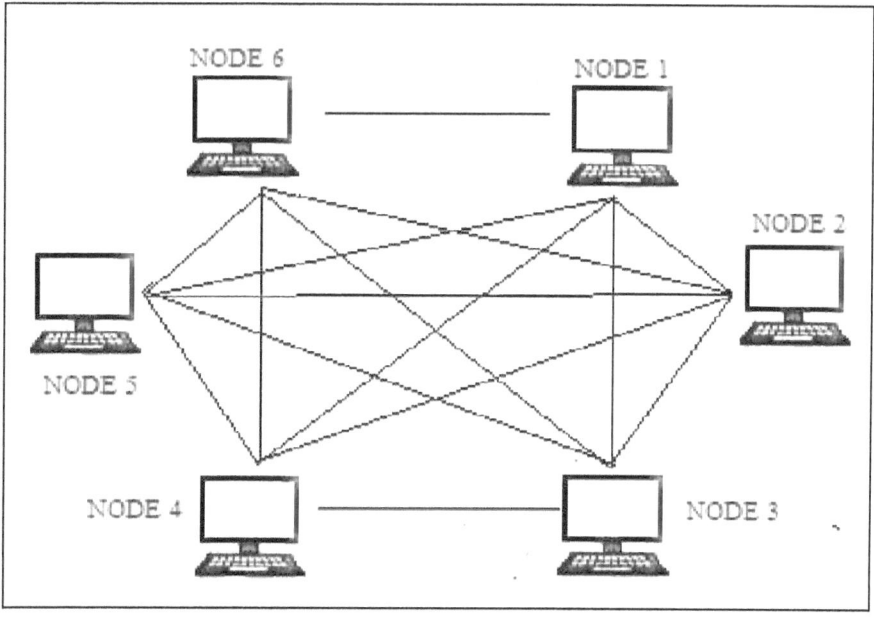

▲ **Figure 6.21:** Mesh topology

Comparison of Network Topologies

The comparisons of Network Topologies are summarized below.

▼ **Table 6.3:** Comparison of Network Topologies

Topology	Characteristics	Problems	Security
Star	All information passes through the central network connection	The central device is a single point of failure	Denial of service attack

Topology	Characteristics	Problems	Security
Bus	It is a network topology in which there is a single line (the bus) to which all nodes are connected and the nodes are connected only to this bus. The cable has a small cap installed at the end, called a terminator. The terminator prevents signals from bouncing back and causing network errors.	If one station experiences a problem, it can negatively affect surrounding computers on the same cable	Not secure due to broadcast
Ring	Information goes in one direction around the ring and passes along the ring until it reaches the correct computer	If one station experiences a problem, it can negatively affect the surrounding computers on the same ring	Has the least security as the information intended for one machine must pass through all the others
Mesh	A network topology in which there are at least two nodes with two or more paths between them	Requires more expense in cabling and extra effort to track down cable faults	A mesh needs secure links, routing, and forwarding
Tree	A bus topology with branches of the main cable	Heavily relies on the main bus cable, a break in the main cable will cripple the entire network. Further adding more nodes and segments makes the maintenance difficult.	Low security as it is physically a star but logically a bus

6.6 MEDIA ACCESS METHODS

A media access method refers to the manner a computer gain and controls access to the network's physical medium. In other way it defines how the network places data on the transmission media and how it takes it off. Common media access methods include the following,

1. Contention systems
    a. CSMA/CD
    b. CSMA/CA
2. Selection systems
    a. Token Passing
    b. Polling
    c. Polling by exception

3. Reservation systems
   a. Frequency Division Multiplexing
   b. Time Division Multiplexing
4. Demand Priority

One of the primary concerns with media access is to prevent packets from colliding as a collision occurs when two or more computers transmit signals at the same time. The Media Access Control (MAC) layer in a communication protocol governs the rules which are a sub layer of Data Link Layer.

CONTENTION SYSTEMS

In contention, any computer in the network can transmit data at any time. In fact it is a first-come-first-served media access method. In a contention-based access method, a node accesses medium when the network is idle. The main contention methods used are Carrier Sense Multiple Access with Collision Detection (CSMA/CD) and   Carrier-Sense Multiple Access with Collision Avoidance (CSMA/CA). Contention is at the heart of these methods which are used in the IEEE 802.3 and the original Ethernet networks.

CSMA/CD: Carrier- Sense Multiple Access with Collision Detection

Carrier Sense means that each station on the LAN continually listens to (test) the cable for the pretense of a signal prior to transmitting. *Multiple Access* means that there are many computers attempting to transmit and compete for the opportunity to send data. *Collision Detection* means that when a collision is detected, the station will stop transmitting and wait for a random length of time before transmitting. CSMA/CD works best in an environment where relatively fewer, longer data frames are transmitted. CSMA/CD is used on Ethernet networks.

CSMA/CD operates as follows:
1. a station that wishes to transmit on the network checks if the cable is free, and if the cable id is free, the station starts transmitting,
2. however, another station may have detected a free cable at the same instant and also start transmitting which result in a collision,
3. once the collision is detected, all stations immediately stop transmitting, and
4. station then wait for a random length of time before checking the cable and then retransmit.

CSMA/CA- Carrier-Sense Multiple Access with Collision Avoidance

It is similar to CSMA/CD. The difference is that the CD (collision detection) is changed to CA (collision avoidance). Instead of detecting and reacting to collisions, CSMA/CA tries to avoid them by having each computer signal and understand the intention to transmit before actually transmitting. CSMA/CA is slower than CSMA/CD, and is used on Apple networks.

The collision detection logic ensures that more than one message on the channel simultaneously will be detected from both ends and eventually stopped. The system is a probabilistic one, since access to the channel cannot be ascertained in advance.

Selection Systems

Token Passing

Collisions are eliminated under token passing because only a computer that possesses a free token (a small data frame) is allowed to transmit. Transmission from a station with higher priority takes precedence over station with lower priority. Token passing works best in an environment where relatively large number of shorter data frames is being transmitted. Token passing is used on Token Ring and ArcNet networks. Token passing works as described below.

1. A station that wishes to transmit on the network waits until the token is free.
2. The sending station transmits its data with the token.
3. The token travels to the recipient without stopping at other stations.
4. The receiving station receives.

Networks that use token passing generally have some provision for setting the priority with which a node gets the token. Higher level protocols can specify that a message is important and should receive higher priority.

Polling

This is one of the simplest master- slave point to point or point to multipoint configuration where the master polls, the slave in predetermined regular intervals and gather the data. Here the master is in total control of the monitoring, decision making and control of the process. The slaves never initiates the data exchange rather wait for the master's request and respond. Essentially it is a half-duplex communication. In case the slave fails to respond timely, then the master retries typically two or three times. If the slave still fails to respond, the master then moves to the next slave in the sequence with a remark that the particular slave is inactive or faulty and requires attention for rectification.

Polling by Exception

On many occasions, the status and the slave may be the same but the polling mechanism gathers the data from the slaves. This unnecessary transfer of data can be minimized or virtually eliminated with a technique called exception reporting. This approach is popular with the CSMA/CD philosophy but it could also offer a solution for the polled approach where there is a considerable amount of data to transfer from each slave.   In exception reporting, the remote station reporting slave devices monitor itself to identify a change of state or data. If there is a change of state, the remote station writes a block of data to the master station when the master station polls the remote. Typical reasons for using polled report by exception include,

- the polling or scanning which is performed with a low data rate due to the channel constraints,
- there is a substantial data being monitored from the remote stations, and
- the number of remote devices connected to master station is reasonably high.

Reservation Systems
Frequency Division Multiple Access (FDMA)

Frequency Division Multiple Access (FDMA) is a channel access technique found in multiple-access protocols as a channelization protocol. FDMA permits individual allocation of single or multiple frequency bands, or channels to the users. FDMA, just like any other multiple access system, harmonizes access between multiple users.

Advantages of FDMA
- It allocates dedicated frequencies to different stations. Moreover there are separate bands for both uplink and downlink. Hence stations transmit and receive continuously at their allocated frequencies.
- It is very simple to implement with respect to hardware resources.
- FDMA is efficient when constant traffic is required to be managed with less number of user populations.

Disadvantages of FDMA
- In FDMA, frequencies are allocated permanently and hence spectrum will be wasted when stations are not transmitting or receiving.
- Network and spectrum planning is cumbersome and time consuming.
- It uses guard bands to prevent interference. This wastes very useful and scarce frequency resources.

- It requires RF filters to meet stringent adjacent channel rejection specifications. This increases the cost of the system.

Time Division Multiple Access (TDMA)

Time Division Multiple Access (TDMA) is a channel access method (CAM) used to facilitate channel sharing without interference. TDMA allows multiple stations to share and use the same transmission channel by dividing signals into different time slots. Users transmit in rapid succession, and each one uses its own time slot. Thus, multiple stations (like mobiles) may share the same frequency channel but only use part of its capacity.

Disadvantages of TDMA: Disadvantage of using TDMA technology is that the users has a predefined time slot. When moving from one cell site to the other, if all the time slots in this cell are full the user might be disconnected. Another problem in TDMA is that it is subjected to multipath distortion.

Demand Priority

This is a new Ethernet media access method that will probably replace popular but older CSMA/CD and by which stations on a 100VG-AnyLan network gain access to the wire for transmitting data. 100VG-AnyLan is a high-speed form of Ethernet based on the IEEE standard 802.12. A 100VG-AnyLan network based on the demand priority access method consists of end nodes (stations), repeaters (hubs), switches, routers, bridges, and other networking devices. A typical 100VG-AnyLan network consists of a number of stations plugged into a cascading star topology of repeaters (hubs). Because of timing, a maximum of five levels of cascading of the physical wiring is permitted. Hubs are connected using uplink ports. Each hub is aware of the stations directly connected to it and the hubs that are uplinked from it. The main features of the demand priority are,

1. used with 100 Mbps Ethernet,
2. requires a *smart* hub,
3. station must require permission from hub before they can transmit,
4. stations can transmit and receive at the same time, and
5. transmission can be prioritized.

Hubs can be thought of as servers and end nodes as clients. With demand priority, before transmitting data, a client must place a request access to the network media. The server processes this request and decides whether to allow the client to access the media. If the server decides to grant permission to the client to access the media, it sends the client a signal informing the decision. The client then takes over the control of the media and transmits its data.

Demand priority is considered as a contention method, but it operates differently from the CSMA/CD access method used in Ethernet networks. Cables in a 100VG-AnyLan network are capable of transmitting and receiving data at the same time using all four pairs of twisted-pair cabling in a quartet signaling method.

6.7 PROS AND CONS OF COMPUTER NETWORKS

Today computer networking has become one of the most effective and successful means of sharing information, where all computers are linked together by a common network using guided or unguided media. However it has certain drawbacks as well. Some of the advantages and disadvantages are briefly described below.

Advantages: Enhances communication and availability of information:

Networking, especially with full access to the web, allows ways of communication that would simply be impossible before it was developed. Instant messaging can now allow users to talk in real time and send files to other people wherever they are in the world, which is a huge boon for businesses. Also, it allows access to a vast amount of useful information, including traditional reference materials and timely facts, such as news and current events.

Allows for more convenient resource sharing: Resource sharing is another benefit which is very important, particularly for larger organisations that really need to produce huge numbers of resources to be shared to all the people.

Makes file sharing easier: Computer networking allows easier accessibility for people to share their files, which greatly helps them with saving more time and effort, since they could do file sharing more accordingly and effectively.

Increases cost efficiency: With computer networking, one can use a lot of software products available on the market which can just be stored or installed in the system or server, and can then be used by various workstations.

Disadvantages

Lacks independence: Computer networking involves a process that is operated using computers, so people will be relying more of computer work, instead of exerting an effort for their tasks at hand. Aside from this, they will be dependent on the main file server, which means that, if it breaks down, the system would become useless, making users idle.

Poses security difficulties: Because there would be a huge number of people who would be using a computer network to get and share some of their files and resources, a certain user's security would be always at risk. There might even be illegal activities that would occur, which one need to be careful about and aware of.

Lacks robustness: if a computer network's main server breaks down, the entire system would become useless. Also, if it has a bridging device or a central linking server that fails, the entire network would also come to a standstill. To deal with these problems, huge networks should have a powerful computer to serve as file server to make setting up and maintaining the network easier.

Security concerns: There would be instances that stored files are corrupt due to computer viruses which may spread easily to other computers, if proper security measures are not implemented. Thus, network administrators should conduct regular check-ups on the system, and the stored files at the same time.

6.8 COMPUTER NETWORKS: PRESENT DAY TECHNOLOGIES

6.8.1 Local Area Networks (LANs)

Local Area Networks (LAN) is a group of computers interconnected one another, located within a limited area such as a room, a house, educational institutions, R & D laboratory, organizations or office building to form a single network. It allows users to share and manage storage devices, printers, applications, data, and other network resources. Usually LAN is limited to a specific geographical area, usually less than two kilometers in diameter. They might use a dedicated backbone to connect multiple sub networks, but they do not use any telecommunication carrier circuits or leased lines. A Local Area Network domain is defined as a sub-network that is made up of servers and clients, each of which are controlled by a centralized database. User approval is obtained through a central server or a domain controller. The term *domain* is referred to descriptors for internet sites, which is a site's web address. Domains on the internet are categorized by levels. Top-level domains or TLDs include. com,. net,. edu and. org. CcTLDs represent top-level domains of countries, while GTLDs are generic top-level domains. Domain names are governed by the internet corporation for Assigned Names and Numbers, which oversees the creation of TLDs as well as their distribution. Domain Name Servers (DNS) maintain a connection between domain names and IP addresses. The DNS system is used by computers to transmit information online.

Advantages and disadvantages of LAN

Local area networks let families, school, businesses and other entities connect their computers, but they are complex. They make administration simple, and are customizable. However, they can be difficult to secure appropriately. They are built on open standards, and the lack of proprietary software and communication standards means that licenses and other costs are not required. Their technology has been proven to be scalable and durable. Their flexibility, however, also makes them complex, and medium and large networks generally require expert maintenance. The setup process is intimidating. LAN technology was designed to scale to large networks, and setting it up requires knowing a number of IT topics that require significant experience. The technology used in LANs is designed to be open, which may leads to security glitches. By default, all network requests are sent, and vulnerable computers are easy to target. If a particular computer on a LAN is not adequately protected, the networking infrastructure find difficult to protect it.

6.8.2 Metropolitan Area Networks (MAN)

A Metropolitan Area Network (MAN) is a network that interconnects users with computer resources in a geographic area or region larger than that covered by Local Area Network (LAN) but smaller than the area covered by a Wide Area Network (WAN). The term is applied to the interconnection of networks in a city into a single larger network. It is also used to mean the interconnection of several local area networks by bridging them with backbone lines.

A MAN is usually a backbone that connects LANs to each other and LANs to WANs, the internet, and telecommunications and cable networks. Presently most of the of MANs are *Synchronous Optical Networks (SONETs)* or *FDDI* rings and Metro Ethernet provided by the service providers. The SONET and FDDI rings cover a large area, and businesses can connect to the rings. In reality, several businesses are usually connected to one ring. SONET is actually a standard for telecommunications transmissions over fiber-optic cables. SONET is *self-healing*, meaning that if a break in the line occurs, it can use a backup redundant ring to ensure transmission continues. All SONET lines and rings are designed for full redundancy. The redundant line waits in the wings in case anything happens to the primary ring. SONET networks can transmit voice, video, and data over optical networks. Slower speed SONET networks often feed into larger, faster SONET networks. This enables businesses in different cities and regions to communicate. MANs can be made up of wireless infrastructures, optical fiber, or Ethernet connections. Ethernet has evolved from just being a LAN technology to being used in MAN environments. Due to its prevalent use within organizations' networks, it is easily extended and interfaced into MAN networks. A service provider commonly uses layer 2 and 3 switches to connect optical fibers, which can be constructed in a ring, star, or partial mesh topology. VLANs are commonly implemented to differentiate between the various logical network connections that run over the same physical network connection.

6.8.3 WAN Technologies

LANs are typically built using affordable hardware like Ethernet cables, network adapters, hubs and switches. A MAN is larger than a LAN, and may consist of smaller LANs spanning several buildings in the same town or city. A WAN is the largest, and may possibly contain several smaller LANs or MANs; it may also be limited to an organization, or publicly accessible. A good example of a global WAN is the Internet. Several varieties of WAN technologies are available to companies today. The information that a company evaluates to decide which is the most appropriate WAN technology for it usually includes functionality, bandwidth demands, service level agreements, required equipment, cost, and what is available from service providers. The following sections go over some of the WAN technologies available today.

Various WAN Technologies and its Characteristics

1. WAN using dedicated line or a leased line is secured because two locations are using the same medium. But are expensive compared to other WAN options.
2. WAN using Frame relay uses packet-switching technology, is a high performance WAN protocol that which works over public networks. Billing is based on bandwidth consumed.
3. High Level Data Link Control protocol (HDLC) is a Layer 2 protocol. It is a simple protocol used to connect point to point serial devices and multipoint communication using data encapsulation method for synchronous serial links.
4. SMDS is a High-speed switching technology used over public networks. It is also obsolete and replaced with other WAN protocols.

5. ATM is another WAN technology using high speed bandwidth switching and multiplexing technology that has relatively low latency.
6. Synchronous Data Link Control (SDLC) is another WAN technology which enables mainframes to communicate with remote offices with polling mechanism.
7. X.25 is the first packet-switching technology developed to work over public networks. X.25 having lower speed than frame relay because of its extra overhead. It also uses SVCs and PVCs. Today it is almost obsolete and replaced with other WAN protocols.

Presently an adequate understanding of WAN technology is an added advantage for a power engineer, as it is extensively deployed in power system SCADA, PMU, WAMS, etc.

LAN and WAN Protocols: Communication error rates are lower in LAN environments than in WAN environments, which make sense when compare the complexity of each environment. WAN traffic may have to travel hundreds or thousands of miles and pass through several different types of devices, cables, and protocols. Because of this difference, most LAN MAC protocols are connectionless and most WAN communication protocols are connection oriented. Connection-oriented protocols provide reliable transmission because they have the capability of error detection and correction.

Advantages and disadvantages of a WAN: WAN cover a wider geographical distance when compared to a LAN. However, they are more expensive and require professional installation. WAN slower than LAN, though it spans a great distance, such as from city to city or even country to country.

6.8.4 Internet

The internet refers to a large computer network that links together other, smaller computer networks. It is a global network of information that is transmitted to billions of computers and devices. These networks are operated from various entities, such as education, government and businesses that follow specific protocols. The Internet was first developed in 1962, but it wasn›t until a few decades later when it became widespread and was adopted among a variety of everyday users. With the accessibility to electronics becoming easier, it makes the internet available to a wider population.

6.8.5 Intranets and Extranets

When an organization uses web based technologies inside its networks, it is using an *intranet*. Organizations set up internal web sites for having centralized business information such as employee details, upcoming events, updates, schedules, directions, circulars, and instructions. Many organizations have implemented web based terminals that enable employees to perform their daily tasks, access centralized databases, make transactions, collaborate on projects, access global calendars, etc. and obtain often available technical or commercial information. Web based clients limit a user's ability to access the computer's system files, resources, and hard drive, access back-end systems, and perform other tasks. The web based client can be configured to provide a GUI with only the buttons, fields, and pages necessary for the users to perform tasks. This gives all users a standard universal interface with similar capabilities. The organizations have web servers and client machines having web browsers, and it uses the TCP/IP protocol suite. The web pages are usually written in HTML and are accessed through HTTP. Using web based technologies has many pluses as they are easy to implement, no major interoperability issues occur, and with just the click of a link, a user can be taken to the requested location. Web based technologies are platform independent, and different types of client workstations can access the with different web browsers.

An *extranet* extends outside the boundaries of the organization's network to enable two or more companies to share common information and resources. Corporate associates commonly set up extranets to accommodate business-to-business communication. An extranet enables corporate associates to work on projects together; share marketing information, communicate and work together on matters, process orders, and share catalogs, pricing structures, and information on upcoming events. Trading partners often use *Electronic Data Transfer (EDT)*, which provides structure and organization to electronic documents, orders, invoices, purchase orders, and a data flow. EDT has evolved into web-based technologies to provide easy access and easier methods of communication. An extranet can be vulnerable if the extranet is not implemented securely and maintained properly. Appropriately configured firewalls need to be in place to control the extranet data flow. Most of the Extranets relies on dedicated lines or highly secure and resilient MPLS-VPN cloud,

6.8.6 Value Added Networks (VAN)

A Value Added Network (VAN) is a private network which is hired by an organization to facilitate Electronic Data Transfer (EDT) or provide other network services. Before the arrival of the World Wide Web, some organizations hired Value Added Networks to move data from their organization to other organizations. With the arrival of the World Wide Web, many companies found it more cost-efficient to move their data over the Internet instead of paying the minimum monthly fees and per-character charges found in typical VAN contracts. In response, contemporary value-added network providers now focus on offering EDT, encryption, secure e-mail, management reporting, and other extra services for their customers.

SUMMARY

This chapter begins with explaining common terminologies of computer networking, and then move on to explain the different network topologies. The OSI reference model has been explained in detail, followed by the TCP/IP and EPA models. The TCP/IP vulnerabilities and attack vectors to different layers and concept of port are also briefly explained.

CHAPTER 07

PROTOCOLS: THE NERVES AND VEINS

7.1 INTRODUCTION

To obtain full functionality in Power System SCADA (PSS), especially in distributed PSS it needs appropriate protocols for transmitting data between its components effectively and securely. SCADA, being the heart and brain of PSS and Smart Grid, standardized and interoperable SCADA protocols are very much needed for their efficient and smooth operation. The old SCADA communication protocols such as Modbus RTU, Profibus and Conitel are widely adopted and used. But they are SCADA vendor specific. Standard protocols which are becoming prevalent today are IEC 61850, IEC 60870-5-101 or 104, and DNP3. These communication protocols are standardized and adopted by almost all major SCADA vendors. Many of these protocols are now considerably modified and contain extensions to operate over TCP/IP as well. However it is a good security engineering practice to avoid connecting SCADA systems to the internet so that the attack surface can be considerably reduced. RTUs and other automatic controller devices were being developed before the advent of industry wide standards for interoperability. This results in multitude of control protocols. This chapter is dedicated to explain various protocols which are relevant to SCADA and DCS environment with an emphasis on cyber-security.

7.2 EVOLUTION OF SCADA COMMUNICATION PROTOCOLS

SCADA protocols evolved out of the need to send and receive data and control information locally and over distances in deterministic time. Deterministic in this context refers to the ability to predict the amount of time required for a transaction to take place when all relevant factors are known and understood. To accomplish communication in deterministic time for applications in ICS and DCS systems, manufactures developed their own protocols and communication bus structure. Profibus of Siemens and Modbus of Schneider Electric, are typical examples. Since all the protocols are proprietary, interoperability became a major challenge. This made the control industries and standards organizations to develop open SCADA protocols for control systems which would be nonproprietary and not exclusive to one manufacturer. Further as the internet gained popularity, manufactures started incorporating the protocols and tools which are developed internet such as TCP/IP series of protocols and internet browsers. In addition, manufactures and open standards organizations modified the highly popular and efficient Ethernet LAN technology for implementing data acquisition and control networks.

De jure standards are developed by national and international standardization development organizations such as ANSI, IEEE, NIST, IEC, etc. Many de facto industrial standards are made de jure, after appropriate evaluation. Modbus is one of the typical SCADA communication protocol extensively deployed today. The following sections discuss a few of the popular communication protocols used in DCS scenario. ICCP (IEC 60870-6) is the international standard for one control center to communicate with another control center of Power System SCADA. For communication from master stations to substations DNP3 is used in North America and IEC T-101 serial and T-104 (TCP/IP) are used in Europe. For communication between field equipment, IEC 61850, DNP3 (IEEE1815) and Modbus are developed.

7.3 SCADA COMMUNICATION PROTOCOLS

As explained above, the use of international open protocol standards are now recognized throughout the industry especially in electric utility as a key to successful integration of the various parts of the electric utility enterprise. Benefits of open systems include longer expected system life, investment protection, upgradeability and expandability, and readily available third party components. The following sections elaborate some of the important communication protocols which are presently used by many industries and power utilities.

7.3.1 Distributed Network Protocol 3 (DNP 3)

DNP3 is extensively used in many industries in automation like electricity, oil and gas, and water. It is popular and extensively used in the United States of America, Canada, South America, Australia, and parts of Asia and Africa. It is a set of an open SCADA protocol that is used for serial or IP communication between control devices, initially developed by Westronics in Calgary,

Alberta, Canada. DNP3 is mainly used between components in process automation systems especially in SCADA systems employed in electric and water companies, but usage in other industries is not common. It has larger data frames and can carry larger RTU messages, and is very much useful for communications between various types of data logging and control devices such as Remote Terminal Units (RTUs), Data Concentrators and Intelligent Electronic Devices (IEDs). It is specifically designed to achieve reliability and efficiency when used in real-time data transfer. Another feature is that it supports time-stamped data communications between a master station and RTUs or IEDs or Phasor Measurement Units (PMU).

DNP3 faces the similar cyber-security problems as IEC 60870-5-104[T-104] is explained in subsequent sections. Scrutiny made by security engineers for integrity in DNP3 revealed that it lacks authentication and encryption. The DNP3 function codes and data types are well known, hence it can be easy to manipulate and compromise a DNP3 communication session.

7.3.1.1 Protocol Architecture of DNP3

DNP3 is based on the Enhanced Performance Architecture (EPA) and uses the frame format FT3 specified by IEC 60870-5. The lower layers of physical and data link defining the communication between devices are similar to IEC 60870-5-101 and the higher levels of data units and functionality are different. DNP3 uses cyclic redundancy check for error detection.

The DNP3 protocol structure uses the basic three layer EPA model with some added functionality. It adds an additional layer named the pseudo-transport layer. The pseudo-transport layer is a combination of network and transport layer of the Open Systems Interconnection (OSI) model and also includes some functions of the data link layer. Network function is concerned with the routing and data flow over the network from sender to receiver. Transport function includes proper delivery of the message from sender to receiver, message sequencing, and corresponding error correction. This function of a transport layer is limited when compared to the OSI layer, and hence the name pseudo-transport layer, as shown in Table 7.1.

▼ **Table 7.1:** DNP3 and OSI model layers

OSI MODEL	DNP 3 (DISTRIBUTED NETWORK PROTOCOL 3)
	USER DEFINED PROCESS APPLICATION FUNCTIONS
Application Layer	Application Layer (ASDUs)

Presentation Layer	
Session Layer	
Transport Layer	PSEUDO TRANSPORT LAYER
Network Layer	
Data Link Layer	Data Link Layer
Physical Layer	Physical Layer

7.3.2 Modbus

Modbus is a request-response serial communications protocol implemented using a master-slave relationship where communication always occurs in pairs. Which means one device (the slave) must initiate a request and then wait for a response, and the initiating device (the master) is responsible for initiating every interaction is shown in Figure 7.1. Typically, the master is a Human Machine Interface (HMI) or SCADA server and the slave is a sensor, RTU, PLC, or Programmable Automation Controller (PAC). The content of these requests and responses, and the network layers, across which these messages are sent, are defined by the different layers of the protocol.

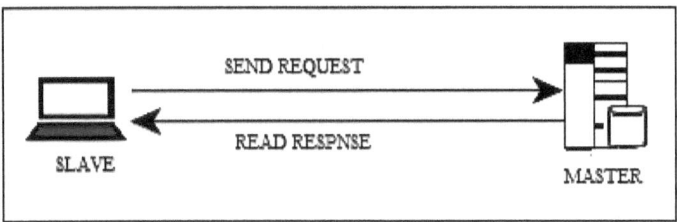

▲ **Figure 7.1:** A Master-Slave Networking Relationship

Modbus is originally published by Modicon which is presently owned by Schneider Electric for use in their PLCs. Modbus being simple and robust, is a de facto standard communication protocol. It is now commonly available and used for connecting industrial electronic devices in ICS. The main reasons for the use of Modbus in the industrial environment are,
1. developed with industrial applications in mind,
2. openly published and royalty-free,
3. easy to deploy and maintain, and
4. moves raw bits or words without placing many restrictions on vendors.

Modbus enables communication among many devices connected to the same network, for example, a system that measures temperature and humidity and communicates the results to a computer. Modbus is often used to connect a supervisory computer with RTUs in SCADA systems. Many of the data types are named from its use in driving relays viz. a single-bit physical output called a coil, and a single-bit physical input called a discrete input or a contact.

The development and update of Modbus protocols has been managed by the Modbus organization, since Schneider Electric transferred rights to that organization. The Modbus organization is an association of users and suppliers of Modbus-compliant devices that seek to drive the adoption and evolution of Modbus. The different versions of the Modbus protocol which are presently used are,
1. Modbus RTU,
2. Modbus ASCII,
3. Modbus TCP/IP or Modbus TCP,
4. Modbus over UDP,
5. Modbus over TCP/IP or Modbus RTU/IP,
6. Modbus over UDP or Modbus Plus,
7. Pemex Modbus, and
8. Enron Modbus.

Among these Modbus RTU, Modbus ASCII, and Modbus TCP/IP or Modbus TCP are the popular versions. A brief description and the frame format of these versions are summarized below.
1. *Modbus RTU*: This is the most common implementation available for Modbus and is used in serial communication by making use of a compact, binary representation of the data for protocol communication. The RTU format follows the commands/data with a Cyclic Redundancy Check (CRC) checksum as an error check mechanism to ensure the reliability of data. A Modbus RTU message must be transmitted continuously without inter-character hesitations. Modbus messages are framed (separated) by idle (silent) periods. A Modbus frame is composed of an Application Data Unit (ADU), which encloses a Protocol Data Unit (PDU).
   - ADU = Address + PDU + Error check,
   - PDU = Function code + Data.

Modbus RTU frame format (primarily used on 8-bit asynchronous lines like EIA-485) and can be represented as shown below.

Start	Address	Function	Data	CRC	End
28 bits	8 bits	8 bits	n x 8 bits	16 bits	28 bits

2. *Modbus ASCII*: This is used in serial communication and makes use of ASCII characters for protocol communication. The ASCII format uses a longitudinal redundancy checksum. Modbus ASCII frame format (primarily used on 7 or 8bit asynchronous serial lines) is given below.

Start	Address	Function	Data	LRC	End
8 bits	16 bits	16 bits	n x 2 bits	16 bits	16 bits

3 *Modbus TCP/IP or Modbus TCP*: This is a Modbus variant used for communications over TCP/IP networks, connecting over port 502. It does not require a checksum calculation, as lower layers already provide checksum protection. Modbus TCP frame format is given below.

Transaction identifier	Protocol identifier	Length field	Unit identifier	Function code	Data bytes
16 bits	16 bits	16 bits	8 bits	8 bits	n bits

One of the other Modbus variants gaining popularity is the Modbus plus. It is a proprietary to Schneider Electric and unlike the other variants, it is a high speed peer to peer network protocol based on token passing communication. It requires a dedicated co-processor to handle fast HDLC-like token rotation. It uses twisted pair at 1 Mbit/s and includes transformer isolation at each node, which makes it transition/edge-triggered instead of voltage/level-triggered. Special hardware is required to connect Modbus Plus to a computer, typically a card made for the ISA, PCI or PCMCIA bus.

In Modbus, data transactions are traditionally stateless, making them highly resistant to disruption from noise. It requires minimal recovery information at either end. Programming operations, on the other hand, expect a connection-oriented approach. This was achieved on the simpler variants by an exclusive login token, and on the Modbus Plus variant by explicit program path capabilities which maintained a duplex association until explicitly broken down. The main reason why the connection-oriented TCP/IP protocol is used is to keep control of an individual transaction by enclosing it in a connection, which can be identified, supervised, and cancelled without requiring specific action on the part of the client and server applications. This gives the mechanism of wide tolerance to network performance changes, and allows security features such as firewalls and proxies to be easily added. The other versions of Modbus are briefly explained below.

1. *Modbus over UDP*: Modbus over UDP on IP networks removes the overheads required for TCP.
2. *Enron Modbus:* This is another extension of standard Modbus developed by Enron Corporation with support for 32-bit integer and floating-point variables and historical and flow data. Data types are mapped using standard addresses.
3. *Pemex Modbus:* This is an extension of standard Modbus with support for historical and flow data. It was designed for the Pemex oil and gas company for use in process control but never gained widespread adoption.

Data model and function calls are identical for the Modbus RTU, Modbus ASCII, Modbus TCP/IP, and Modbus over UDP variants, but the encapsulation is different. However the variants are not interoperable.

7.3.2.1 Modbus Limitations

1. Modbus protocol does not provide any security against unauthorized commands.
2. Modbus was designed and developed initially for the communication with the PLC. Hence the number of data types was limited to those which understood by PLCs at that time.
3. No standard way exists for a node to find the description of a data object, such as determining whether a register value represents a temperature between T_1 and T_2 degrees.
4. Since Modbus is a master-slave protocol, there is no way for a field device to *report by exception*. As a result the master node must routinely poll each field device and look for changes in the data. This consumes considerable bandwidth and network time in applications.
5. Modbus is restricted to addressing 254 devices on one data link, which limits the number of field devices that may be connected to a master station.
6. Modbus transmissions must be contiguous, which limits the types of remote communications devices to those that can buffer data to avoid gaps in the transmission.

7.3.2.2 Attacks on DNP3 and Modbus

As the original focus while developing these was not security rather efficiency, Modbus and DNP3 virtually have no security measures incorporated. The lack of authentication in Modbus means that remote terminals accept commands from any machine that appears to be a master. The lack of integrity checking or encryption allows messages to be intercepted, changed and forwarded. In fact it is susceptible to Man In The Middle attack (MITM) which is most perilous in ICS and DCS which control Critical Infrastructure. For DNP3, message from an outstation can easily be spoofed, making it appear to be unavailable to the master. It is also common that passwords may be sent across the network in clear text. Though the DNPSecure has been developed, deployment is difficult as most of the systems using DNP3 are in the Critical Infrastructure, where downtime is very crucial and almost not feasible.

Modbus protocol is highly vulnerable to attacks like response and measurement injection, and command injection. A scrutiny on Modbus protocol with various levels of injection attacks ranging from naïve injection to complex injections targeting specific fields and values based on domain knowledge, reveals the possible consequences, such as sporadic sensor measurements, altered system control schemes and altered actuator states. This can result in partial communication disruption to complete shutdown of the device.

7.3.3 Profibus

Profibus is an open standard, serial, smart field-bus technology widely used in time-critical control and data acquisition systems. It can provide data transmission rates of 31 Kbps, 1Mbps, and 2,5 Mbps in the physical layer. It provides determinism for real-time control applications and supports multimaster communication networks.

Profibus uses the bus topology where a central line, or bus, is wired throughout the system. Devices are attached to this central bus. One bus eliminates the need for a full-length line going from the central controller to each individual device. In the past, each Profibus device had to be connected directly to the central bus. Technological advancements, however, have made it possible for a new *two-wire* system. In this topology, the Profibus central bus can connect to a ProfiNet Ethernet system. In this way, multiple Profibus buses can connect to each other.

Profibus devices which are connected to a central line can communicate information in an efficient manner which can go beyond the automation messages. Profibus devices can also participate in self-diagnosis and connection diagnosis. At the most basic level, Profibus benefits from superior design of its OSI layers and basic topology. There are three versions of Profibus, which are summarized below.

7.3.3.1 Profibus Process Automation (PA)

Profibus PA is a protocol designed for Process Automation. In fact, Profibus PA is a type of Profibus DP (Decentralized Peripherals) application profile. It connects data acquisition and control devices on a common serial bus and supports reliable, intrinsically safe implementations. It also provides power to field devices through the bus. Profibus PA standardizes the process of transmitting measured data. It does hold a very important and unique characteristic. Profibus PA was designed specifically to use in hazardous environments. Profibus uses the basic functions and extensions available in Profibus DP.

In most environments, Profibus PA operates over RS485 twisted pair media. This media, along with the PA application profile supports power over the bus. In explosive environments, though, that power can lead to sparks that induce explosions. To handle this, Profibus PA can be used with Manchester Bus Powered technology (MBP). The MBP media was designed specifically to be used in Profibus PA. It permits transmission of both data and power. The stepping down of power, reduces, or nearly eliminates, the possibility of explosion. Buses using MBP can reach 1900 meters and can support branches.

7.3.3.2 Profibus Factory Automation (Decentralized Peripherals-DP)

The second type of Profibus is more universal which is referred to as Profibus DP, for Decentralized Periphery, this new protocol is much simpler and faster. It uses different physical layer standards than those employed by Profibus PA. Application profiles allow users to combine their requirements for a specific solution. Profibus DP itself has three separate versions. Each version, from DP-V0 to DP-V1 and DP-V2, provides newer, more complicated features such as diagnostics, alarm messaging, and parameterization.

7.3.3.3 Profibus Fieldbus Message Specification (FMS)

The initial version of Profibus was Profibus FMS, Fieldbus Message Specification. Profibus FMS was designed to communicate between Programmable Controllers and PCs, sending complex information between them. Unfortunately, being the initial effort of Profibus designers, the FMS technology was not as flexible as needed. This protocol was not appropriate for less complex messages or communication on a wider and complicated network. Though it offers a large number of functions and is, generally, more complicated to implement than Profibus PA or Profibus DP. New types of

Profibus would satisfy those needs. Profibus FMS is still in use, though the vast majority of users find newer solutions to be more appropriate.

A comparison of the three Profibus versions is given in Table 7.2.

▼ **Table 7.2:** Profibus Versions

Profibus PA	Profibus DP	Profibus FMS
intrinsically safe, reliable, bus powered, process applications	high speed, decentralized applications	general automation, large number of applications

7.3.3.4 Communication Architecture of Profibus

Table 7.3 illustrate the communication architecture of the Profibus versions and shows their relationships in the OSI seven layer model.

▼ **Table 7.3:** Profibus PA, DP, and FMS layered protocols.

OSI MODEL	Profibus PA	Profibus DP	Profibus FMS
Application Layer	Application layer not used	Application layer not used	Application layer field bus message specification
Presentation Layer			
Session Layer			
Transport Layer			
Network Layer			
Data Link Layer	Data link layer IEC interface	Data link layer fieldbus data link	Data link layer fieldbus data link
Physical Layer	Physical layer IEC 61158-2	Physical layer EIA-485, fiber optics, radio waves	Physical layer EIA-485, fiber optics, radio waves

Profibus systems can have three types of physical media. The first is a standard twisted-pair wiring system, in this case RS485. Two more advanced systems are also available. Profibus systems can now operate using fiber-optic transmission in cases where that is more appropriate. A safety-enhanced system called Manchester Bus Powered (MBP), is also available in situations where the chemical environment is prone to explosion. The physical layers can also use either the EIA-485 standard or the IEC 61158-2 standard. If desired, all three Profibus versions can use the same bus line if they employ EIA-485 in the physical layer. However, if the application requires intrinsically safe circuitry, IEC 61158-2 operates at 31.25 Kbps.

7.3.4 IEC 60870-5-101/103/104

IEC 60870 was introduced by the IEC Technical Committee 57 and widely used in ICS and DCS communication. It is mainly popular in Europe and China and suitable for controlling electric power transmission grids and other geographically widespread ICS. This is an open protocol originally written for serial communication and was released in 1995. The IEC 60870-5-104 standard, released in 2000, present a combination of the application layer of IEC 60870-5-101 and the transport functions provided by TCP/IP. Presently this is widely used in SCADA and DCS communication protocol for monitoring and controlling of remote field locations of DCS especially in electrical grids. The structure of IEC 60870-5 standard is hierarchical and has six parts, each having different sections, and has four companion standards. Main part of the standard defines the fields of application, whereas the companion standards elaborate the information regarding the application field by specific details. These companion standards may be referred to as T-101, T-102, T-103, and T-104, where T stands for Tele-control. The five documents specify the base IEC 60870-5 Tele-control equipment and systems. The six sections of Part 5 of the IEC 60870-5 are described below.

1. IEC 60870-5-1 Transmission Frame Formats.
2. IEC 60870-5-2 Data Link Transmission Services.
3. IEC 60870-5-3 General Structure of Application Data.
4. IEC 60870-5-4 Definition and Coding of Information Elements.
5. IEC 60870-5-5 Basic Application Functions.
6. IEC 60870-5-6 Guidelines for conformance testing for the IEC 60870-5 companion standards.
7. IEC TS 60870-5-7 Security extensions to IEC 60870-5-101 and IEC 60870-5-104 protocols by applying IEC 62351.

7.3.5 IEC 60870-5-101 [T-101]

An Enhanced Performance Architecture (EPA) based protocol released by IEC in the beginning of 90s, which found extensive acceptance in power system SCADA. It is a master slave communication for multi-drop or bus topology. It polls data by cyclic polling technique using the data link layer and use parity check as well as checksum error detection techniques. The data error is decreased by these checks in IEC 60870 protocols. This standard is widely used in power system communications for tele-control, teleprotection, and associated telecommunications. The application function of T-101 includes station initiation, parameter loading, data acquisition, cyclic data transmission, clock synchronization, etc. for a remote substation. This is completely compatible with IEC 60870-5-1 to IEC 60870-5-5 standards and uses standard asynchronous serial tele-control channel interface between master and slave. The standard is suitable for multiple configurations like point-to-point, star, multidropped etc. The specific features of IEC 608705-101 are described below.

1. Both master initiated and master/slave initiated modes of data transfer are available,
2. It functions based on Link address and Application Service Data Unit (ASDU) bit,
3. ASDU addresses are provided for classifying the end station,
4. Data can be classified into different information objects and each information object is provided with a specific address,
5. Capability to assigning priority to the data before transmitting the data,
6. Option of classifying the data into 16 different groups to get the data according to the group, by issuing specific group interrogation commands from the master,
7. Capability for time synchronization, and
8. Various schemes for data transfer are available which is very much useful in IED's to transfer the SoE and disturbance files for fault analysis.

7.3.6 IEC 60870-5-103 [T-103]

This is a master-slave standard protocol based on the EPA architecture and is mainly used to couple the central unit to several protection devices and is primarily used in the energy sector. The standard is especially designed for the communication with protection devices and therefore difficult to adapt it to other applications. It defines a companion standard that enables interoperability between protection equipment and devices of a control system in a substation. It handles protection functions as status indications of circuit breakers, types of fault, trip signals, auto-reclosure, relay pickup, etc. The device complying with this standard can send the information using two methods for data transfer, either using the explicitly specified ASDU or using generic services for transmission of all the possible information. The standard supports some specific protection functions and provides the vendor a facility to incorporate its own protective functions on private data ranges. IEC 60870-5-103 is mostly used for comparatively slow transmission media with RS232 and RS485 interfaces. Connection via optical fiber is also covered by the standard. The transmission speed in general is specified with a maximum of 19200 Baud.

7.3.7 IEC 60870-5-104 [T-104]

IEC 60870-5-104 [T-104] protocol is a standard for tele-control equipment and systems with coded bit serial data transmission in TCP/IP based networks for monitoring and controlling geographically widespread processes. It is an extension of IEC 101 protocol suitable for networked circumstances with the changes in transport, network, link and physical layer services to suit the complete network access. The standard uses an open TCP/IP interface network to have connectivity to the Local Area Network (LAN) and routers with different facility used to connect to the Wide Area Network (WAN). Within TCP/IP various network types can be utilized including X.25, Frame Relay, ATM, ISDN, Ethernet and serial point to point (X.21), Application layer of IEC 104 is preserved same as that of IEC 101 with some of the data types and facilities which are not used. There are two separate link layers defined in the standard, which is suitable for data transfer over Ethernet and serial line Point-to-Point Protocol (PPP). The control field data of IEC104 contains various types of mechanisms for effective handling of network data synchronization.

Unfortunately, by design itself the security of IEC 104 is problematic, same as many of the other contemporary SCADA protocols developed. IEC Technical Committee(TC) 57 have published a security standard IEC 62351, which implements end-to-end encryption which would prevent such attacks as replay, man-in-the-middle and packet injection. However due to the increase in complexity and cost, vendors are reluctant to include this security features on their ICS networks.

7.3.7.1 Protocol Architecture of IEC 60870-5

Protocol Architecture of IEC 60870 is based on Enhanced Performance Architecture (EPA) model. The EPA model has three layers viz. Physical, Data Link and Application layers. An user layer is added to the top of the EPA model to provide the interoperability between equipements in a tele-control system. This four layer model is used for T-101, and T-103 companion standards. For companion standard T-104, which is the network adaptation, some additional layers are included from the OSI model. These are network and transport layers that are essential for the networked architecture. This networked architecture is useful for the transportation of data and messages over the network. Thus a non networked version is used for T-101, T-103, and the networked version for T-104, as shown in Figure 7.2. It may be noted that the lower four layers of T-104 are now the TCP/IP suite for networking applications.

OSI model	T-101, T-103 (EPA with user defined process functions)	T-104- EPA with network and transport with user defined process functions
	User defined process application functions	User defined process application functions
Application layer	Application layer(ASDUs)	Application layer(ASDUs)
Presentation layer		
Session layer		
Transport layer		TCP/IP transport and network protocol suite
Network layer		TCP/IP transport and network protocol suite
Data link layer	Data link transmission procedures and frame formats	TCP/IP transport and network protocol suite
Physical layer	Physical interface specification	

▲ **Figure 7.2:** Communication layers of IEC 60870-5

7.3.7.2 Attacks on IEC 60870-5

SCADA StrangeLove is an independent group of cyber-security engineers founded in 2012, focused exclusively on security assessment of ICS and SCADA. They could successfully prove detection of an IEC 60870 device on a SCADA network. They also released python scripts which can identify and return the common address of an IEC 60870 device. The common address is an address used for all data contained within the IEC 60870-5 packet, used to identify the physical device. With the help of these existing scripts it is possible to scan an ICS/DCS network for any specific IEC 60870 hosts. Once investigation became successful, the scripts could be used to detect possible targets for a Man in the Middle (MITM) attack. Further improperly terminated VPN running on IEC 60870-5 protocols is highly susceptible to DoS attacks, hence end node security issue has to be sorted out with utmost care, else can be catastrophic.

7.3.8 IEC 61850

Multiple protocols exist for electrical substation automation, which include many proprietary protocols with custom communication links. Interoperation of devices from different vendors is a need of the hour in power system automation, as it would be indeed an advantage to the designers of substation automation. This brought a need for a robust interoperable standard to serve the power system SCADA. This requirement has been attended by a group of about 60 members from different countries worked in three IEC working groups from 1995 and created IEC 61850 which accomplished the following objectives.

1. A single protocol for complete substation automation considering all different data transfer requirement,
2. Definition of basic services required to data transfer,
3. Provides high inter-operability between systems from different vendors,
4. A common method/format for storing complete data, and
5. Defines complete testing required for the equipment which conforms to the standard.

IEC 61850 is a layered architecture standard with full OSI layers that separates the functionality required for electric utility applications from the lower level networking tasks. The layered architecture illustrating the separation of functions is shown in Figure 7.3.

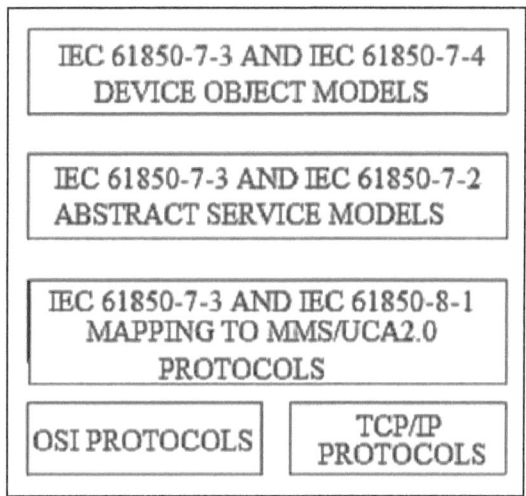

▲ **Figure 7.3:** IEC 61850 Layered Architecture

IEC 61850 is an object oriented substation automation standard which defines how to describe the devices in an electrical substation and how to exchange the information about these devices. The information model of IEC 61850 is based on two main levels of modeling. The breakdown of physical device into logical device, and the breakdown of logical device into logical nodes, data objects and data attribute. The approach of IEC 61850 is to decompose the application function into the smallest entities which are used to exchange information.

While implementing the PSS, the data exchange between the process level devices and substation is necessary. In order to achieve this all participating devices must be compatible with this protocol. As the IEC 61850 standard has been developed with a clear vision to fully incorporate the IED interoperability, irrespective that the IEDs belong to different vendors. In fact the IEC 61850 compliance of devices takes the substation automation to a next level. IEC 61850 offers three types of communication models.

1. Client/server type communication services model,
2. A publisher subscriber model, and
3. Sample Value model for multicast measurement values.

Generic Object Oriented Substation Events (GOOSE): Client server communication is preferred for managing the process level equipments at substation because of its low latency. Generic Object Oriented Substation Events (GOOSE) is a controlled model mechanism in which any format of data like Status Values (SVs) is grouped into a data set and transmitted within a time period of 4 milliseconds. In GOOSE the critical commands of a power system such as a trip or interlocking are directly mapped to the link layer. The message structure of GOOSE supports the exchange of a wide range of possible common data organized by a dataset. Being the GOOSE message a multicast, it is received by all the IEDs which have been connected and configured to subscribe it. GOOSE message usually carries the information of a status change and time of the last status change. In other words, GOOSE time stamped events from IEDs are appropriately communicated to the subscribers. The message frame consists of the destination or source Media Access Control (MAC), addresses, tag protocol identifies, length, two reserved fields and application protocol data unit. Another feature of the GOOSE message is that it repeats until the subscriber receives it.

7.3.8.1 Comparison of DNP3 and IEC-61850 GOOSE

IEC 61850 is not just a communication protocol as it not only defines how the data is transmitted and received but also describes how data is executed and stored. This makes data model of IEC61850 very different to the OSI reference model. This difference in the data model requires IEC61850 to work over a real set of communication protocols such as DNP3 or IEC T[101/3/4], requiring the data model of IEC 61850 to be mapped on to one of the above mentioned protocols. It's more advantageous if used only within a substation and its specification states that connection to the remote control centers such as remote MCC is beyond its scope. The use of IEC 61850 would be most appropriate if it is used in a substation environment where a number of IEDs interact with the SCADA master. In fact it is referred to as substation protocol.

7.3.8.2 Attacks on IEC 61850 Protocol

Though IEC 61850 based automated substations can provide various advantages over traditional substations, the power distribution utilities are very much cautious about its implementation due to security concerns. Security Engineers have identified a number of security vulnerabilities and flaws in the IEC 61850 protocol such as the lack of encryption used in the GOOSE messages, lack of Intrusion Detection System (IDS) implementation in IEC 61850 networks, and no firewall implementation inside IEC 61850 substation network. If these vulnerabilities are deviously exploited, a number of security attacks can be launched on IEC 61850 substation network.

One of the vulnerabilities within the GOOSE communication of IEC 61850 that can be exploited is by sending GOOSE frames containing higher status numbers. It prevents genuine GOOSE frames from being processed. This effectively causes a hijacking of the communication. This attack could be used to implement a Denial of Service (DoS) attack. This weakness in GOOSE can be exploited to insert spoofed messages with incorrect data between each valid message. This can be used to demonstrate using Scapy, which is a Python program that enables the user to sniff, dissect, forge, and send network packets. This attack is possible due to unencrypted and unauthenticated nature of the GOOSE message.

7.3.9 ICCP TASE 2(IEC 60870-6)

The Inter Control Center Communications Protocol (ICCP or IEC 60870-6/TASE.2) is the protocol used by various power utilities throughout the world to provide data exchange over Wide Area Networks (WANs) between utility control centers, power pools, regional control centers, and independent power producers. In fact today ICCP is the international standard adopted by IEC as ICCP Tele-control Application Service Element 2 (TASE2) and is an essential protocol in Power System Automation. Usually a typical national power grid includes a hierarchy of control centers to manage the generation, transmission, and distribution of power throughout the grid. The grid is controlled by one or more hierarchical control centers, which are responsible for scheduling of power generation to meet customer demand, and for managing major network outages and faults.

7.3.9.1 ICCP Functionalities

The basic functionality of the ICCP is to establish an appropriate link between the other control centres by managing and configuring it for proper information exchange. Generally ICCP establish link with the following control centers.
1. Generation control centers, responsible for managing the operation of generating plants such as coal-fired, natural gas, nuclear, solar, wind, etc. and for adjusting the power generated according to the requirements of the system control center,
2. Transmission control centers, responsible for the transmission of power from generating stations to network distributors, and
3. Distribution control centers, responsible for the distribution of power from the transmission networks to individual consumers.

7.3.9.2 Protocol Architecture of ICCP TASE 2

At present, the ICCP TASE 2 protocol is the internationally recognized standard for communications between electrical utility control centers. ICCP uses the Manufacturing Messaging Specification (MMS) for the messaging service needed by ICCP. It is based on client/server principles. Consequently data transfers initiated with a request from one control center (client) to another control center (server). Control centers may be both clients and servers. ICCP TASE 2 operates at the application layer in the OSI model. Any physical interfaces transport and network services that fit this model are supported. However TCP/IP over Ethernet seems to be the most common. ICCP may operate over a single point-to-point link between two control centers. The logical connections or *associations* between control centers are completely general. A client can establish association with more than one server and a client can also establish more than one association with the same server. Multiple associations with same server can be established at different levels of Quality of Service (QoS) so that high priority real-time data is not delayed by lower priority or non-real-time data transfers.

7.3.9.3 Implementation Issues and Interoperability

ICCP is a standard real-time data exchange protocol. It provides numerous features for the delivery of data, monitoring of values, program control and device control.

All the protocol specifics needed to ensure interoperability between different vendor's ICCP products have been included in the specifications. The ICCP specifications, however, do not attempt to specify other areas that will need to be implemented in an ICCP software product but that do not affect interoperability. These areas are referred to as local implementation issues in the specification. Some of the local implementation issues in the specification are listed below.
1. The API through which local applications interface to ICCP to send or receive data,
2. A user interface to ICCP for user management of ICCP data links,
3. Management functions for controlling and monitoring ICCP data links,
4. Failover schemes where redundant ICCP servers are required to meet stringent availability requirements, such as those typically experienced in an EMS/SCADA system environment, and
5. How data, programs or devices will be controlled or managed in the local SCADA/EMS to respond to requests received via an ICCP data link.

The wide acceptance of ICCP by the utility industry has resulted in several ICCP products are available on the market. Extensive interoperability testing between products of some of the major vendors has been a feature of ICCP protocol development. An ICCP purchaser must define functionality required in terms of conformance blocks and the objects within those blocks. Application profiles for the ICCP client and server conformances must match if the link is to operate successfully. Interoperability among ICCP of different vendors for the power grid is crucial for achieving the benefits of standardization such as application evolution, open architecture and scalability, plug and play capability of components and services, reliability and service orientation.

7.3.9.4 ICCP-Product Differentiation

ICCP is a real-time data exchange protocol providing features for data transfer, monitoring and control. For a complete ICCP link there need to be facilities to manage and configure the link and monitor its performance. The ICCP standard does not specify any interface or requirements for these features that are necessary but nevertheless do not affect interoperability. Similarly failover and redundancy schemes and the way the SCADA responds to ICCP requests is not a protocol issue, hence is not specified. These non-protocol specific features are referred in the standard as local implementation issues. ICCP implementers are free to handle these issues as per their requirements. Local implementation means that developers have to differentiate their product in the market with added features and values. The money spent for a product with appropriately developed maintenance and diagnostic tools will be valued, during its life expectancy, only if the ICCP connection grow and adapt the changes.

7.3.9.5 ICCP- Product Configurations

Commercial ICCP products are generally available for one of three configurations,
1. as a built in protocol embedded in the SCADA host,
2. as a networked server, and
3. as a gateway processor.

As an embedded protocol the ICCP management tools and interfaces are all part of the complete suite of tools for the SCADA. This configuration offers maximum performance because of the direct access to the SCADA database without requiring any intervening buffering. This approach may not be available as an addition to a legacy system. The ICCP application may be restricted to accessing only the SCADA environment in which it is embedded. A networked server making use of industry standard communication networking to the SCADA host may provide, performance approaching of an embedded ICCP application. On the application interface side the ICCP is not restricted to the SCADA environment but is open to other systems such as a separate data historian or other databases. Security may be easier to manage with the ICCP server segregated from the operational real-time systems. The gateway processor approach is similar to the networked server except it is intended for legacy systems with minimal communications networking capability and so has the lowest performance. In the most minimal situation the ICCP gateway may communicate with the SCADA host via a serial port in a similar manner to the SCADA RTUs.

7.4 OTHER SG PERTINENT STANDARDS

Certain other standards which are deployed in Power System SCADA (PSS) and Smart Grid and under development are briefly described below.

7.4.1 IEEE C37.118.1synchrophasor Standard

This standard defines synchrophasor measurements. It also defines a data communication protocol, including message formats for communicating the data in real-time. Mainly intended to take measurements at substations, real-time data sent to control center and collect and align data, for sending on to applications or higher level processing.

A little more elaborated, this standard defines synchrophasors, frequency, and Rate of Change of Frequency (ROCOF) measurement under all operating conditions. It specifies methods for evaluating these measurements and requirements for compliance with the standard under both steady-state and dynamic conditions. Time tag and synchronization requirements are included. Performance requirements are confirmed with a reference model, provided in detail. This document defines a Phasor Measurement Unit (PMU), which can be a stand-alone physical unit or a functional unit within another physical unit. This standard does not specify hardware, software, or a method for computing phasors, frequency, or ROCOF.

7.4.2 IEC 61968 Standard

IEC 61968 is a series of standards that will define standards for information exchanges between electrical distribution systems. These standards are being developed by Working Group 14 (WG 14) of Technical Committee 57(TC 57) of the IEC. IEC 61968 is intended to support the inter-application integration of a utility enterprise that needs to collect data from different applications that are legacy or new and each has different interfaces and runtime environments. IEC 61968 defines interfaces for all the major elements of interface architecture for Distribution Management Systems (DMS) and is intended to be implemented with middleware services that broker messages among applications.

7.4.3 IEC 61970 Standard

The IEC 61970 series of standards deals with the application program interfaces for Energy Management Systems (EMS). The series provides a set of guidelines and standards to facilitate the following.
1. The integration of applications developed by different suppliers in the control center environment,
2. The exchange of information to systems external to the control center environment, including transmission, distribution and generation systems external to the control center that need to exchange real-time data with the control center,
3. The provision of suitable interfaces for data exchange across legacy and new systems.

7.4.4 IEC 62325 Standard

IEC 62325 is a set of standards related to deregulated energy market communications, based on the Common Information Model (CIM). IEC 62325 is a part of the IEC Technical Committee 57 (TC57) reference architecture for electric power systems, and is the responsibility of Working Group 16(WG16) which works on Standards related to energy market communications.

IEC 62325-301 specifies the common information model for energy market communications. The CIM is an abstract model that represents all the major objects in an electric utility enterprise typically involved in utility operations and electricity market management. By providing a standard way of representing power system resources as object classes and attributes, along with their relationships, the CIM facilitates the integration of Market Management System (MMS) applications developed independently by different vendors, between entire MMS systems developed independently, or between an MMS system and other systems concerned with different aspects of market management, such as capacity allocation, day-ahead management, balancing, settlement, etc.

Note: *IEC Technical Committee 57(TC 57) is one of the technical committees of the IEC. TC 57 is responsible for development of standards for information exchange for Power System SCADA, distribution automation and teleprotection.*

7.4.5 IEC 61508 Standard

IEC 61508 is an international standard published by the IEC and is titled Functional Safety of Electrical/Electronic/Programmable Electronic Safety-related Systems. This protocol is intended to be a basic functional safety standard applicable to all kinds of industry. It defines functional safety as part of the overall safety relating to the Equipment Under Control (EUC) and the EUC system which depends on the correct functioning of the E/E/PE safety-related systems, other technology safety-related systems and external risk reduction facilities. The standard covers the complete safety life cycle, and may need interpretation to develop sector specific standards. It has its origin in the process control industry. The safety life cycle of IEC 61508 has 16 phases which can be categorized into three groups as follows.
1. Phases 1-5 address analysis,
2. Phases 6-13 address realization, and
3. Phases 14-16 address operation.

All the phases are concerned with the safety function of the system.

IEC 61508:2010 sets out the requirements for ensuring that systems are designed, implemented, operated and maintained to provide the required Safety Integrity Level (SIL). Four SILs are defined according to the risks involved in the system application, meant to protect against the highest risks. The standard specifies a process that can be followed by all links in the ICS so that information about the system can be communicated using common terminology and system parameters. The standard is in eight parts which are described below.

1. IEC 61508-0, Functional safety and IEC 61508
2. IEC 61508-1, General requirements
3. IEC 61508-2, Requirements for E/E/PE safety-related systems
4. IEC 61508-3, Software requirements
5. IEC 61508-4, Definitions and abbreviations
6. IEC 61508-5, Examples and methods for the determination of safety integrity levels
7. IEC 61508-6, Guidelines on the application of IEC 61508-2 and IEC 61508-3
8. IEC 61508-7, Overview of techniques and measures

IEC 61508 has been adopted in the UK as BS EN 61508, with the EN indicating adoption also by the European Committee electrotechnical standardization (CENELEC). Other standards are being produced for the application of the 61508 approach to particular sectors. Sector specific standards related to IEC 61508 include,

1. IEC 61511 Process industries,
2. IEC 61513 Nuclear power plants,
3. IEC 62061 Machinery sector, and
4. IEC 61800-5-2 Power drive systems.

7.4.6 IEC 62351 Security Standard

IEC 62351 is an industry standard developed for improving security in automation systems especially in the PSS domain. It comprises provisions to ensure the integrity, authenticity and confidentiality for different protocols used in PSS.

IEC 62351 is developed by WG15 which works on Data and Communication Security of IEC TC57 mainly for handling the security of TC 57 series of protocols including IEC 60870-5 and its derivatives, IEC 60870-6 (TASE.2), and IEC 61850. In addition, security through network and system management has to be addressed. These security standards have been developed in such a way that it should meet different security objectives for the different protocols, which vary depending upon how they are used. Some of the security standards can be used across a few of the protocols, while others are very specific to a particular profile. The different security objectives include authentication of data transfer through digital signatures, ensuring only authenticated access, prevention of eavesdropping, prevention of playback and spoofing, and intrusion detection. The IEC 62351 standards consist of introduction and overview, glossary of terms, security for profiles including TCP/IP, security for IEC 60870, security for IEC 61850, objects for network and system management, role based access control, key management, security architecture guidelines, and security for XML files.

7.4.7 IEC 62056 Electricity Metering Data Exchange Standard

IEC 62056 is a set of standards for electricity metering data exchange by International Electrotechnical Commission. The IEC 62056 standards are the International Standard versions of the DLMS/COSEM specification. DLMS or Device Language Message Specification (originally Distribution Line Message Specification), is the suite of standards developed and maintained by the DLMS User Association (DLMS UA) and has been adopted by the IEC TC13 WG14 into the IEC 62056 series of standards. The DLMS UA maintains a liaison with IEC TC13 WG14 responsible for international standards for meter data exchange and establishing the IEC 62056 series. In this role, the DLMS UA provides maintenance, registration and compliance certification services for IEC 62056 DLMS/COSEM.

COSEM or Companion Specification for Energy Metering includes a set of specifications that defines the Transport and Application Layers of the DLMS protocol. The DLMS UA defines the protocols into a set of four specification documents namely Green Book, Yellow Book, Blue Book and White Book. The Blue book describes the COSEM meter object model and the Object Identification System (OBIS), the Green book describes the Architecture and Protocols, the Yellow book treats all the questions concerning conformance testing, the White book contains the glossary of terms. If a product passes the conformance test specified in the Yellow book, then a certification of DLMS/COSEM compliance is issued by the DLMS UA.

The IEC TC13 WG14 groups the DLMS specifications under the common heading: *Electricity metering data exchange-the DLMS/COSEM suite*. DLMS/COSEM protocol is not confined to electricity metering, rather it is used for gas, water and heat metering.

7.4.8 IEC 62056-21 Standard

IEC 61107, the current IEC 62056-21, is an international standard for a computer protocol to read utility meters. It is designed to operate over any media, including the internet. A meter sends ASCII or High level Data Link Control (HDLC) data to a nearby Hand-Held Unit (HHU) using a serial port. The physical media are usually either modulated light, sent with an LED and received with a photodiode, or a pair of wires, usually modulated by a 20mA current loop. The protocol is usually half-duplex.

7.5 SECURE COMMUNICATION (SCOMMUNICATION)

The main difference between DCS networks and IT networks is mainly the control part and which is not at all a surprise. As an example a compromised power system SCADA is an unacceptable threat to the reliability of the Bulk Electric System (BES). All software can be hacked, including firewalls, hence best-practices and unconditional security standards are recommended. Obviously ICCP communication meets all of these requirements, it does not provide authentication or encryption. These services are normally provided by lower protocol layers. ICCP uses bilateral tables to control access. A bilateral table represents the agreement between two control centers connected with an ICCP link. The agreement identifies data elements and objects that can be accessed via the link and the level of access permitted. Once an ICCP link is established, the contents of the bilateral tables in the server and client provide complete control over what is accessible to each party. There must be matching entries in the server and client tables to provide access to data and objects.

7.6 SELECTING THE RIGHT PROTOCOL FOR SCADA

There are so many protocols available as both proprietary and open, many factors are to be considered when choosing the protocol for SCADA. The following points are useful while designing the SCADA.

1. Determine the system area with which the SCADA may become concerned, e.g.
   - the protocol from a SCADA master control station to the SCADA RTUs,
   - a protocol from substation IEDs to an RTU or a PLC, or a LAN in the substation.

2. As the technology is changing so fast that the timing of utility installation can have a great impact on the protocol which is selected. Hence the installation period must be determined most appropriately.

If new IEDs are implemented in the substation and scheduled to be in service within six months, protocols may be selected accordingly as Modbus and Modbus Plus are suitable at present in Indian scenario. But if the period of installation and intended applications is for a long time, then consider IEC 61850 and UCA2 MMS as the protocol.

If the timeframe is around one year, make protocol choices from implementing agency that acts as the industry initiatives and incorporates this technology into their product's migration paths. This helps to protect the investment from becoming obsolete by allowing incremental upgrades to new technologies.

In the design phase, protocol choices differ with the application areas. Different application areas are in different stages of development. An awareness of development stages will help to determine realistic plans and schedules for specific projects. Earlier when SCADA were designed, information and system security was not a priority. Most of the SCADA were designed as proprietary, stand-alone systems, and their security resulted from their physical and logical isolation and controlled access to them.

With the advancement of Information and Communication Technology (ICT), SCADA and DCS began to accept open standards and advanced networking technologies especially the internet technology. Suppliers acquired the capability to implement Web-based applications to perform monitoring, control, and remote diagnostics. Obviously this introduces control system cyber-vulnerabilities. In addition to traditional IT vulnerabilities, SCADA specific vulnerabilities become the real threat which triggers unique cyber-security requirements to protect ICS against these SCADA specific attacks and vulnerabilities. Further it is a fact that present day technology may become obsolete tomorrow as the pace of the technological advancement is so fast, it seems the time between the present and the future is shrinking. Hence it is most important that one must evaluate not only the vendor's or implementation agency's present products but also their future product development and implementation strategies.

SUMMARY

This chapter begins with the evolution of communication protocols and then move on to explain the various communication protocols used today such as DNP3, Modbus, Profibus, IEC 60870, IEC 61850, and ICCP TASE 2 (IEC 60870-6). Other relevant SG and PSS protocols such as IEEE C37.118.1 Synchrophasor Measurement Standard, IEC 61968, IEC 61970, IEC 62325, IEC 61508, IEC 62351, IEC 62056 and IEC 62056-21 are also fleetingly presented. Finally the significant points to select the right protocols are also discussed.

CHAPTER 08

PHYSICAL-CYBER SECURITY OF SCADA

8.1 INTRODUCTION

In recent years, with the integration of Information and Communication Technology (ICT), dependency of the electrical power grid on cyber-space grew, cyber-threats have materialized as new vulnerabilities, and the resilience and security of the power grids have become a major concern. The power system automation and integration of Smart Grid technologies transforms the power generation and flow on the Nation's electricity grid, with remote controlling mechanisms by means of various ICT technologies. Further new intelligent technologies and Intelligent Electronics Devices (IEDs) utilizing bi-directional communications and other digital advantages are being introduced and optimized with internet connectivity. Modernization of many Industrial Control Systems (ICS) particularly, the Supervisory Control and Data Acquisition (SCADA) system also has resulted in tremendous dependency on the internet and public networks. In fact these dependencies made the cyber-security of the automated power grid, one of the prime focus areas which need serious efforts to protect its integrity.

The advancement in unguided communication media and the Distributed Control System (DCS) technology made the remote access secure and reliable especially in the Smart Grid or SCADA DMS area. Hence engineers working in the power system automation must consider and address the attributes such as interoperability, extent of openness, scalability, simplicity and security, while considering the geographical, operational, and logistical constraints of Smart Grid and Power System SCADA. To achieve these requirements a valid and secure remote access policy is essential. But the implications of communications crossing the physical confines of their immediate network is the major challenge while accomplishing these benefits, and need additional meticulous security requirements. The increasing frequency of cyber intrusions on Industrial Control Systems (ICS) of critical infrastructure certify these requirements. Further the recent reports of intrusions into ICS clearly indicate that the attackers are with sound technical capability, having the capability of taking down the control systems that operate on the power grids, and other critical infrastructure. Today this is a nerve-wracking and challenging issue for SCADA and DCS engineers. In fact communication is an essential component of the DCS but security is a crucial concern with highest priority and not at all an add on.

If the system collapses under a cyber-attack it creates hazard and inconvenience to consumers and utilities alike. Hence while implementing the Smart Grid or power system SCADA, the important areas which need considerable special attention are, secured remote access and End Node Security (ENS), control center architecture and security, AMI and secured communication, firewall deployment and HIPS, and threats, vulnerabilities and solutions.

This chapter discusses the difference between IT security and SCADA security, various security concerns, solutions and standards such as remote access techniques, authentication techniques, NERC CIP standards etc. pertaining to power system automation.

8.2 IT SECURITY AND SCADA SECURITY

The cyber-space which is the present warfront is described as the interdependent network of information technology infrastructures, which includes the internet, communications networks, computer systems, and embedded processors and controllers of critical industries. Today this interdependent cyber-space is a crucial component of National Critical Infrastructure which needs distinct protective measures. Cyber-space is used to exchange information, buy and sell products and services, and enable many online transactions across a wide range of sectors, both nationally and internationally. As a result, a secure cyber-space becomes most critical to the economy and security of any Nation.

Without developing an appropriate cyber-security policy, cloud based process automation can be disastrous. Hence to protect the Nation's critical information infrastructures, from risks such as online fraud, identity theft, and misuse of information online, a well defined cyber-security and mitigation policy in accordance with the utility's safety and security has to be devised with utmost care.

Today the rapid growth of ICTs and social inter-dependency has changed the perception of Critical Information Infrastructure threats and, as a consequence, cyber-security has become international agenda. It is crucial to understand the risks that accompany new technologies in order to maximize the benefits. Growing threats to security, at the level of the individual, the firms, government and critical infrastructures, make security as responsibility of public. It is important to understand and keep conversant contours of fast changing challenges. Present power system automation mainly based on the SCADA technology. The security requirement of the SCADA is quite different from IT security as it involves mission critical processes. The basic difference of the IT security and SCADA security are briefly described below.

Generally in SCADA systems, or Industrial Control Systems, the fact that any logic execution within the system has a direct impact in the physical world warranting safety to be paramount. The field devices being on the first frontier to directly face human lives and ecological environment, the field devices in SCADA systems are deemed with no less importance than central hosts. Further certain operating systems and applications running on SCADA systems, which are unconventional and proprietary to IT personnel's, may not work correctly with commercial off-the-shelf IT cyber-security solutions. Also, factors like the continuous availability demand, time-criticality, constrained computation resources on edge devices, large physical base, wide interface between digital and analog signals, social acceptance including cost effectiveness, user reluctance to change, legacy issues, etc. make SCADA system an extraordinary security engineering task.

SCADA systems are hard real-time systems as the completion of an operation after its deadline is considered useless and potentially can cause cascading effect in the physical world. The operational deadlines from event to system response impose stringent constraints as a missing deadline can cause a complete failure of the system. Latency is very important and can be destructive to SCADA system performance. If the system does not react within a certain time frame can cause great loss in safety, such as damaging the surroundings including fatal accidents. It's not the length of time frame but whether a particular operation meeting its deadline is vital in SCADA. Soft real-time systems, may tolerate certain latency and respond with decreased service quality (graceful degradation) which can be tolerated.

Non-major violation of time constraints in soft real-time systems leads to degraded quality rather than system failure. Furthermore due to the physical nature, tasks performed by SCADA system and the processes within each task are often needed to be interrupted and restarted. The timing aspect and task interrupts can prevent the use of conventional encryption block algorithms. Vulnerability of SCADA rises from the fact that memory allocation is more critical in a Real-Time Operating System (RTOS), than in other operating systems. Many field level devices in SCADA system are embedded systems such as RTUs, IEDs, etc. which run years without rebooting but accumulating fragmentation. Thus, buffer overflow is more problematic in SCADA than in traditional IT.

8.3 REMOTE ACCESS AND OPEN COMMUNICATION SYSTEMS

For any Power system SCADA and Smart Grid implementations, communication among various automation components is critical. Power measurement devices must talk to real-time control components across the entire power generation, transmission and distribution systems. All automation components must be connected to the SCADA system and these SCADA system must be linked to one another. All of these connections and linkages require open communication systems, often based on Ethernet and the internet. open systems are preferred today as they reduce communication system costs in the following way as listed below.

1. *Hardware and software are relatively inexpensive:* Open systems cut purchase costs because communications hardware and software based on Ethernet and the internet are much less expensive than their alternatives.
2. *Installation relies on familiar tools and techniques:* Installation is eased because of a widespread familiarity with these types of systems among implementation agencies.
3. *Existing communications system can be used:* Present communications infrastructure can be used in many cases, dramatically reducing installation and other related costs.
4. *Open protocols cut integration costs:* Integration expenses for connecting different Smart Grid components are reduced because Ethernet is used as a common communications hardware protocol.
5. *Skilled manpower is extensively available:* On-going maintenance and operation costs are reduced because many in the industry are familiar with Ethernet and the internet.

Open communication systems keep costs down, but these systems are much more vulnerable to cyber-attack than their proprietary and more closed alternatives because of the following reasons.
1. Large number of interconnections creates multiple vulnerabilities,
2. Armies of professional hackers are familiar with open system protocols,
3. Browser-based internet servers and clients create entry points,

4. Windows-based systems invite attack,
5. Vulnerable TCP/IP software stacks are used across multiple platforms, and
6. Older closed protocols lack security when ported to open protocols like TCP/IP.

Proprietary systems not only have fewer connections to other systems, they are also less familiar to professional hackers, creating a possible *security through obscurity* defense. On the other hand communication systems based on Ethernet, TCP/IP protocols, internet and popular operating systems such as Windows, invite attacks from millions of hackers worldwide.

Power system SCADA and Smart Grid with open communication systems are the technologies to stay as it is economical and affordable to many power utilities. Hence cyber-threats to these systems and their underlying power generation, transmission and distribution assets will become a real challenge. These cyber-threats have to be thwarted with a well-defined cyber-security policy and solutions. Communication equipment such as Ethernet switches, firewalls and gateway controllers are the cyber-security gatekeepers to the field devices. In addition to mitigating cyber-attacks, utilities must meet regulatory complainces and requirements preverably the NERC CIP regulations.

Most cyber-security regulations are just reaching the point of implementation in the utility industry, and many utilities are struggling to comprehend and implement to achieve these compliance. Some utilities are forging ahead with cyber-security plans, while others are taking a wait-and-see approach. Watching and waiting may sound sensible, but can open a utility to violations and fines. Fortunately, many consultants, implementing agencies and suppliers serving the power utility industry are helping to fill the knowledge void with a variety of hardware and software tools that comply with existing and anticipated standards, while at the same time effectively protecting against cyber-attacks. There are three main security objectives to be considered while migrating to Power System SCADA or Smart Grid and they are,
1. availability of uninterrupted and quality power supply,
2. integrity of the data communicated, and
3. Confidentiality of data.

Though the power system automation and Smart Grid evolution are revolutionizing the electrical grid, many devices, especially in the customer domain, are ill-equipped and lack sufficient and efficient security measures and policy enforcement. Some of these can be attributed to lack of customer awareness, while in other instances, vendors are not obligatory to build their devices with the best security features and standard. Further the implementing agencies may not ensure the required security standards and configurations, exploiting the lack of technical awareness of power utilities and of course the incompetent consultants to certain extent. The remote access problem is exacerbated in the Smart Grid's distributed architecture where devices on which its operations are unmanned and not necessarily continually monitored and serviced. Because of the rising security issues many power utilities started preferring their own communication network without even sharing the bandwidth.

8.4 VPN AND MPLS IN POWER SYSTEM AUTOMATION

Power system automation is a typical example of DCS spanning across a large geographical area in which secure communication is very important. Many utilities do not have their own full-fledged communication network such as optical, copper, or PLCC connectivity to link all remote sites within the automation network. In *such* cases MPLS/VPN is an ideal solution if implemented properly.

A Virtual Private Network (VPN) is a secure, private connection through an untrusted network such as the internet normally established between a client device and a server device. It is considered as a private connection in a public network or in an untrusted network because the encryption and tunneling protocols are used to ensure the confidentiality and integrity of the data in transit. It is important to remember that VPN technology requires a tunnel to work and it assumes encryption.

This is very much useful for establishing secured remote accessing/connectivity in Smart Grid. Here normally, a VPN client might be a Remote Terminal Unit (RTU) and the VPN server might be a server in the critical control network. Typically the client is the one that initiates the connection, and the server accepts and authenticates incoming connection requests from one or more clients. Once a VPN connection is established between a client and a server, the networks upstream of the client and the server are connected together such that network traffic may pass between them.

In the case of the RTU client as aforementioned, the RTU would appear as if it was actually plugged into the network upstream of the VPN server. As such, it would receive a new virtual IP address suitable for local network and could access other devices just as if it was directly connected to the network.

When using VPNs, it's critical to remember that the VPN only secures the tunnel and not the client or server. To ensure network security, it's critical that the VPN is seamlessly integrated into a suitable firewall.

During exigent situations, remote VPN access provides secured way of maintaining the operational continuity and support. Remote VPN access permits bulk power organizations to keep personnel away from sites during dangerous weather conditions. This reduction of travel risks during such conditions permits the power utilities to protect some of their skilled manpower, the most critical assets. Remote VPN access allows bulk power organizations to limit the number of personnel at their facilities during periods of heightened physical security threats. During increased security risks, bulk power utilities should limit the number of personnel entering their facilities with greater scrutiny screening. This helps in reduction of operational staffs at risk to a physical attack on the facility.

8.4.1 VPN Architecture

For many years the de facto standard VPN software was Point-To-Point Tunneling Protocol (PPTP), which was made most popular when Microsoft included it in its Windows products. Since most internet-based communication first started over telecommunication links, the industry needed a way to secure PPP connections. The original goal of PPTP was to provide a way to tunnel PPP connections through an IP network, but most implementations included security features also since protection was becoming an important requirement for network transmissions at that time.

The two methods of VPN architecture today used are site-to-site VPN and remote user access VPN. Both these architecture helps the organization to replace long distance dial-up or leased line with local dial-ups or leased line to Internet Service Provider (ISP).

8.4.2 Security Issues of VPN

The advantage of using secure remote access sometimes makes the network susceptible to security breaches. The possibility of accessing enterprises network with laptops having unsecured External Access Points (EAP) cannot be ruled out. If the laptop device or a device connected to the EAP has a virus or some other malicious software, it can spread it to the enterprises network.

Though a properly installed VPN can prevent some of the performance issues associated with supporting multiple protocols and data transmission mediums, VPNs are only as fast as the slowest internet connection between the two endpoints. In addition, most IP applications were designed for low-latency and high reliability network environments. This means that network performance issues will become more pronounced with the increasing use of real-time and interactive applications. While some applications can be reprogrammed or reconfigured to work with increased latency, getting this workaround to work with some applications can be challenging, if not impossible.

8.4.3 Remote Access VPN

A remote access VPN allows individual users to establish secure connections with a remote computer network. Those users can access the secure resources on that network as if they were directly plugged into the network's servers. An example of a company that needs a remote-access VPN is a large firm with hundreds of salespeople in the field. Another name for this type of VPN is virtual private dial-up network (VPDN), acknowledging that in its earliest form, a remote-access VPN required dialing into a server using an analog telephone system.

There are two components required in a remote access VPN. The first is a Network Access Server (NAS) also called a media gateway or a Remote Access Server (RAS). A NAS might be a dedicated server, or it might be one of multiple software applications running on a shared server. It's a NAS that a user connects to from the INTERNET in order to use a VPN. The NAS requires that user to provide valid credentials to sign into the VPN. To authenticate the user's credentials, the NAS uses either its own authentication process or a separate authentication server running on the network.

8.4.4 VPN Termination in Remote Access

VPN terminating node in remote access is the node where the VPN protocols actually ends. Here the payload is exposed and unwrapped. Data is no longer cypher rather the plain text. Extra care must be given to decide on VPN termination node else it can be a point of vulnerability. It is a common mistake in substation automation, terminating VPN at routers having multiple access point with connected devices have security holes.

As technology advances, there is an exponential growth of mobile, wireless and widely distributed networks which presents a vastly greater potential for unauthorized remote access. Hence it is better to secure all remote access over VPN with proper End Node Security (ENS) using point-to-point IPSec or clientless Secured Socket Layer (SSL) technology.

8.4.5 Site-To-Site VPN

Site-to-site VPN is a type of VPN connection that is created between two separate locations. It provides the ability to connect geographically separate locations or networks, usually over the public internet connection or a WAN connection. Site-to-site VPN typically creates a direct, unshared and secure connection between two end points. Site-to-site VPN can be intranet based or extranet based. Intranet based site-to-site VPN is created between an organization's propriety networks, while extranet-based site-to-site VPN is used for connecting with external partner networks or an intranet. The connection in a site-to-site VPN is generally enabled through a VPN gateway device.

Traditional VPN rely on internet Protocol Security (IPSec) to tunnel between the two endpoints. IPSec works on the Network Layer of the OSI Model, securing all data that travels between the two endpoints without an association to any specific application. When connected on an IPSec VPN the client computer is *virtually* a full member of the corporate network which is able to see and potentially access the entire network.

The majority of IPSec VPN solutions require third party hardware or software. In order to access an IPSec VPN, the workstation or device in use must have an IPSec client software application installed. This has both an advantage and disadvantage. The advantage is that it provides an extra layer of security, if the client machine is running the right VPN client software to connect to the IPSec VPN with proper configuration. These are additional hurdles that a hacker would have to get over before gaining access to the customers network.

The disadvantage is that it can be a financial burden for the utilities to maintain the licenses for the client software and for the technical support to install and configure the client software on all remote machines. It is this disadvantage is well projected as one of the negatives by the rival Secure Sockets Layer(SSL) VPN solutions. SSL is a common protocol and most web browsers have SSL capabilities built in. Therefore almost every computer in the world is already equipped with the necessary *client software* to connect to an SSL VPN. Another advantage of SSL VPN is that they allow more precise access control. Primarily, they provide tunnels to specific applications rather than to the entire corporate LAN. So, users on SSL VPN connections can only access the applications that they are configured to access rather than the whole network. Secondly, it is easier to provide different access rights to different users and have more granular control over user access.

A disadvantage of SSL VPN is that the customers are accessing the applications through a web browser which means that they exclusively work for web based applications. It is possible to web enable other applications so that they also can be accessed through SSL VPN, but doing so adds more complexity to the solution and eliminates some of the advantages.

Having direct access only to the web enabled SSL applications also means that the users don't have access to network resources such as printers or centralized storages and are unable to use the VPN for file sharing or file backups.

SSL VPN have been gaining in prevalence and popularity, however they are not the right solution for every instance. Likewise, IPSec VPN is not suited for every instance either. Vendors are continuing to develop ways to expand the functionality of the SSL VPN and it is a technology that a costumer should watch closely if the customer is in the market for a secure remote networking solution. Presently, it is important to consider carefully the needs of the customer's remote uses and consider the pros and cons of each solution to determine what works best for them.

Point-to-Point Tunneling Protocol (PPTP) is a network protocol used in the implementation of VPN. Newer VPN technologies like OpenVPN, L2TP, and IPSec may offer better network security support, but PPTP remains a popular network protocol especially on Windows computers. PPTP uses a client server design that operates at Layer 2 of the OSI model. PPTP VPN clients are included by default in Microsoft Window and also available for both Linux and Mac OS X. PPTP is most commonly used for VPN remote access over the internet. In this usage, VPN tunnel are created through the following two step process.

1. The user launches a PPTP client that connects to their internet provider, and
2. PPTP creates a TCP control connection between the VPN client and VPN server. The protocol uses TCP port 1723 for these connections and General Routing Encapsulation (GRE) to finally establish the tunnel.

PPTP also supports VPN connectivity across a local network. Once the VPN tunnel is established, PPTP supports two types of information flow which is mentioned below.
- *Control messages* for managing and eventually tearing down the VPN connection. It passes directly between VPN client and server, and
- *Data packets* that pass through the tunnel, to or from the VPN client.

Layer2TunnelingProtocol (L2TP) is another tunneling protocol used to support VPN or as part of the delivery of services by Spit does not provide any encryption or confidentiality by itself. Rather, it relies on an encryption protocol that it passes within the tunnel to provide privacy. L2TP combines the features of PPTP and Cisco's Layer 2 Forwarding (L2F) protocol. L2TP tunnels Point-to-Point Protocol (PPP) traffic over various network types such as IP, ATM, X.25. In fact it is not restricted to IP networks as PPTP. PPTP and L2TP have very similar focuses, which is to get PPP traffic to an end point that is connected to some type of network that does not understand PPP. Like PPTP, L2TP does not actually provide much protection for the PPP traffic, but it integrates with protocols that do provide security features. L2TP inherits PPP authentication and integrates with IPSec to provide confidentiality, integrity, and potentially another layer of authentication.

8.4.6 Difference between IPSec VPN and SSL VPN

1. Generally, to start the IPSec VPN secure connection, the user has to start the application by installing IPSec 3rd party client application or hardware in client PC. This incur financial burden to utilities, as they have to buy licenses for VPN clients. But for SSL VPN, it is not necessary to install separate applications. Almost all the modern standard web browsers can use SSL connections.
2. In IPSec communication, once the client is authenticated to the VPN, then the client has the full access of the private network, which may not be necessary, but in SSL VPNs, it provides more precise access control. At the beginning of the SSL authentication, it creates tunnels to specific applications using sockets rather than to the whole network. Also, this enables to provide role based access having different access rights for different users.
3. One of the disadvantages of SSL VPN is that, it is only used for web based applications. For other applications, though it is possible to use by web-enabling it adds some complexity for the applications.
4. Since it provides access only for Web Enabled Applications (WEA), SSL VPN is difficult to use with applications like file sharing and printing, but IPSec VPNs provide highly reliable printing and file sharing facilities.
5. SSL VPNs are becoming more popular due to ease of use and reliability but, as mentioned above, it is not reliable with all the applications. Therefore, selection of the VPN (SSL or IPSec) totally depends on the application and requirements.

IPSec VPN gateways are usually implemented on the perimeter firewall, and permit or deny remote host access to center private subnets. SSL VPN gateways are usually deployed behind the perimeter firewall with ruleset that permit or deny access to application services or data. The Figure 8.1 shows that SSL users have access to their own mailboxes on a private network and to a subset of URLs hosted on a Web Server, Java Application Servers, etc.

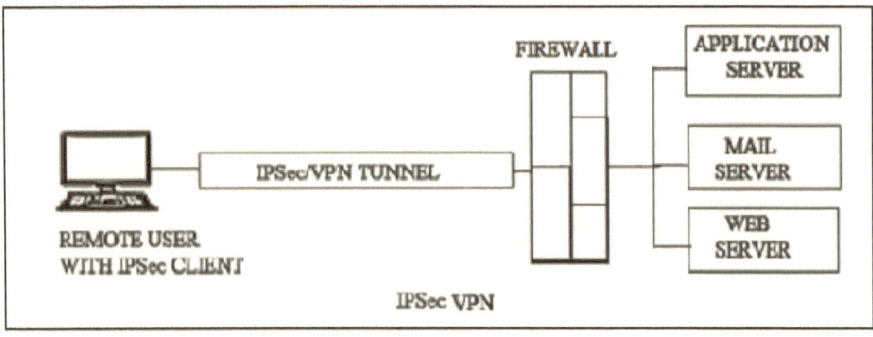

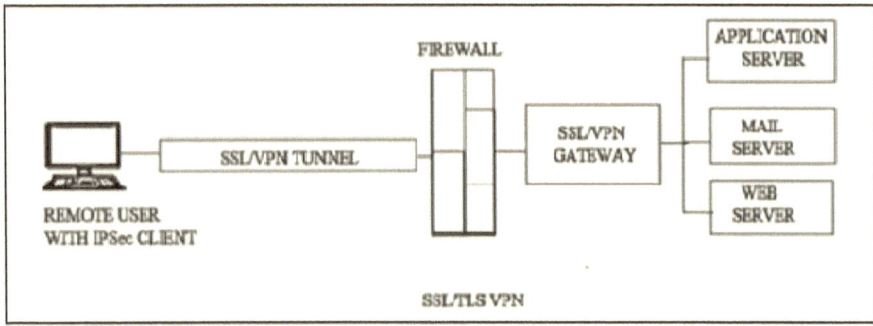

▲ **Figure 8.1:** IPSec and SSL VPN

An appropriate summary of Tunneling Protocols are briefly described below.
- PPTP operates in client/server architecture capable of extending and protecting PPP connections. It is a data link layer protocol works and transmits over IP networks only.
- L2TP is a hybrid of Layer 2 Forwarding (L2F) and PPTP. It extends and protects PPP connections. It also operates at the data link layer. It works and transmits over multiple types of networks, joint with IPSec for security.
- IPSec is capable of handling multiple VPN connections at the same time, provides secure authentication and encryption, operates only on IP networks. It mainly focuses on LAN-to-LAN communication rather than user-to-user and operates on the network layer, and provides security on top of IP.

- SSL operates on the transport layer and protects mainly web-based traffic. Most of the web browsers already embedded SSL as it is for deployment. However it can only protect a small number of protocol types, thus is not an infrastructure-level VPN solution. It has granular access control, configuration and better security features as SSL VPNs are closer to the application layer. However, there are a smaller number of traffic types that can be protected through this VPN type.

One VPN solution is not necessarily better than the other, they just have their own focused purposes and are described below.
- PPTP is used when a PPP connection needs to be extended through an IP based network.
- L2TP is used when a PPP connection needs to be extended through a non IP based network.
- IPSec is used to protect IP based traffic and is commonly used in gateway-to-gateway connections.
- SSL VPN is used when a specific application layer traffic type needs protection.

The predecessor to SSL is Transport Layer Security (TLS), which is an actual standard. SSL is a de facto standard and is more popular than TLS because of its large deployment base. Attackers commonly encrypt their attack traffic so that countermeasures which are put into place to analyze traffic for suspicious activity are not effective. Attackers can use SSL, TLS, or PPTP to encrypt malicious traffic as it transverses the network. When an attacker compromises and opens up a back door on a system, the attacker will commonly encrypt the traffic and that will then get exchanged between his system and the compromised system. It is important to configure security network devices to only allow approved encrypted channels.

8.4.7 Deploying VPN

To deploy an efficient and operational VPN, several different elements are needed at various steps along the path, starting from the client, through the cloud, to the network boundary and into enterprise networks. Within the enterprise network, VPN can terminate at Communication Front End (CFE) or at the server, by properly ensuring the secure data transfer. The basic VPN necessities and components are briefly described below.
1. Client VPN software to make a secure remote connection,

2. VPN aware routers and firewalls which permit authentic unobstructed VPN traffic, and
3. VPN applications, concentrators or servers to handle and manage incoming VPN traffic and to establish and manage VPN sessions and their access to network resources.

8.4.8 MultiProtocol Label Switching (MPLS)

Multi Protocol Label Switching (MPLS) was originally presented as a way of improving the forwarding speed of routers based on short path labels rather than longer network addresses but is now emerging as a crucial standard technology that offers new capabilities for large scale IP networks. Directing the data from one network node to the next based on short path labels rather than long network addresses, avoids complex lookups in a routing table. This technique saves significant time over traditional IP-based routing processes. In fact, MPLS is a protocol for speeding up and shaping network traffic flows. Furthermore, MPLS is designed to handle a wide range of protocols through encapsulation. Thus, the network is not limited to TCP/IP and compatible protocols. This enables the use of many other networking technologies, including T1/E1, ATM, Frame Relay, SONET, and DSL. MPLS got its name because it works with the internet Protocol (IP), Asynchronous Transfer Mode (ATM) and Frame Relay network protocols. Any of these protocols can be used to create a Label Switched Paths (LSPs). It was created initially to save router's idling time by avoiding stop and look up routing tables. A common misconception is that MPLS is only used on private networks, but the protocol is used for all service provider networks including internet backbones. Today, Generalized Multi Protocol Label Switching (GMPLS) extends MPLS to manage Time Division Multiplexing (TDM), Lambda Switching and other classes of switching technologies beyond packet switching.

MPLS allows most packets to be forwarded at Layer 2 (the switching level) rather than having to be passed up to Layer 3 (the routing level). Each packet gets labeled on entry into the service provider's network by the ingress router. The label determines the pre-determined path LSP which the packet has to follow. LSPs, allow service providers to decide the best way for certain types of traffic to flow within a private or public network ahead of time. All the subsequent routing switches perform packet forwarding based only on those labels. These labels never look the IP header until they are badly required. Finally, the egress router removes the labels and forwards the original IP packet towards its final destination.

Service providers can use MPLS to improve Quality of Service (QoS) by defining LSPs that can meet specific Service Level Agreements (SLAs) on traffic latency, jitter, packet loss and downtime. For example, a network might have three service levels and they are,
1. level for voice,
2. level for time sensitive traffic, and
3. level for best effort traffic.

MPLS also supports traffic separation and the creation of virtual private networks (VPNs), Virtual Private LAN Services (VPLS) and Virtual Leased Lines (VLLs).

8.4.9 Choosing MPLS VPN Services

While choosing MPLS VPN service for an utility, one must be clearly aware of the requirements, network design and options of service providers. It is better to keep the following points in mind while selecting the MPLS VPN.
1. Carefully evaluate the needs and optimized requirements,
2. Gather the service providers offers in the near geographic area,
3. Compare between the requirements and options of service providers, which matches the best to the requirements of the utility, and
4. Recommendations from an experienced consultant is always better particularly, when the utility is selecting the MPLS services for the first time.

In many cases, finding the best MPLS VPN service need to combine multiple services. For example, many enterprises use Layer 3 MPLS VPNs for smaller sites, pseudo wires for point-to-point links between data centers and Virtual Private LAN Service (VPLS) for sites that need high availability to control convergence speed and routing protocol behavior. If the customer is planning for an MPLS VPN connectivity, it would be better to consider the following points, while finalizing the vendor evaluation.
1. *Internet access:* Most vendors allow customer to connect their MPLS VPN directly to the internet through a shared network firewall. However, some of them restrict the outbound traffic, while others allow establishing an IPSec tunnel to the network firewall and then hopping into customer's network. Still others allow inbound access through an encapsulated GRE tunnel that dumps off in front of another firewall in control, and

2. *The full mesh:* While MPLS technology typically facilitates a full mesh of connectivity among all of the customer's sites, this requires a single MPLS network. Some service providers have split their MPLS networks into geographic regions, and customer has to pay more to get connectivity from one region to another. Without this, traffic from one location to another may be forced through a third site acting as a hub. This can unnecessarily complicate the routing and make it inefficient.

Keep the point in mind that MPLS VPNs are not encrypted, rather they logically separate the customer data from other customer's data. The data shares the same physical path with other customers of the service provider, just like Frame Relay or any other WAN. Some vendors may offer additional services that allow the customer to encrypt their traffic. In fact, the customer may want to explore the possibility of using their existing IPSec VPN equipment to create permanent tunnels between sites over a new high speed MPLS backbone to get the best of both worlds.

8.4.10 Advantages and Disadvantages of MPLS VPNS

The main advantages of MPLS VPN are briefly described below.

Many MPLS VPNs offer more cost effective price points with more flexibility than other WAN technologies. The label switching technology offers Class of Service (CoS) and Quality of Service (QoS) capabilities. As an encapsulation and VPN mechanism, MPLS brings many benefits to the IP networks which are described below.

1. *Faster packet processing with MPLS when compared to IP:* MPLS was designed to provide faster packet processing when compared to IP based lookup. As MPLS works with label switching operation instead of IP destination based lookup, it provides faster processing and provides performance benefit.
2. *Border Gateway Protocol (BGP) Free Core with MPLS:* If BGP is running on a network, without MPLS, it needs to run on every device on the path. MPLS eliminates this need which reduces the requirements of protocols, making the network simpler and easy to maintain.
3. *Hiding service specific information from the core of the network:* When MPLS is used on the network, only the edge devices has to keep the customer specific information such as MAC address, VLAN number, IP address, etc. Core of the network only provides reachability between the edges.

4. *Improve scalability:* Does not have service specific information on the core of the network which provides better scalability. From the point of view of CPU and memory utilization and with the routing protocol updates, link state changes and many other problems become immaterial from the core of the network by using MPLS.
5. *MPLS Layer 2 and Layer 3 VPNs:* Probably the most important reason and main benefit of MPLS is MPLS VPNs. MPLS is most suitable to create point-to-point, point-to-multipoint and multipoint-to-multipoint type of MPLS layer 2 VPN and MPLS layer 3 VPNs. By using BGP, LDP or RSVP protocols, VPNs can be created.
6. *Traffic Engineering:* MPLS with the Resource Reservation Protocol - Traffic Engineering (RSVP-TE) Provides traffic engineering capability which allows better capacity usage and guaranteed SLA for the desired service.
   a. *Fast Reroute:* With RSVP-TE, MPLS provides MPLS Traffic Engineering Fast Reroute Link and Node Protection. RSVP-TE is one of the options but possible with Label Distribution Protocol (LDP), Loop Free Alternate (LFA) and Remote LFA. This can be setup if RSVP-TE is not used in the network. MPLS Traffic Engineering Fast Reroute can protect the important service in any kind of topology. On the other hand, IP Fast Reroute (IP FRR) mechanisms require highly meshed topology to provide full coverage in the case of failures.
   b. MPLS doesn't bring security by default. If security is needed then IPSec should run on top of MPLS. MPLS is used mainly for the Wide Area Network (WAN).

Though MPLS VPN offers many advantages, it has certain disadvantages and the communication engineer or power system engineer should be aware of these drawbacks while opting for MPLS/VPN connectivity. With MPLS VPN, service providers run the core of user's network, which presents several disadvantages which are described below.
- User routing protocol choice might be limited,
- User's end-to-end convergence is controlled primarily by the service provider,
- The reliability of user's L3 MPLS VPN is influenced by the service provider's competence level,
- Deciding to use MPLS VPN services from a particular service provider also creates a very significant lock-in. It's hard to change the provider when it's operating user's network core, and
- An additional layer is added and the router has to understand MPLS.

8.5 CRITICAL INFRASTRUCTURE PROTECTION AND NERC CIP

Certain infrastructure which are most essential for the normal function of a Nation such as energy and electric power, banking and finance, transportation, etc. are to be protected at any cost, else no Nation can survive. The following sessions briefly describes the critical infrastructure and its protection.

8.5.1 Critical Infrastructure

Critical Infrastructure is defined as the system that compose the assets, systems, and networks, whether physical or virtual, and/or the computer programs, computer data, content data and/or traffic data so vital to a country that the incapacitation or destruction or interference with such systems and assets would have a debilitating impact on security, national or economic security, national public health and safety, or any combination thereof.

Every day, products and services that support the way of life flow, almost flawlessly in areas of energy and electric power, banking and finance, transportation, Information and Communications Technology (ICT), water systems, Government and private emergency services. All these are made possible with the proper integration of the systems and networks such as the roads, airports, power plants, and communication facilities. If just one of these systems in the infrastructure is disrupted there could be awful consequences. These infrastructures that are essential for operations of the economy and government are generally termed as Critical Infrastructure of a Nation. Today these critical sectors whose operations greatly depend on ICT and therefore it become inevitable to protect these sectors from cyber-threat.

8.5.2 Critical Infrastructure Protection (CIP)

As explained the operational stability and security of Critical Infrastructure is vital for economic security of a Nation and hence its protection must be given paramount importance. Many Nations consider power sector as Super Critical Infrastructure (SCI) as it is the back bone of almost all industry. One of the purposes of Critical Infrastructure protection is to establish a real-time ability for all sectors of the critical infrastructure community to share information on the current status of infrastructure elements. Ultimately, the goal is to protect the Critical Infrastructure by eliminating the known vulnerabilities and develop a proactive defense mechanism to counter cyber-attacks to Critical Infrastructure. Thus the need of the hour is to chalk out a national policy for Critical Infrastructure Protection (CIP), created through a partnership between the government and private industry.

8.5.3 Critical Information Infrastructure Protection

Within the last two decades, advances in information and communications technologies have revolutionized Government, scientific, educational, and commercial infrastructures. Higher processing power of end devices, miniaturization, reducing memory storage cost, wireless networking technologies capable of supporting high bandwidth and widespread use of internet have transformed stand-alone systems and predominantly closed networks into a virtually seamless fabric of interconnectivity. ICT or information infrastructure enables large scale processes throughout the economy, facilitating complex interactions among systems across global networks. Their interactions propel innovation in industrial design and manufacturing, e-commerce, e-governance, communications, and many other economic sectors. The information infrastructure provides for processing, transmission, and storage of vast amounts of vital information used in every domain of society, and it enables Government agencies to rapidly interact with each other as well as with industry, citizens, state and local governments, and the Governments of other Nations. information infrastructure also encompass interconnected computers, servers, storage devices, routers, switches and other related equipments increasingly support the functioning of such critical national capabilities. Thus ICT has become an integral part of the Critical Information Infrastructure as well and need to be protected from cyber-attack.

8.5.4 NERC CIP and Bulk Electric System (BES)

It is high time for the critical sectors such as defense, finance, energy, transportation and telecommunications of a Nation to realize that they need to depend on improved and secured software development, systems engineering practices and the adoption of strengthened security models and best practices. This prompts the enterprises and utilities to adhere to international/national standards and regulations. Various standards and regulations are put up for each and every area of Critical Infrastructure Protection and are in practice.

The North American Electric Reliability Corporation's (NERC) current set of standards, Critical Infrastructure Protection (CIP) Version 5 standards are unique and address the issues of power sector automation very efficiently.

Today NERC CIP standards are becoming mandatory security compliance for Bulk Electric System (BES) while migrating to automated system. The adequacy of current standards depends where and whether the utilities do apply with failsafe manner is the real issue of achieving the NERC CIP compliance. The set of standards put forward by NERC to secure bulk electric systems provide physical-cyber network security administration and supporting best-practice industry processes. The NERC CIP plan consists of a set of standards having 45 requirements covering the security of electronic perimeters and the protection of critical cyber-assets as well as personnel and training, security management and disaster recovery planning and they are elaborated in subsequent chapters.

The current set of NERC's standards which almost cover all cyber-issues of a Bulk Electric System (BES) is flexible and acceptable. Presently it is observed that the power utilities move towards an active consideration of NERC CIP version 5 system security as an effective security implementation especially after the cyber-attack on Ukraine power sector

8.6 SECURITY CONCERNS IN SUBSTATION AUTOMATION

Substation Automation (SA) is the need of the hour as substations form the base of the most important functions of power utilities. Any major breakthrough in a substation technology is seen as one of the foremost aspects of the Smart Grid revolution. SA not only revolutionized the mode of operation and maintenance, rather it introduced a new culture of managing the utility network. Anticipating the upcoming Smart Grid revolution, many transmission and distribution utilities are explicitly targeting to complete substation automation in a time bound manner. The two main issues are some of the substations are partially automated while the remaining is not. The profusion of the substation and feeder automation equipments from different manufactures, which ranges from SCADA to alarm processing, retrofitting is relatively an easy job today. But an essential knowledge of ICT becomes mandatory without which substation automation can be a bottleneck for implementing Smart Grid.

However many power utilities, mainly due to economic reasons, adopt conventional substation design. Though it caters the present requirements, migration to Smart Grid may become burdensome. Further in the long run O & M costs exceeds relatively high when compared to one-time investment for automated digital substation, as it requires minimal human intervention and has enhanced operation efficiency which saves money.

A block diagram of a typical automated substation which is shown in Figure 8.2 is generally preferred today as it can accommodate the Smart Grid revolution with relatively lesser investment.

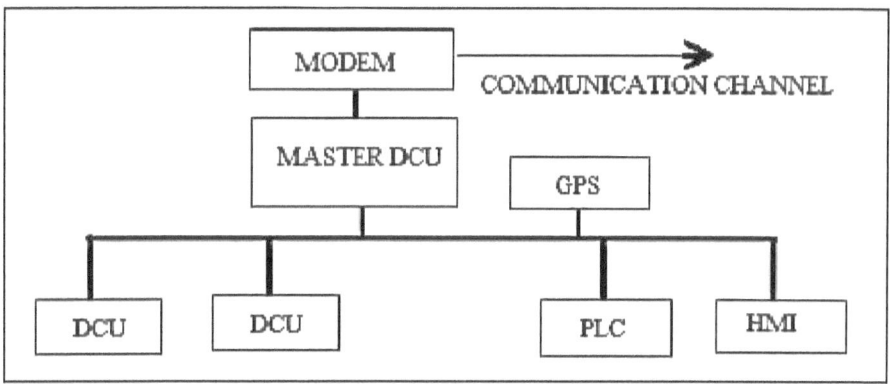

▲ **Figure 8.2:** Block diagram of an automated substation

Generally the data concentrator is an integral part of the RTU, which collect the information, and send the data to the control center through a communication channel. Since the substations are remote sites, method like dial-up modem, wireless GSM/CDMA, or dedicated RS232/RS485 serial are to be used. This introduces threat points from hackers as well as insiders. The attack points in a conventional substation with RTU and Human Machine Interface (HMI) are shown in Figure 8.3.

Hence care must be taken to protect the system with proper encryption and authentication between each and every communication nodes. Ensuring the NERC CIP compliance or international equivalent standards is becoming mandatory in power sector today to secure not only the Bulk Electric Supply (BES) but also in all other areas of power generation, transmission and distribution.

If a local HMI is provided in the substation with connectivity to local data concentrator, it is most critical that the communication between the data concentrator and HMI must be appropriately authenticated and secured by required level of encryption. Today the local HMI or Local Data Monitoring System (LDMS) which are directly linked with the data concentrator or RTU for local monitoring are discouraged due to internal threat, unless the connectivity and communication security is ensured. The other solution is to provide a separate communication channel from the control center to the substation for local monitoring but is expensive and requires recurring cost.

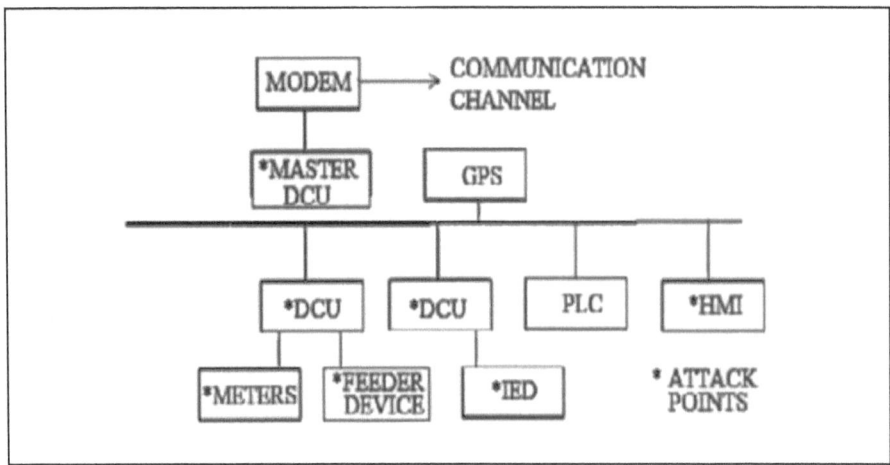

▲ **Figure 8.3:** Attack points in an automated substation with DCU and HMI

The security threats can be from intruders and from insiders of the corporate network. External intruders can gain access to the system through the SCADA communication link, modem, and any dial up lines to certain systems like IEDs. The lines can be privately owned or leased from a carrier and are vulnerable to eavesdropping and intrusions that can corrupt the data. The intruders could be members of general public or cyber-criminals. Internal intrusion can happen inside the corporate WAN, as many unauthorized users can gain access to the data unless the utility is very strict with the password policy, privileges, and authentication measures. Employees or suppliers of substation equipment may gain access to critical data and can extract crucial information or cause damage unless prevented effectively from doing so.

Encryption is one of the methods that can effectively handle many of the attacks discussed. Inserting IPSec protocol in IP level is a typical example. But this encryption requires RTUs/IEDs with processors of higher computing capability and makes the system very expensive. There is a grave requirement to accelerate the implementation of cyber-security measures in Smart Grid and power system SCADA.

8.6.1 Attack Vector through Substation HMI

Today, within the various SCADA solutions, the Human Machine Interface (HMI), especially the HMI at remote locations for local data monitoring and configuring is the most favorite and most preferred target for attackers.

The HMI at the control center acts as a unified hub for managing Critical Infrastructure. If an attacker succeeds in compromising the HMI, nearly anything can be done to the infrastructure itself, causing even physical damage to SCADA equipment. Even if attackers decide not to disrupt operations, they can still exploit the HMI to gather information about a system or disable alarms and notifications meant to alert operators of danger to SCADA equipment.

Many cyber-security professionals found that most HMI vulnerabilities fall into four categories viz. memory corruption, credential management, lack of authentication/authorization and insecure defaults, and code injection. All of which are preventable through secure development practices. It is also observed that the average time needed between disclosure of a Zero Day Initiative (ZDI) of a SCADA vendor and the time for releasing a patch takes up to 150 days. This delay in releasing patches must be minimized by the HMI vendors by giving special attention and respond accordingly. Also the ZDI developers should start with basic fuzzing techniques to find new vulnerabilities in HMIs.

Software developers should also look for new file associations during installation to aid in fuzzing, as many of the file formats are wide open. Developers of HMI and SCADA solutions would be well advised to adopt the secure life cycle practices implemented by OS. By taking simple steps such as auditing for the use of banned APIs, vendors can make their products more resilient to attacks. SCADA developers also should expect that their products may be used in manners that they did not intend to and developers have to indeed assume that their products and solutions will be connected to a public network. Hence developers must have a mindset that assumes the worst-case scenario while developing the applications. It is better that if developers implement more defense-in-depth strategy to enhance protection.

The SCADA HMI malware specifically targeting ICS aggressively aim HMIs. The ZDI program encourages researchers to find and report the bugs associated with HMI and other SCADA systems to the bounty program. By working together, the developers receive compensation for their work while the vendors receive valuable data for improving their products. Present scenario indicates that bugs in SCADA systems will likely be present for many years to come. Again by working together, the security of these systems will continue to improve though a completely secure system will never be created. Implementing strong research and development tactics will be the best chance to keep the lights on as long as needed else HMI may become a Hacker Machine Interface.

8.6.2 Security Concerns of SCADA Control Center

The control center architecture must be designed with utmost security and with proper disaster recovery center. Secured ingress and egress must be ensured with the field devices, IEDs, Smart Meters, other control centers, etc. Usually the field devices and Smart Meters communicate with the Communication Front End (CFE) server through VPN preferably SSL/VPN. Obviously the CFE should have the capability of handling very large data, and certain situation act as a Data Management Server with Web server capabilities. Front End Processor (FEP) of CFE server should have very high processing capability and must be hot redundant. Today it is a general practice that the control center architecture is logically segmented to various zones and critical networks are isolated. The data transfer between these zones is generally through firewalls with proper configuration. One of the usual recommended practice is to separate the SCADA network from the corporate network.

The nature of network traffic on these two networks should be different. Internet access, FTP, e-mail, and remote access will typically be permitted on the corporate network but should not be on the SCADA network. Rigorous change control procedures for network equipment, configuration, and software changes may not be in place on the corporate network. If SCADA network traffic is carried on the corporate network, it could be intercepted or be subjected to a Denial of Service (DoS) attack. Having separate networks, security and performance problems on the corporate network will not be able to affect the SCADA network. Typical control center architecture of Power System SCADA having Firewall with De Militarised Zone (DMZ) between corporate network and control network are shown in Figure 8.4.

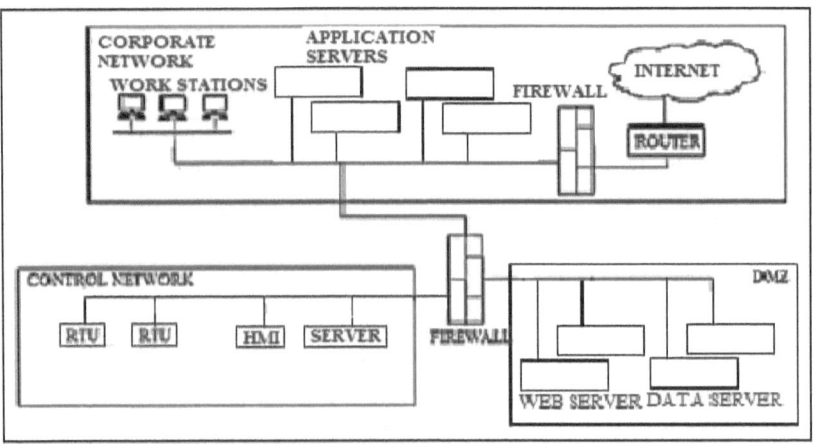

▲ **Figure 8.4:** Firewall with DMZ between Corporate and Control Networks

Using firewalls with the ability to establish a DMZ between the corporate and control networks are preferred today in MCC architecture. Each DMZ holds one or more critical components, such as the data historian, the wireless access point, or remote and third party access systems. In effect, the use of DMZ capable firewalls allows the creation of an intermediate network.

Creating a DMZ requires the firewall to offer three or more interfaces, especially when linking public and private interfaces. One of the interfaces is connected to the corporate network, the second to the control network, and the remaining interfaces to the shared or insecure devices such as the data historian server or wireless access points on the DMZ network.

Non-firewall based solutions will not provide suitable isolation between control networks and corporate networks. The two zone solutions especially without DMZ are marginally acceptable but should be only be deployed with extreme care. The most secure, manageable, and scalable control network and corporate network segregation architectures are typically based on a system with at least three zones, incorporating a DMZ.

8.6.3 Defense-in-Depth Architecture

Another SCADA control center architecture which gained much popularity with the incorporation of modern firewall technology is the defense-in-depth architecture. A single security product, technology or solution cannot adequately protect a SCADA by itself. A multiple layer strategy involving two or more different overlapping security mechanisms is desired so that the impact of a failure in any one mechanism is minimized. A defense-in-depth architecture strategy includes the use of firewalls, the creation of demilitarized zones, intrusion detection capabilities along with effective security policies, training programs and incident response mechanisms. In addition, an effective defense-in-depth strategy requires a thorough understanding of possible attack vectors on a SCADA. These include,
1. backdoors and holes in network perimeter,
2. vulnerabilities in common protocols,
3. attacks on field devices,
4. database attacks, and
5. communications hijacking and *Man-In-The-Middle (MITM)* attacks.

Figure 8.5 shows a PSS defense-in-depth architecture strategy that has been developed with improved control systems cyber-security. The control systems cyber-security using *defense-in-depth strategies* for organizations use control

system networks which maintain a multi-tier information architecture which requires,
1. maintenance of various field devices, telemetry collection, and/or industrial level process systems,
2. access to facilities via remote data link or modem, and
3. public facing services for customer or corporate operations.

This strategy includes firewalls, the use of demilitarized zones and intrusion detection capabilities throughout the ICS architecture. The use of several demilitarized zones provides the added capability to separate functionalities and access privileges and has proved to be very effective in protecting large architectures comprised for networks with different operational mandates. Intrusion detection deployments apply different rule-sets and signatures unique to each domain being monitored.

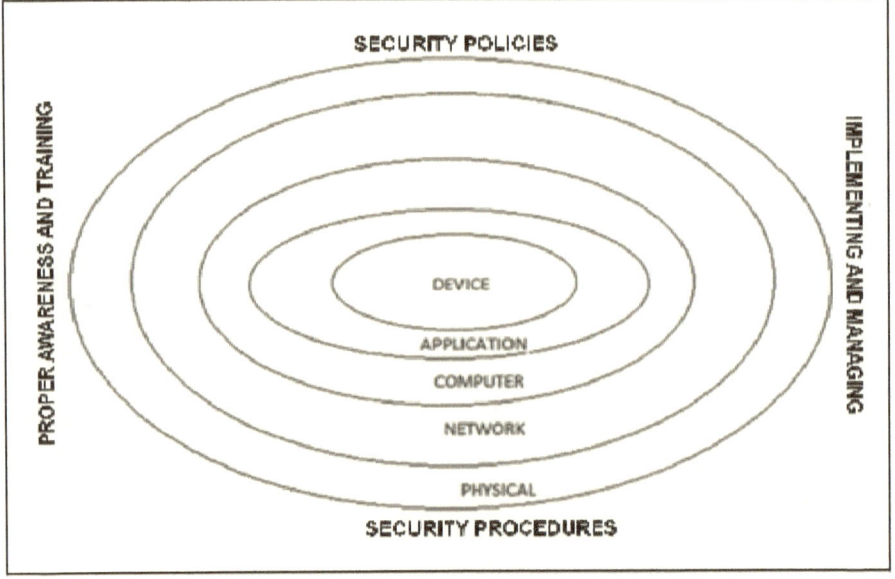

▲ **Figure 8.5:** Defense-in-Depth Architecture

8.6.4 Firewall deployment and Policies

Once the defense-in-depth architecture is in place, the security engineers have to determine exactly the traffic that should be allowed through the firewalls. Configuring the firewalls to deny all except, for the traffic absolutely required for business needs of every organization. Cyber-security engineers of the utility must be aware of the data flow requirements of business and the security impacts of allowing that traffic through. Typical example is, many organizations considered allowing SQL traffic through the firewall as required for business for many data historian servers. Unfortunately, SQL was also the vector for the Slammer worm. Many important protocols used in the industrial world, such as HTTP, FTP, OPC/DCOM, Ethernet/IP, and MODBUS/TCP, have significant security vulnerabilities.

While deploying the firewalls in SCADA which is the heart of Smart Grid, the following points are most important from the point of security. When installing a single two-port firewall without a DMZ for shared servers particular care needs to be taken with the ruleset design. The best practice is to keep rules which should be stateful with IP address and port specific. The address portion of the rules should restrict incoming traffic to a very small set of shared devices on the control network from a controlled set of addresses on the corporate network. Allowing any IP addresses on the corporate network to access servers inside the control network is not recommended. In addition, the allowed ports should be carefully restricted to relatively secure protocols such as Hypertext Transfer Protocol Secure (HTTPS). Allowing HTTP, FTP, or any unencrypted SCADA protocol to cross the firewall is a security risk due to the potential for traffic sniffing and modification. Rules should be added to deny inbound communication with the control network.

Rules should allow internal devices in the control network to establish connections outside the control network in a most secured manner. On the other hand, if the DMZ architecture is being used, then it is possible to configure the system so that traffic will not pass directly between the corporate network and the control network. With a few special exceptions which are mentioned below, all traffic from either side can terminate at the servers in the DMZ. This allows more flexibility in the protocols allowed through the firewall. A typical example, MODBUS/TCP might be used to communicate from the RTUs/IEDs to the data historian, while HTTP might be used for communication between the historian and enterprise clients. Both protocols are inherently insecure, yet in this case they can be used safely because neither actually crosses between the two networks. An extension to this concept is the idea of using *disjoint* protocols in all control networks to corporate network communications. That is, if a protocol is allowed between the control network and DMZ, then it is explicitly not allowed between the DMZ and corporate network. This design greatly reduces the chance of a worm such as Slammer actually making its way into the control network, since the worm would have to use two different exploits over two different protocols.

One area of considerable variation in practice is the control of outbound traffic from the control network, which could represent a significant risk if unmanaged. Typical example is Trojan horse software that uses HTTP tunneling to exploit poorly defined outbound rules. Thus, it is important that outbound rules be as stringent as inbound rules. A summary of the preferred outbound and inbound rules for the secure data flow are described below.

1. Inbound traffic to the control system should be blocked. Access to devices inside the control system should be through a DMZ.
2. Outbound traffic through the control network firewall should be limited to essential communications only.
3. All outbound traffic from the control network to the corporate network should be source and destination-restricted by service and port.

In addition to these rules, the firewall should be configured with outbound filtering to stop forged IP packets from leaving the control network or the DMZ. In practice this is achieved by checking the source IP addresses of outgoing packets against the firewall's respective network interface address. The intent is to prevent the control network from being the source of spoofed communications, which are often used in DoS attacks. Thus, the firewalls should be configured to forward IP packets only if those packets have a correct source IP address for the control network or DMZ networks. Finally, internet access by devices on the control network should be strongly discouraged. Usually the firewalls deployed in PSS come with the default configuration as described below. However the installation and security engineers do confirm the ruleset of the firewalls without any compromise, else can be catastrophic.

1. The base rule set should be deny all, permit none.
2. Ports and services between the control network environment and the corporate network should be enabled and permissions granted on a specific case-by-case basis. There should be a documented business justification with risk analysis and a responsible person for each permitted incoming or outgoing data flow.
3. All permit rules should be both IP address and TCP/UDP port specific, and stateful if appropriate.
4. All rules should restrict traffic to a specific IP address or range of addresses.
5. Traffic should be prevented from transiting directly from the control network to the corporate network. All traffic should terminate in the DMZ.

6. Any protocol allowed between the control network and DMZ should explicitly NOT be allowed between the DMZ and corporate networks (and vice-versa).
7. All outbound traffic from the control network to the corporate network should be source and destination-restricted by service and port.
8. Outbound packets from the control network or DMZ should be allowed only if those packets have a correct source IP address that is assigned to the control network or DMZ devices.
9. Control network devices should not be allowed to access the internet.
10. Control networks should not be directly connected to the internet, even if protected via a firewall.
11. All firewall management traffic should be carried on either a separate, secured management network or over an encrypted network with two-factor authentication. Traffic should also be restricted by IP address to specific management stations.

These are only guidelines; hence a vigilant assessment of each control environment is required before finalizing and implementing the firewall ruleset.

8.6.5 Design Considerations of PSS Security

The design objective of any cyber-security system comprises of mainly four components and they are prevention, detection, response, and disaster recovery which are briefly explained below.

Prevention: Actions taken and measures put in place for the repeated assessment and readiness of necessary actions to reduce the risk of threats and vulnerabilities, to intervene and stop an occurrence, or to mitigate effects.

Detection: Devise methods to recognize abnormal data flow, behaviors, discover intrusions, detect malicious code, and other activities or events that can disrupt electric power grid operations, as well as techniques for evidence collecting.

Response: Activities that address the short term, direct effects of an incident, including immediate actions to save lives, protect property, and meet basic human needs. Response also includes the execution of emergency operations plans and incident mitigation activities designed to limit the loss of life, personal injury, property damage, and other unfavorable outcomes.

Disaster Recovery: Development, coordination, and execution of service and site restoration plans for affected facilities and services, reconstitution of

Smart Grid operations and services through individual, private-sector, non-governmental, and public sector actions.

Presently there is no single, canonical security architecture for Smart Grid. Hence whichever security architecture is chosen, it has to be appropriately customized to the utilities requirements by recognizing the gaps and limitations. These considerations can bring dramatic effect on the automation of extensive security data collection and analytics and the design of Public Key Infrastructure (PKI) for device identification.

Both IEC and IEEE have developed a set of standards related to Smart Grid. NISTIR 7628 is a conceptual model as the context for these standards ensuring a fast, efficient and dependable communication infrastructure. In the 62351 series, IEC TR 62351-10 Security Architecture Guidelines provides useful guidance related to security controls relevant to Smart Grid. Though the NISTIR 7628 has been obtained wide acceptance, it has certain drawbacks as well which are briefly described below.

1. Thorough mapping of risk and attack scenarios to the architecture are insufficient.
2. Lack of physical security issues, lack of specifications regarding privacy, and insufficient strategies for detecting, analyzing and responding to target attacks.
3. Though it concentrates on definitions and management, a detailed architecture for a specific Smart Grid is absent. In fact it is only a guideline which helps in designing Smart Grid architecture.

Nevertheless it is extremely important, before moving to a detailed design and implementation, establish an architecture that provides a longer-term and durable understanding of the system.

8.6.6 PSS Potential Risks

As already described, with the evolution of the Smart Grid, there are many potential risks associated which should be seriously addressed. Some of them are described below.

1. Greater complexity increases exposure to the potential attackers and unintentional errors,
2. Smart Grid networks link more frequently to other networks which introduce common vulnerabilities that increase the potential for cascading failures. Presence of more interconnections increases opportunities for Denial of Service(DoS) attacks, introduction of

malicious code or compromised hardware, and related types of attacks and intrusions,
3. As the number of network nodes increases, the number of EAP and paths which potential enemies might exploit also increases, and
4. Extensive data gathering and bi-directional information flows may broaden the possibility for compromises of data confidentiality, compromises of personal data and intrusions of customer privacy.

SUMMARY

This chapter gives a description about how the SCADA security is different from IT security and its importance. The requirement of open communication system and standardization are discussed with emphasis on security. The VPN and MPLS technology with their advantage and disadvantages, selection criteria are also discussed. The critical infrastructure protection requirements as well as the NERC CIP standard are also described in this chapter. Security concerns of the substation automation and control center architecture are discussed with solutions. Malware threats especially the attack of the lethal malware Stuxnet is a nightmare for power system SCADA implementing agencies because of the ZDVs of the Windows OS. These attack vectors and proposed solutions are discussed in this chapter. Threats and vulnerabilities of ICS and power system SCADA with various types of attacks and mitigating techniques are also discussed.

CHAPTER 09

CYBER RISKS AND MITIGATION STRATEGIES

9.1 INTRODUCTION

Presently one of the ICS architecture which gained much popularity with the incorporation of modern firewall technology is the defense-in-depth architecture. A single security product, technology or solution cannot adequately protect a SCADA by itself. A multiple layer strategy involving two or more different overlapping security mechanisms is desired so that the impact of a failure in any one mechanism is minimized.

A defense-in-depth architecture strategy includes the use of firewalls, the creation of demilitarized zones, intrusion detection capabilities along with effective security policies, training programs and incident response mechanisms. In addition, an effective defense-in-depth strategy requires a thorough understanding of possible attack vectors on a ICS. These include,
1. backdoors and holes in network perimeter,
2. vulnerabilities in common protocols,
3. attacks on field devices,
4. database attacks, and
5. communications hijacking and *Man-In-The-Middle (MITM)* attacks.

The control systems cyber-security using *defense-in-depth strategies* for organizations use control system networks which maintain a multi-tier information architecture which requires,
1. maintenance of various field devices, telemetry collection, and/or industrial level process systems,
2. access to facilities via remote data link or modem, and
3. public facing services for customer or corporate operations. This strategy includes firewalls, the use of demilitarized zones and intrusion detection capabilities throughout the ICS architecture.

This chapter presents the *defense-in-depth strategies starting from the Purdue Reference Architecture, and explain the five level securities needed to secure the*

ICS viz. Physical Security, Network Security, Computer Security, Application Security and Device Security.

9.2 COMMON SCADA NETWORK SECURITY ATTACKS

1. DOS and DDOS attack

There are many cases where a website's server gets overloaded with traffic and simply crashes. But more commonly, this is what happens to a website during a DoS or DDoS attack. When a website has too much traffic, it's unable to serve its content to visitors.

A DoS attack is performed by one machine and its internet connection, by flooding a website with packets and making it impossible for genuine users to access the content of flooded website.

A distributed denial-of-service attack, is similar to DoS, but is more forceful. It's harder to overcome a DDoS attack. It's launched from several computers, and the number of computers involved can range from just a couple of them to thousands or even more. Since it's likely that not all of those machines belong to the attacker, they are compromised and added to the attacker's network by malware. These computers can be distributed around the entire globe, and that network of compromised computers is called botnet. Since the attack comes from so many different IP addresses simultaneously, a DDoS attack is much more difficult for the victim to locate and defend against.

2. Phishing

Phishing is a method of gathering personal information using misleading e-mails and websites. A method of a social engineering with the goal of obtaining sensitive data such as passwords, usernames, credit card numbers mainly using deceptive e-mails. The recipient of the email is tricked into opening a malicious link, which leads to the installation of malware on the recipient's computer. It can also obtain personal information by sending an email that appears to be sent from a bank, asking to verify the identity and stealing away the private information.

3. Rootkit

Rootkit is a collection of software tools that enables remote control and administration level access over a computer or computer networks. Once remote access is obtained, the rootkit can perform a number of malicious actions as it come equipped with key loggers, password stealers and antivirus disablers. Rootkits are installed by hiding in legitimate software when the user give permission to that software to make changes to the OS, the rootkit installs itself in the computer and waits for the hacker to activate it. Other ways of rootkit distribution include phishing emails, malicious links, files, and downloading software from suspicious websites.

4. SQL Injection Attack

Many servers store data for websites using SQL. SQL Injection Attack is a common attack vector that uses malicious SQL code for backend database manipulation to access information that was not intended to be displayed. This information may include any number of items, including sensitive company data, user lists or private customer details. With the advancement of technology, network security threats become sophisticated, making threat of SQL injection common and dangerous to privacy issues and data confidentiality.

5. Man-In-the-Middle Attacks

Man-in-the-middle attacks are cyber security attacks that allow the attacker to eavesdrop on communication between two targets. It can listen to a communication which should, in normal settings, be private. As a typical example, a man-in-the-middle attack happens when the attacker wants to intercept a communication between person A and person B. Person A sends their public key to person B, but the attacker intercepts it and sends a forged message to person B, representing themselves as A, but instead it has the attackers public key. B believes that the message comes from person A and encrypts the message with the attackers public key, sends it back to A, but attacker again intercepts this message, opens the message with private key, possibly alters it, and re-encrypts it using the public key that was firstly provided by person A. Again, when the message is transferred back to person A, they believe it comes from person B, and this way, it has an attacker in the middle that eavesdrops the communication between two targets. Some of the various types of MITM attacks are mentioned below.
- DNS spoofing
- HTTPS spoofing *
- IP spoofing
- ARP spoofing
- SSL hijacking
- Wi-Fi hacking

6. Computer Virus

A computer virus is a type of computer program that, when executed, replicates itself by modifying other computer programs and inserting its own code. When this replication succeeds, the affected areas are then said to be *infected* with a computer virus for everyday Internet users, computer viruses are one of the most common threats to cyber security. Statistics show that approximately 33% of household computers are affected with some type of malware, more than half of which are viruses.

Computer viruses are pieces of software that are designed to be spread from one computer to another. They're often sent as email attachments or downloaded from specific websites with the intent to infect the computer and other computers on contact list by using systems on the user network. Viruses are known to send spam, disable the security settings, corrupt and steal data from the computer including personal information such as passwords, even going as far as to delete everything on the hard drive.

7. Rogue security software

Rogue security software poses a growing threat to computer security. It is malicious software that mislead users to believe there is a computer virus installed on their computer or that their security measures are not up to date. Then they offer to install or update users' security settings. They'll either ask the user to download their program to remove the alleged viruses, or to pay for a tool. Both cases lead to actual malware being installed on the computer.

8. Trojan horse

Trojan horse or Trojan refers to tricking someone by inviting into a securely protected area to deceive. In computing, it holds a very similar meaning. It is a malicious bit of attacking code or software that tricks users into running it willingly, by hiding behind a legitimate program. The spread of Trojan horse is often by phishing email. When user click on the email and its included attachment, malware got downloaded immediately to the computer. Trojans also spread when click on a false advertisement. Once inside the computer, a Trojan horse can record the user passwords by logging keystrokes, hijacking the webcam, and stealing any sensitive data from the computer.

9. Adware and spyware

Adware is deceptive software that earns its creators money through fraudulent user clicks. Fortunately, it's one of the most detectable types of malware. Adware may slow down your computer and affect browsing experience. It could also add vulnerabilities to the computer that could be exploited, and at times, it can collect and send the browsing history to third parties without the user consent. It can also trick you into installing a real malware using its ad network.

Spyware works similarly to adware, but is installed on the computer without the user's knowledge. It can contain key loggers that record personal information including email addresses, passwords, even credit card numbers, making it dangerous because of the high risk of identity theft.

10. Computer worm

Computer worms are pieces of malware programs that replicate quickly and spread from one computer to another. A worm spreads from an infected computer by sending itself to all of the computer's contacts, then immediately to the contacts of the other computers.

A worm spreads from an infected computer by sending itself to all of the computer's contacts, then immediately to the contacts of the other computers. Interestingly, they are not always designed to cause harm; there are worms that are made just to spread. Transmission of worms is also often done by exploiting software vulnerabilities.

9.2.1 Threat Sources to PSS

A *threat* can be defined as any person, circumstance or event with the potential to cause loss or damage. Vulnerability can be defined as any weakness that can be exploited by an opponent. Both are evaluated based on the consequences and the amount of loss or damage occurred from a successful attack. Smart Grid is a typical Cyber Physical Systems (CPS), which tightly integrates a physical power transmission system with the cyber process of network computing and communication at all scales and levels. Obviously it is susceptible to cyber threats and vulnerabilities and can be exploited by different attack groups. They are mainly classified as described below.

1. *Threat from crackers:* who break into computers for profit or bragging rights. Internet increases availability of hacker tools along with information about infrastructures and control systems,
2. *Ransomware threats*: who breaks into computers with a strong financial motive for cyber-crime to exploit vulnerabilities,
3. *Threat from insiders*: who disrupt their corporate network, sometimes an accident, often for revenge, *Threat from terrorists*:who attack systems for cause or ideology, and

4. *Threat from hostile countries:* countries which attack computers and servers of the enemy countries.

9.2.2 Power System SCADA Vulnerabilities, Threats and Attacks

With the integration of ICT, transition to a smarter electrical grid is very much optimistic and promising. However it introduces cyber-vulnerabilities to the Grid. The major points which lead Smart Grid vulnerable to cyber-attacks are described briefly below.

1. *Bi-directional communication:* Though bi-directional communication provides great benefits to the utility and the customer with the capability to communicate and share information, it makes the system vulnerable to cyber-attacks.
2. *Customer data privacy:* The information shared over the Smart Grid is intrinsically sensitive, requires high level privacy and personal security.
3. *AMI:* Various manufactures provides devices with different security features with different security levels built in, making it a challenge to standardize security practices.
4. *Distributed connectivity:* Smart Meters are part of the Neighborhood Area Network (NAN) in Smart Grid which is not confined to a specific geographic area. Hence the boundaries of the network will expand and become more difficult to secure.
5. *Authentication and access controls:* As the number of customers, suppliers and contractors increases, it becomes difficult to gain access to the network resulting in identity theft.
6. *Proper employee training and awareness*: Without proper training and awareness, there are chances of increasing the insider threats and lapses.
7. *Guidelines, standards and interoperability:* This may pose the potential for gaps in visibility, defense and recovery.

These threats are real that many Nations have declared its digital infrastructure as strategic asset and made cyber-security a national agenda with highest priority. They are setting up security policies to force power utilities responsible for protecting the critical electrical infrastructure. Many Nations entrusted their secret agencies to monitor for hackers attempting to infiltrate the power sector. But government regulation and pressure won't be enough to safeguard the Critical Infrastructure. Hence utilities have to take spontaneous and significant steps to secure their networks. The techniques which attackers use to gain control of PSS or cause different levels of damage are mostly similar to those in case of IT. But they also possess certain explicit techniques and some of the techniques used are briefly described below.

1. *Spreading of malwares:* Malwares developed by an attacker can infect the AMI, RTU, PLC or control center servers or utility's corporate servers. Malwares can replace or alter the device functions or a system including sending false commands and sensitive information.
2. *Backdoor entry through communication devices:* A cyber-attacker may compromise some of the communication devices such as modems, routers, etc. and infiltrate the system using it as a backdoor to launch attacks.
3. *Accessing and manipulating the database:* PSS events and data are stored in a database on the control center server network and then mirror the logs into the business network. If the database management systems are not properly configured, an adroit cyber-attacker can gain access to the business network database, and then exploit the control system network.
4. *False Data Injection:* An attacker can send false packets of information into the network, such as wrong Smart Meter data, false price tariffs, fake emergency event, etc. Fake information can cause huge financial impact on the electricity markets. These types of attacking technologies are advancing much faster than security patch ups to control it. Hence it is very essential that the end node security aspects of Smart Grid especially the Smart Meter should ensure that it has communication capabilities that meet all basic integrity and confidentiality criterions. Unfortunately the Smart Meters today do not meet the required protection against false data injection. These facts highlight a much larger potential issues with data integrity throughout the Smart Grid infrastructure.
5. *DoS attacks network availability:* DoS attacks might attempt to delay, block, or corrupt information transmission in order to make Smart Grid resources unavailable. As the Smart Grid uses IP protocol and TCP/IP stack, it becomes subject to all the vulnerabilities inherent in the TCP/IP stack and hence the DoS attacks.
6. *Modbus security issue:* Modbus protocol is widely used in Power System SCADA especially for the communication between the IEDs and RTUs. Hence all Modbus security issues are applicable to the SCADA system or all the facility-based processes such as Smart Grid processes.
7. *Eavesdropping and traffic analysis:* An enemy can obtain sensitive information by monitoring network traffic. It can be the utilities business strategy, future price information, control structure of the grid, and power usage. Later these data can be used for hostile deeds.

9.2.3 Alarming Power System SCADA Threats

The present day Smart Grid threats are much advanced technically and the implemented Smart Grid and PSS are secured just because of *security through obscurity*. A brief description of the technically advanced lethal threats and vulnerabilities are explained in the following sections.

9.2.3.1 Zero Day Vulnerabilities

The term zero day implies that the developer does not get enough time to develop and deploy a patch to overcome the flaw. Before that, an attacker exploits the flaw and/or creates and deploy malwares to attack the SCADA system. There are many zero-day flaws that may affect a SCADA system. Stack overflow is one of them. This attack can occur on the field devices as well as the servers. The stack buffer in the memory can be corrupted by a malicious player, leading to injection of dangerous executable code into the running program and thus usurping the control of the industrial process. The vulnerability reported in China is a well-known example. Zero Day Attacks can also occur in the form of DoS attacks that overload computer resources.

Stuxnet is a lethal computer worm which uses four zero-day Windows vulnerabilities. It was primarily written to target Iranian nuclear centrifuges. Its final goal is to disrupt ICSs by modifying programs implemented on PLCs to make them work in a manner that the attacker intended and to hide those changes from system operators. It is believed that Stuxnet is introduced to a computer network through an infected removable drive. To hide itself while spreading across the network and realizing the final target, the virus installs a Windows rootkit by exploiting four zero-day vulnerabilities. The success of this virus in penetrating the PLC environment shows that traditional security measures are not sufficient for the complete protection of safety-critical infrastructures.

9.2.3.2 Non-Prioritization of Tasks

This is a serious flaw in real-time operating systems of many Industrial Control Systems (ICS). In certain embedded operating systems, there may not have the feature of prioritization of tasks. Memory sharing between the equally privileged tasks lead to serious security issues. The features such as the accessibility to create Object Entry Point (OEP) in the kernel domain can lead to loopholes in security. Non-kernel tasks may be protected from overflows using guard pages. But the guard pages may be small and cannot provide stringent protection.

9.2.3.3 Database Injection

Detrimental query statements can be injected to exploit the vulnerabilities in a PSS especially when the client inputs are not properly filtered. This is widely reported for SQL-based databases. Here the attacker sends a command to SQL server through the web server and attempt to reveal critical authentication information.

9.2.3.4 Communication Protocol Issues

Today with the developments in encryption and authentication, IT security is capable of encountering the sophisticated cyber-attacks and threats. But they are not adopted in an adequate manner in PSS and Smart Grid especially when the process is controlled with the client server architecture. Earlier SCADA security was not a major concern and hence communication protocols did not give sufficient importance to authentication. This does not mean that authentication and encryption methods cannot be used with these systems. It should be noted that encryption is effective only in an authenticated communication between entities. To have a secure TCP/IP communication, internet Protocol Security (IPSec) framework has to be employed. It will help to create a secure channel of communication for ICS.

IPSec uses two protocols for authentication and encryption viz. Encapsulating Security Payload (ESP) and Authentication Header (AH). Advanced Persistent Threat (APT) attacks which monitor network activity and steal data for a future attack, can be effectively dealt with protocols like Syslog that keeps security logs which provide a means for detecting stealthy attempts to gather information prior to building sophisticated attacks by malicious players.

9.2.3.5 Stealthy Integrity Attack

Stealth attacks are targeted to disrupt the service integrity, and make the networks to accept false data value. Security experts reports that powerful adversaries equipped with in-depth knowledge, disclosure resources, and disruption capabilities who are capable to perform stealthy attacks which partially or totally bypass traditional anomaly detectors. The detectability of an attack strategy depends on the capabilities of adversaries to coordinate attack vectors on control signals and sensor measurements.

9.2.3.6 Replay Attack

These are the network attacks in which an attacker spies the communication between the sender and receiver and takes the authenticated information

e.g. stealing the key and then contact the receiver with that key. In replay attack the attacker gives the proof of his identity and authenticity. The negative effect of a replay attack on a feedback control system has been found that this attack strategy is carried out in two steps which are mentioned below.
- The hacker records sensor measurements for a certain window of time before performing the attack.
- The hacker replaces actual sensor measurements with previously recorded signals while modifying control signals to drive system states out of their normal values.

A replay attack is capable of bypassing the classical detectors.

9.2.3.7 False Data Injection Attack

Smart Grid may operate in very hostile environments. AMI components lacking tamper-resistance hardware increases the possibility to be compromised. The attacker may inject false measurement reports to disrupt the Smart Grid operation through the compromised meters and sensors. The objective of the attacker is to fool the state estimator by carefully injecting a certain amount of false data into sensor measurements. Those attacks are denoted as false data injection attacks. It can upset the grid system and lead to a false state estimation resulting to the disruption of the energy distribution. Security experts are of the opinion that the false data injection attack is a discrete-time state-space model driven by Gaussian noises. A Kalman filter is generally used to perform state estimation, and a failure detector is employed to detect abnormal situations.

9.2.3.8 Zero-Dynamics Attack

In PSS Zero-Dynamics Attack (ZDA) is one of the hardest attacks to defend, especially in closed-loop feedback system which possesses an unstable zero, such as an unbounded actuator or sensor. These attacks cannot be observed by the monitoring data. The ZDA can be easily implemented in the cyber space or injected into the communication links by an attacker who has a proper understanding of Data Acquisition System (DAS). As modern control systems are implemented on digital computers using sample and hold mechanisms, where the controllers can be dealt with in a Sampled Data (SD) framework, can generate vulnerability to stealthy attacks due to the unstable sampling zeros in the SD system.

9.2.3.9 Covert Attack

This attack is a targeted attack, but it is constructed and deployed using public tools. These are custom-made and minimally equipped. The strategy of covert attack consists of coordinating control signals and sensor measurements into a concerted malicious attack. This attack is executed in two steps as described below.
- The state attack vector can be chosen freely based on malicious targets and available resources.
- The sensor attack vector is designed in such a way that it can compensate for the effects of the state attack vector on the sensor measurements.

The covert attack strategy can be considered as the worst case attack because it has the capability to bypass traditional anomaly detectors. However, the covert attack needs to compromise numerous sensors to assure its stealth. Therefore, PSS defenders can remove a covert attack by protecting some critical sensors or deploying secure sensors.

9.2.3.10 Surge Attack, Bias Attack, and Geometric Attack

The three types of stealthy attacks, viz. the surge attack, the bias attack, and the geometric attack are also important and are to be addressed by a cyber-security expert. The surge attack seeks to maximize the damage as soon as possible, while the bias attack tries to modify the system by small perturbations over a long period. The geometric attack integrates the surge attack and the bias attack by shifting the system behavior gradually at the beginning and maximizing the damage at the end.

9.2.4 Dreadful PSS Malwares

As PSS and Smart Grid improve the efficiency and performance of the power grid, they also increase the grid vulnerability to potential cyber-attacks. Black Energy, Stuxnet, Havex, Duqu, and Sandworm are all recent examples of malwares targeting PSS. The AMI components especially the Smart Meters and increase in External Access Points (EAP) added with the integration of Renewable Energy Sources (RES) introduced new additional areas through which a potential cyber-attack may be launched on the grid. The present malware intrusions have resulted in a significant disruption of grid operations like what it had occurred to Ukraine by the dreadful malware Black Energy. The following sections are dedicated to describe some of the SCADA malwares.

9.2.4.1 BlackEnergy

BlackEnergy is a Trojan horse malware program which infects the SCADA systems. Though it has been detected in 2007, the vehemence of destruction is realized only in 2015 with the Ukraine power sector attack. Till the Ukraine power sector attack, it has been believed that BlackEnergy was designed only for nuisance spam attacks and not targeted PSS or critical energy infrastructure.

Today, BlackEnergy is a special concern for Critical Infrastructure companies especially to PSS because the software being used is in an Advanced Persistent Threat (APT) form, apparently to gather information. BlackEnergy specifically targets HMI software which is typically running 24/7, with the provision of remote access. It is rarely updated, thus making it a favorite target for opportunistic hackers. While no attempts to damage, modify, or otherwise disrupt the victim systems' control processes immediately after infection of BlackEnergy, indicates that the APT variant of BlackEnergy is a special concern as it is a modular malware capable moving through network files.

9.2.4.2 Ukraine Incident

The December 23rd 2015, a cyber-attack has been carried out by the Black Energy malware which resulted in disrupting the PSS network almost completely in a brilliant sabotage operation which is briefly described below.

The hackers penetrated the PSS networks through the hijacked VPNs and sent commands to disable the UPS systems they had already reconfigured. Then they issued commands to open breakers. But before they did, they launched a Denial-of-Service (DoS) attack against customer call centers to prevent customers from calling in to report the outage. DoS attacks sent a flood of data to the web servers and as a result, the phone systems of the control center were flooded with thousands of bogus calls, in order to prevent genuine callers from getting through.

This clearly demonstrates a high level of complexity, cleverness and organization of the attackers. Expert cybercriminals and even Nation sponsored attackers often fail to anticipate in all likelihoods. However, regarding BlackEnergy attack, it has been observed that the attackers put very concentrated effort to make sure that they are covering all aspects without any lapse so that nothing could go wrong. The move certainly required considerable time to the attackers to complete their operation. But by the time the operators realized this adverse situation that their machine has been hijacked, a number of substations had already been taken down and the situation had been totally slipped out of their control. In fact the operators become silent witness of this attack.

By gaining the remote control and opening the breakers, the attackers switched off a series of substations from the grid. They carried out these most clever operations in fact paralyzed the grid. They overwrote the firmware of the substation serial-to-Ethernet converters, replacing genuine firmware with their malicious firmware and rendering the converters thereafter inoperable and unrecoverable, unable to receive commands. As a result, these gateways are blown and cannot be recovered until they got new devices and integrate them.

Once the attackers have completed all of these, they used a malware called KillDisk to wipe out files from operator stations to make them inoperable. KillDisk wiped or overwrote the data in essential system files, causing computers to crash. Because it also overwrote the master boot record, the infected computers could not reboot. All these happened during the beginning of night peak, and the power utilities posted a note to their web sites acknowledging that power was out in certain regions and reassuring that they are working vehemently to identify the cause of the crisis. But within half an hour, the KillDisk completed its dirty action and left power operators without any doubt what caused the blackout. The utilities then posted another note to customers intimating the cause of the outage was hackers.

In effect, up to 95% of daily electricity consumption in Ukraine was not supplied. Though the cyber-attacks on the energy distribution companies has been attributed to the Russian APT group, the Black-Energy cyber-attack on Ukrainian power sector not only wrecked the Nation economically, but also worsened the energy customers confidence in the Ukrainian power companies and government.

9.2.4.3 Stuxnet

Stuxnet, the Nation sponsored world's first lethal digital weapon, was unlike any other virus or worm that came before. Rather than simply hijacking targeted computers or stealing information from them, it escaped the digital realm to wreak physical destruction of equipments which are controlled by the computers. Initially it designed specifically to infect the Siemens SIMATIC WinCC and S7 PLC products, either installed as part of a PCS 7 system, or operating on their own. It starts operation by taking advantage of vulnerabilities in the Windows operating systems and Siemens products. Once it detects a suitable victim, it modifies control logic in of PLCs or RTUs. Undoubtedly the objective is to sabotage a specific industrial process using the vendors' variable-frequency drive controllers, along with a supervising safety system for the overall process. Though there has been much speculation on Stuxnet's intended target, recent information suggests it was Iran's nuclear program and more specifically, its uranium enrichment process. Stuxnet is capable of infecting both unsupported/legacy and current versions of Windows including Windows 2000, Windows XP, Windows Server 2003, Windows Vista, Windows Server 2008 and Windows 7. It also infects the Siemens STEP 7 which is one of the world's best known and most widely used engineering software in ICS in such a way that it automatically executes when the STEP 7 project is loaded by an uninfected Siemens system. Some of the important characteristics of the worm are,

1. It propagates cleverly between targets, typically via USB flash drives and other removable media,
2. Once migrated the target, It propagates quickly within the target via multiple network pathways,
3. It very cunningly searches for numerous vendors' anti-virus technologies installed on machines and modifies its behavior to avoid the detection,
4. It contacts a command and control server on the internet for instructions and updates,
5. It establishes a peer-to-peer network to propagate instructions and updates within a target, even to equipments without direct internet connectivity,
6. It modifies PLC or RTU program logic, causing physical processes to malfunction,
7. It hides the modified PLC or RTU programs from control engineers and system administrators who are trying to understand the reason for the malfunctioning of the system,
8. It is signed with certificates stolen from major hardware manufacturers, so that no warnings are raised when the worm is installed, and
9. If a particular machine is not the intended target, the worm removes itself from the machine after it has replicated itself to other vulnerable media and machines.

9.2.4.4 Iranian Experience

In 2012 it has been confirmed that the Iranian nuclear facilities were attacked and infiltrated by the first *cyber-weapon* or the *digital missile* of the world known as the Stuxnet. It is believed that this attack was initiated by a random worker's USB drive. One of the affected nuclear facilities was the Natanz nuclear facility. The first signs that an issue existed in the nuclear facility's computer system was in 2010. Inspectors from the International Atomic Energy Agency (IAEA) visited the Natanz facility and observed that a strange number of uranium enriching centrifuges were breaking. The cause of these failures was unknown at that time. Later Iranian technicians contacted computer security specialists in Belarus for examining their server and network systems which control the facilities. This security firm eventually discovered multiple malicious files on the Iranian computer systems. It has subsequently revealed that these malicious files were the Stuxnet worm. Although Iran has not released specific details regarding the effects of the attack, it is currently estimated that the Stuxnet worm destroyed 984 uranium enriching centrifuges. By current estimations this constituted a 30% decrease in enrichment efficiency.

9.2.4.5 Spreading of Stuxnet

As already explained Stuxnet is one of the most complex, lethal and well-engineered 500kilobyte computer worms the world has ever seen. It took advantage of at least four Zero Day Vulnerabilities (ZDV) with remarkable sophistication. The worm propagates using three totally diverse mechanisms as described below.

1. Via infected removable drives (such as USB flash drives and external portable hard disks)
2. Via Local Area Network communications (such as shared network drives and print spooler services), and
3. Via infected Siemens project files (including both WinCC and STEP 7 files).

Within these three, it uses the following vulnerability exploitation techniques for spreading to new computers in a system. The worm initially exploits a Zero Day Vulnerability in Windows Shell handling of LNK files which is a vulnerability present in all versions of Windows since at least Windows NT 4.0. Then uses several techniques to copy itself to all accessible network and spread from there to all possible locations. It also copies itself to printer servers using zero-day vulnerability, The Conficker RPC vulnerability is also used to propagate through computers which are not properly patched up. If any Siemens WinCCSQLServer database servers are present, then the worm installs itself on those servers via database calls and puts copies of itself into Siemens STEP 7 project files to auto-execute whenever the files are loaded. Certain versions of the worm used a variant of the old *autorun.inf* trick to propagate via USB drives.

In addition to the propagation techniques described above, the worm uses two ZDVs to escalate privilege on targeted machines. This provided the worm with system access privileges so that it could copy itself into system processes on compromised machines. When first installed on a computer with any software, Stuxnet attempts to locate Siemens STEP 7 programming stations and infect these. If it succeeds, it replaces the Dynamic Link Library (DLL). Mostly Stuxnet spreads through the infected removable devices especially the USB. However it is a misconception that it spreads through only removable devices and an effective disabling of the storage device access can prevent the spreading of Stuxnet.

9.2.4.6 Havex

The modified version of Havex malware mainly targets the energy sector. Originally, Havex was distributed via spam email or spear-phishing attacks. The new version of Havex appears to have been designed as a Trojan horse specifically to infiltrate and modify legitimate software from ICS and SCADA suppliers, adding an instruction to run, code containing the Havex malware. In the instance discovered, Havex malware was used as a Remote Access Tool (RAT) to extract data from Industrial Control System (ICS) related software used for remote access. The cyber-attack leaves the company's system in what appears to be a normal operating condition, but a backdoor has been opened by the attacker to access and control the utilities ICS or SCADA operations. The Havex malware possibly enter the control systems of targeted utilities using one or multiple levels of attack as described below.
1. Top level management are targeted with malicious PDF attachments.
2. Websites are likely to be visited by people working in the energy sector. Such websites can be infected and visitors are redirected to another compromised legitimate website hosting an exploit kit. Using this exploit kit, the RAT may be installed.
3. Through software downloads from ICS related vendors which they include the RAT malware.

Havex is also called as *Backdoor. Oldrea* or the *Energetic Bear RAT* as it contains the malware known as *Kragany*. Havex is a product of the Dragonfly group, which appears to be a state-sponsored undertaking focused on espionage with sabotage as a definite secondary capability. The malware allows attackers to upload and download files from the infected computer and run executable files.

It was also reported to be capable of collecting passwords, taking screenshots and cataloguing documents.

9.2.4.7 Sandworm

Sandworm is a type of Trojan horse, focused on exploiting vulnerability in the Windows operating system. USB storage devices with automatically run files, carrying the malware is used for the attack. The primary mode of Sandworm attack is spear phishing. Using well written emails with topics of interest has been sent to the target. The malware contains an attachment that exploits the vulnerability to deliver variants of the BlackEnergy Trojan.

Various reports released regarding the Sandworm team and investigations of the malware samples and domains, realized that Sandworm team is targeting SCADA centric victims especially who are using GE Intelligent Platform's CIMPLICITY HMI solution suite. The HMI can be viewed as an operator console that is used to monitor and control devices in an industrial environment. Sandworm can potentially have greater impacts on an enterprise, as the malware could be transferred to other corporate business systems.

It is important to note that CIMPLICITY is mainly used as an attack vector by Sandworm. However, there is no sign that this malware is manipulating any actual SCADA systems or data. Since HMIs are located in both the corporate and control networks, this attack could be used to target either network segment, or used to cross from the corporate to the control network.

9.2.4.8 Duqu and Flame

Duqu and Flame are computer malwares that were discovered on 1st September 2011 and 28th May 2012, respectively. Duqu is almost identical to Stuxnet but with a different tenacity. Flame is also known as Flamer or Sky-wiper. The goal of Duqu is to collect information that could be useful in launching an ICS attack later. Flame like Stuxnet and Duqu, uses rootkit functionality to evade information security methods. Unlike Stuxnet, which was designed to sabotage ICSs, the target of Flame is to gather technical diagrams such as AutoCAD drawings, PDFs, and text files. Though Duqu and Flame were not designed to target ICSs directly, their penetration and acquiring information about the systems, is an indication of a targeted stealthy attacks in the future.

9.2.5 Flash Drive Usage and End Node Security (ENS)

USB attacks are becoming more sophisticated, affecting all classes of USB device instead of just storage. As USBs have become a common method for easily sharing information locally between devices, they have become a common source of information system cyber-compromise. As per NERC CIP guidelines, use of USB or USB type ports are strongly discouraged because a USB port is not immune to protection from *unauthorized access*. It would be helpless against connecting modems, network cables that bridge networks or insertion of an infected USB pen drive. Cyber protection for USB ports can be enforced, however, it is often cost prohibitive and is not one hundred percent effective. There is no essential requirement for using a USB instead of other standard and more secure interfaces such as Ethernet and serial ports. Some of the common methods to protect the USB ports are,

1. disabling (via software) the physical ports,
2. prominent physical port usage discouragement such as, a port cover plate or tamper tape, and
3. physical port obstruction using removable locks.

These measures are examples of defense-in-depth methods, but the CIP guidelines acknowledge that these control approaches can be easily circumvented. It is also not uncommon for an employee or authorized contractor to inadvertently compromise a device simply by plugging in an infected smart phone to charge the battery. USB flash drives pose two major challenges to critical infrastructure cyber-security viz.

1. ease of data theft owing to their small size and transportability, and
2. system compromise through infections from computer viruses, malware and spyware.

It is a well-accepted fact that a USB supported portable peripheral device can trigger a massive cyber-attack, even when the computer system targeted is isolated and protected from the outside with firewalls and other types of security devices.

9.2.5.1 BadUSB

BadUSB is an USB which includes firmware in addition to disk space. It is inherently a microcontroller with writable storage memory registers. This firmware however can be embedded with executable codes which cannot be verified by third party security software applications since the firmware is not an open source. This flaw in USBs opens the door to modification of USB firmware, which can easily be done from inside the operating system, and hide the malware in a way that it becomes almost impossible to detect. The flaw is even more potent because complete formatting or deleting the content of a USB device won't eliminate the malicious code, since it is embedded in the firmware. Patches made for BadUSB have been largely ineffective and a fix is years away. In fact till date, there has been no practical defensive solution against BadUSB attacks and it exposes the fundamental vulnerabilities of unconstrained privileges in USB devices. This being the situation, PSS and Smart Grid design and implementation must be in such a manner that it completely eliminates the need for a USB port which is advised and recommended in the interest of reliable and safe operations.

9.2.5.2 Cyber Incidents Using USB

1. Two US based power plants were infected with malware after using USB drives in their PSS. At one of the plants this resulted in downtime and delayed the plant's restart by three weeks. This caused considerable financial loss.
2. Another cyber incident of malware wreaking havoc includes the recent widespread infections at Saudi Aramco and Ras-Gas, where malware were planted by USB drives. At Saudi Aramco approximately 40,000 computer hard drives were completely wiped off. As none of the infected power plants had updated anti-malware softwares, the infected computers were totally incapable of detecting the malware on the inserted USB drives.

Usually, almost every vendors or implementation agencies offer or sign the contract stating that the control system shall be protected by an automatically updated antivirus system. The agreement is reasonable, but it may be very difficult to comply with. The PSS industry is very much aware of this problem. Different solutions have been tried, including the use of glue guns to disable USB ports and physical USB locks.

9.3 PURDUE REFERENCE ARCHITECTURE FOR ICS

Purdue model is an Enterprise Reference Architecture developed by Theodore J.Willaims and colleagues of Purdue University for ICS and was adopted model by ISA-99 as a concept model for ICS network segmentation. This reference model is a resource for segmenting the modern ICS architecture and also help to understand the Industrial Cyber Security Landscape.

The Purdue model divides the ICS architecture into three zones and they are, Enterprise zone, Industrial Demilitarized zone and Industrial or process zone. Process zone is further divided into four levels and they are,
- Level 3: Site operations- managing production work flow to produce the desired products. Batch management, manufacturing execution/operations management systems (MES/MOMS), laboratory, maintenance and plant performance management systems, data historians and related middleware. Time frame: shifts, hours, minutes, and seconds.
- Level 2: Area supervisory control- supervising, monitoring and controlling the physical processes. Real-time controls and software; DCS, human-machine interface (HMI), supervisory and data acquisition (SCADA) software.
- Level 1: Basic control- sensing and manipulating the physical processes. Process sensors, analyzers, actuators and related instrumentation.
- Level 0: The process- defines the actual physical processes.

The enterprise zone has been divided into two levels viz. enterprise network and site business and logistics. These levels manage the business-related activities of the manufacturing operation. ERP is the primary system; establishes the basic plant production schedule, material use, shipping and inventory levels. Time frame: months, weeks, days, shifts. In fact Purdue model Increases resiliency by segmenting the OT network.

The enterprise zone

The enterprise zone is the part of the ICS where business systems such as ERP and SAP typically live. Here, tasks such as scheduling and supply chain management are performed. The can be subdivided into two levels:
- Level 5: Enterprise network
- Level 4: Site business and logistics

Level 5 - Enterprise network

The systems on the enterprise network normally sit at a corporate level and span multiple facilities or plants. They take data from subordinate systems out in the individual plants and use the accumulated data to report on the overall production status, inventory, and demand. Technically not part of the ICS, the enterprise zone relies on connectivity with the ICS networks to feed the data that drives the business decisions.

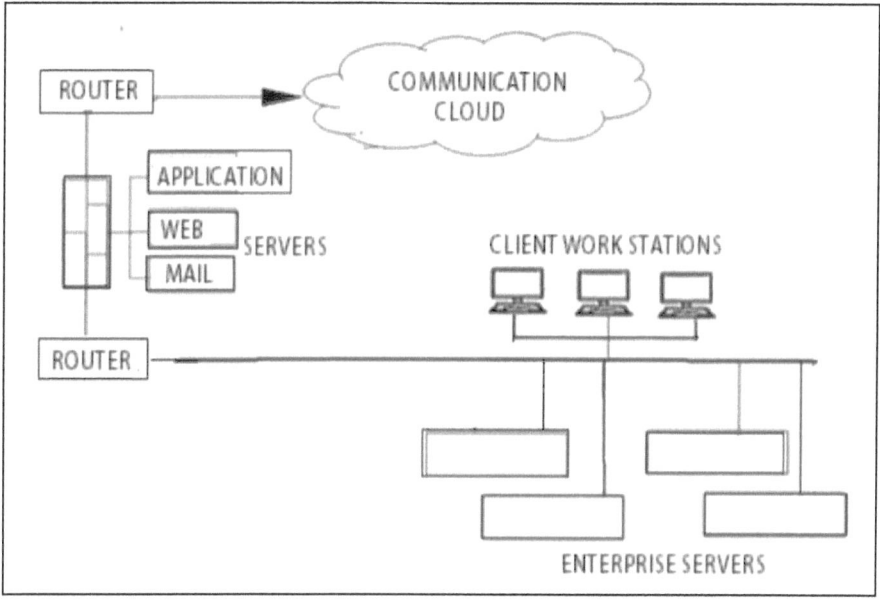

▲ **Figure 9.1:** Enterprise zone

Level 4 - Site business planning and logistics
Level 4 is home to all the **Information Technology (IT)** systems that support the production process in a plant of a facility. These systems report production statistics such as uptime and units produced for corporate systems and take orders and business data from the corporate systems to be distributed among the Operation Technology (OT) or ICS systems.

Systems typically found in level 4 include database servers, application servers (web, report, MES), file servers, email clients, supervisor desktops, and so on.

Industrial Demilitarized Zone (IDMZ)
Usually the Industrial Demilitarized Zone (IDMZ) lies between the enterprise zone and the Industrial zone as shown in Figure 9.3. Like the traditional (IT) DMZ, the OT-IDMZ allows to securely connect networks with different security requirements. Here different make firewalls at the two different levels are recommended for improves security.

DMZs are in essence a network between networks, and in the industrial security context, an added network layer between the OT, ICS, or SCADA, network and the less-trusted IT or enterprise network. Deploying the DMZ between two firewalls means that all inbound network packets are screened using a firewall or other security appliance before they arrive at the servers the organization hosts in the DMZ. There are many organizations and standard bodies that recommend segmenting the enterprise zone from the industrial zone by utilizing an industrial demilitarized zone (IDMZ). It acts as a zone and conduit system protecting physical processes, separating networks according to their different purposes, requirements and risks.

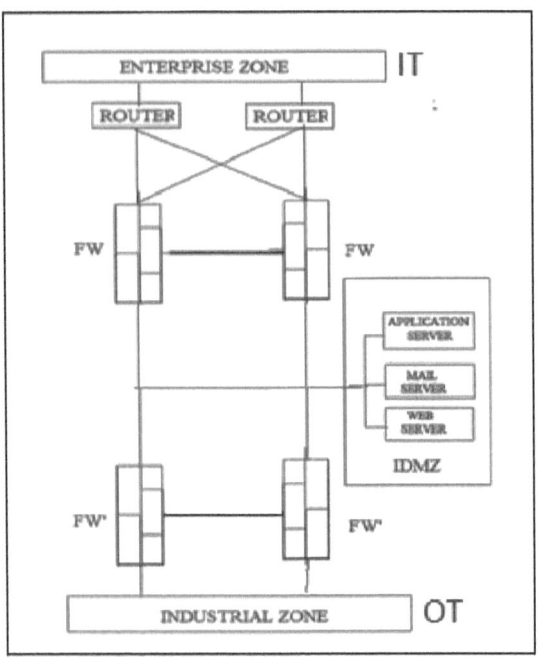

▲ **Figure 9.2:** Industrial Demilitarized Zone (IDMZ)

In the Purdue model, the Industrial DMZ is an information sharing layer between the business or IT systems in levels 4 and 5 and the production or OT systems in levels 3 and lower created as per security standards such as the NIST cyber security Framework and NERC CIP. Direct communication between IT and OT systems is prevented and having a proxy service in the IDMZ relay add an extra layer of separation and scrutiny. Systems in the lower layers are not directly exposed to attacks or compromise. If something were to compromise a system at some point in the IDMZ, the IDMZ is configured such a manner that it will automatically shut down without compromising and the production will be continued. Systems typically kept in the Industrial Demilitarized Zone are WEB servers, Microsoft domain controllers, Mail servers, etc.

The manufacturing zone

The manufacturing zone is where the action is; it is the zone where the process lives, by all means, this is the core of the. The manufacturing zone is subdivided into four levels viz. Level 3: Site operations, Level 2: Area supervisory control, Level 1: Basic control and Level 0: The process.

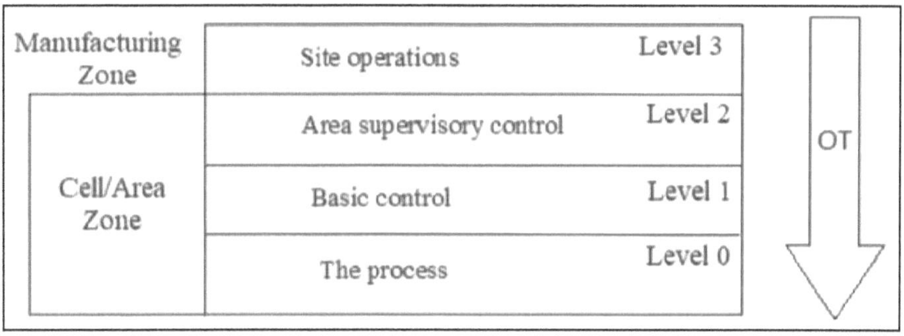

Manufacturing Zone	Site operations	Level 3	
Cell/Area Zone	Area supervisory control	Level 2	OT
	Basic control	Level 1	
	The process	Level 0	

▲ **Figure 9.3:** Manufacturing Zone

Level 3 - Site operations

Level 3 is where systems that support plant wide control and monitoring functions reside. At this level, the operator is interacting with the overall production systems. Think of centralized control rooms with HMIs and operator terminals that provide an overview of all the systems that run the processes in a plant or facility. The operator uses these HMI systems to perform tasks such as quality control checks, managing uptime, and monitoring alarms, events, and trends.

Level 3, site operations, is also where the OT systems that report back up to IT systems in level 4 live. Systems in lower levels send production data to data collection and aggregation servers in this level, which can then send the data to higher levels or can be queried by systems in higher levels (push versus pull operations).

Systems typically found in level 3 include database servers, application servers (web and report), file servers, Microsoft domain controllers, HMI servers engineering workstations, and so on.

Level 2 - Area supervisory control

Many of the functions and systems in level 2 are the same as for level 3 but targeted more toward a smaller part or area of the overall system. In this level, specific parts of the system are monitored and managed with HMI systems. Think along the lines of a single machine or skid with a touch screen HMI to start or stop the machine or skid and see some basic running values and manipulate machine or skid-specific thresholds and set points.

Systems typically found in level 2 include HMIs (standalone or system clients), supervisory control systems such as a line control PLC, engineering workstations, and so on.

Level 1 - Basic control

Level 1 is where all the controlling equipment lives. The main purpose of the devices in this level is to open valves, move actuators, start motors, and so on. Typically found in level 1 are PLCs, Variable Frequency Drives (VFDs), dedicated proportional-integral-derivative (PID) controllers, and so on. Although you could find a PLC in level 2, its function there is of supervisory nature instead of controlling.

Level 0 - Process

Level 0 is where the actual process equipment that we are controlling and monitoring from the higher levels lives. Also known as Equipment Under Control (EUC), level 1 is where we can find devices such as motors, pumps, valves, and sensors that measure speed, temperature, or pressure. As level 0 is where the actual process is performed and where the product is made, it is imperative that things run smoothly and uninterrupted. The slightest disruption in a single device can cause mayhem for all operations.

9.4 PHYSICAL SECURITY

Physical security is the first line of defense against environmental risks and fickle human behavior. It is the protection of physical property, encompasses both technical and nontechnical components. Most of the information security experts often-overlooked physical security as they do about information and computer security and the associated hackers, ports, viruses, and technology-oriented security countermeasures. But information security without proper physical security could be very dangerous. The physical threats that an organization faces fall into the following categories viz.
- Natural disaster like floods, earthquakes, fires etc.,
- Strikes, riots, terrorist attacks etc,
- Location and layout of building infrastructure,
- Power supply failure, communications interruptions, water supply interruption etc.,

- Unauthorized access and damage by disgruntled employees, employee errors and accidents, fraud, theft etc.,
- Unsecure network devices used.

In all situations, the primary consideration, above all else, is that nothing should obstruct life safety goals as it has given the highest priority. A wise planning can balance life safety and other security measures.

The physical security of computers and their resources in the decades back was not as challenging as it is today because computers were mostly mainframes that were locked away in server rooms, and only a few people knew what to do with them. Presently, a computer is available with almost every desk in every utility, and access to devices and resources is spread throughout the environment. Organizations have server rooms, and remote access, resources out of the facility. Properly protecting these computer systems, networks, facilities, and employees has become an overwhelming task to many companies.

Theft, fraud, sabotage, and accidents are raising costs for many companies because environments are becoming more complex and dynamic. Security and complexity are at the opposite ends of the spectrum. As environments and technology become more complex, more vulnerabilities are introduced that allow for compromises to take place. Most companies have had memory or processors stolen from workstations, while some have had computers and laptops taken. Even worse, many companies have been victims of more dangerous crimes, such as robbery at gunpoint, a shooting rampage by a disgruntled employee, anthrax, bombs, and terrorist activities.

Many companies may have implemented security guards, closed-circuit TV (CCTV) surveillance, intrusion detection systems (IDSs), and requirements for employees to maintain a higher level of awareness of security risks. These are only some of the items that fall within the physical security boundaries. If any of these does not provide the necessary protection level, it could be the weak link that causes potentially dangerous security breaches.

From a holistic view of physical security, there are so many components and variables such as secure facility construction, risk assessment and analysis, secure data center implementation, fire protection, IDS and CCTV implementation, personnel emergency response and training, legal and regulatory aspects of physical security, etc. Each has its own focus and skill set, but for an organization to have a solid physical security program, all of these areas must be understood and addressed.

Many thefts and deaths could be prevented if all organizations were to implement physical security in an organized, mature, and holistic manner.

When security professionals look at *information* security, they think about how someone can enter an environment in an unauthorized manner through a port, wireless access point, or software exploitation. When security professionals look at *physical* security, they are concerned with how people can physically enter an environment and cause an array of damages.

Physical security must be implemented based on a *layered defense model*, which means that physical controls should work together in a tiered architecture. The concept is that if one layer fails, other layers will protect the valuable asset. Layers would be implemented moving from the perimeter toward the asset.

A physical security program should comprise safety and security mechanisms. Safety deals with the protection of life and assets against fire, natural disasters, and devastating accidents. Security addresses vandalism, theft, and attacks by individuals. Many times an overlap occurs between the two, but both types of threat categories must be understood and properly planned for.

Mitigation Strategies

From a holistic view of physical security, there are so many components and variables such as secure facility construction, risk assessment and analysis, secure data center implementation, fire protection, IDS and CCTV implementation, personnel emergency response and training, legal and regulatory aspects of physical security, etc. Each has its own focus and skill set, but for an organization to have a solid physical security program, all of these areas must be understood and addressed. Presently the physical security is implemented based on a layered defense model and it generally falls in to the following categories.
- Building Location And Layout
- Building Infrastructure.
- Utilities Such as Power, Water, Fire Suppression, etc and
- Network Devices Used.

Layered security includes
- Fencing
- Reinforced Barricades
- Walls
- Gates/Entry Points
- Vehicle Barriers
- On-Site Security Guards
- CCTV Cameras

- Motion Detectors
- Intrusion Detection Systems
- Vibration Detectors
- Secured Cabling

Control room should be highly secured, such as the control room access may be depend on time, employment status, work assignment, level of training etc. The authentication for the access to the control room should be reliable such as biometric devices, cipher locks or access cards. The devices in the control room must be fixed permanently. Adequate lighting should be provided. Regular auditing of access logs must be done. Natural disasters must not affect the proper functioning of the firm. Field devices must be protected from the intruders using proper alarms, cipher locks with proper authentication.

9.5 NETWORK SECURITY

The ICT network security is the most important as it connects the field devices, corporate networks, SCADA networks, DR network, etc. A proper zone base segmentation and security based on functionalities using firewalls, gateways, data diodes, etc are the present techniques employed for ensuring network security.

From a mitigation perspective, simply deploying IT security technologies into an ICS may not be a viable solution. Although modern control systems use the same underlying protocols that are used in IT and business networks, the very nature of control system functionality may make even proven security technologies inappropriate. Some sectors, such as energy, transportation, and chemical, have time sensitive requirements, so the latency and 'throughput' issues associated with security strategies may introduce unacceptable delays and degrade or prevent acceptable system performance. Due to these facts, currently the main mitigation strategy is a security conscious segmentation. A network segment is also known as a network security zone which is a logical grouping of information and automation systems in an ICS network. Usually a Industrial Control Network has been segmented into four security zones with different trust levels as described below.

- Enterprise Zone-   Low Trust Level
- Industrial Demilitarised Zone-   Medium Trust Level
- Industrial Zone-   High Trust Level
- Cell Area Zone-   High Trust Level

Understanding attack vectors is essential to building effective security mitigation strategies. The degree of understanding of the control system by the security engineers regarding these vectors are most essential to mitigate these vulnerabilities effectively and efficiently. Effective security depends on how well the security engineers and vendors understand the ways that architectures can be compromised. Critical cyber security issues that need to be addressed include those related to,
- Backdoors and holes in network perimeter
- Vulnerabilities in common protocols
- Attacks on Field Devices
- Database Attacks
- Communications hijacking and *Man-in-the-middle* attacks

Demilitarized Zone (Level 3.5). This first line of defense in isolating the OT network from IT network. This is a critical segmentation because IT network are generally targeted before OT network. Manufacturing Zone (Level 3). This segmentation protects each ICS system / remote sites / factories. The purpose of this segmentation is to keep this site operational even if other ICS systems / sites comes under attack. Cell Zone (Level 2). To further increase resiliency of each ICS environment, each functional cell or production line within the ICS network is further segmented. This will ensure that if a functional cell is attacked, other adjacent functional cell are still functioning. A SCADA network security engineer must perform the following tasks without any compromise.
- Ensure the firewalls and Intrusion Prevention System(IPS) are properly placed in the network with proper configuration as per the organisation's security policy and standards adopted such as NERC CIP,IEC62443 etc.
- Security engineer must be aware of the port scan by the external attackers and their capability of exploiting vulnerabilities.
- Must be aware of internet connections, remote access capabilities, layered defenses and placements of hosts on networks.
- Aware of the interaction of the security devices installed in the network such as firewalls, IPSecs, antivirus etc. and their routable protocols if supports.
- Must be aware of the vulnerable protocols such as SSL.
- Unprotected ports which are commonly attacked.
- Network monitoring and maintenance.

For nearly three decades, digital network DMZs or *demilitarized zones* have been used as a data protection strategy in IT networks to broker access to information by external untrusted networks. This IT DMZ model and approach is well established, but is not necessarily an effective security measure when applied to OT networks.

In ICS network, DMZs is a network between networks, an added network layer between the OT, ICS, or SCADA, network and the less-trusted IT or enterprise network. If correctly implemented, no TCP or any other connection exchanging messages should ever traverse between IT and OT; through DMZ, it is a place where information originates from, or terminates completely.

For industrial purposes, what should be placed inside the DMZ is all of the applications and servers that have TCP connections out to the IT network. In practice, this can be a whole range of things; so when designing IT/OT network integration architectures, one must generally see customers deploy an intermediate system to aggregate OT data which needs to be shared with the enterprise. The most common aggregator is one of the many process historians, or one of the many variants of OPC server.

The modern industrial DMZ acts as a zone and conduit system protecting physical processes, separating networks according to their different purposes, requirements and risks. The best practice for implementing an IT/OT DMZ is to put two firewalls around the DMZ, one between the OT network and DMZ network, the other between the DMZ and the IT network. This practice supports the assumption that two firewalls reduce the likelihood that single firewall software vulnerability will open an attack pathway straight in to control system networks. In addition, all three networks must be on separate domains with their own authentication systems and no sharing of domain credentials.

Network segmentation has traditionally been accomplished by using multiple routers. Firewalls should be used to create DMZs to protect the control network. Multiple DMZs could also be created for separate functionalities and access privileges, such as peer connections, the data historian, the Inter Control Center Communications Protocol (ICCP) server in SCADA systems, the security servers, replicated servers, and development servers. All connections to the Control System LAN should be routed through the firewall, with no connections circumventing it. Network administrators need to keep an accurate network diagram of their control system LAN and its connections to other protected subnets, IDMZs, the corporate network, and the outside.

Generally it is observed that, most sites deploy only a single firewall with three ports- one connected to IT, one connected to OT and one to the DMZ network- meaning single vulnerabilities are again a concern. More fundamentally, modern attacks don't exploit software vulnerabilities, modern attacks exploit permissions. From an attack perspective, exploiting vulnerabilities involves a lot of work and code writing – unless of course someone else has already done the work and released an attack tool to the public. Exploiting permissions, on the other hand, can be as easy as stealing the firewall password or stealing a password on the IT network that the OT network then trusts or allows an attacker to go into a historian or other system in the DMZ right through the firewall.

9.6 COMPUTER/SERVER SECURITY

When consider the computer/server security, one of the important task is the patch management. Keeping regular IT systems and applications with the latest firmware, software, and patch management is a daunting task especially in the industrial zone of the ICS network.

Uptime requirements for the critical ICS computer systems do not allow them to reboot after updates. For those critical systems that are allowed to be altered, a different strategy to protect them is better and generally employed. For systems that can be updated and patch especially in the field operation zone, a readily available, update and convenient patching solution should be provided. Typical examples are windows server updates services and system center configuration manger of Microsoft.

End node protection is another important task in the computer system. This defensive control in the form of end node protection software and can be installed locally with remote administrative capability. Some of the generally available end node protection software are:
- Host Based Firewalls,
- Anti Malware software, and
- Application Whitelisting software.

9.7 APPLICATION SECURITY

Application security is the process of making applications more secure by finding, fixing, and enhancing the security of apps. Much of this happens during the development phase, but it includes tools and methods to protect apps once they are deployed. This is becoming more important as hackers increasingly target applications with their attacks. SCADA systems need to have a strategy that supports not only management of knowledge and training but security knowledge that include policy, standards, design and attack patterns, threat models, code samples, reference architecture and security framework. Application security can help organizations protect all kinds of applications

used by internal and external stakeholders including customers, business partners and employees. Vulnerabilities in SCADA applications include the following key components.
- Input Validation Vulnerabilities
- Software Tampering
- Authorisation and Authorisation Vulnerabilities
- Configuration Vulnerabilities
- Session Management Vulnerabilities
- Parameter management Vulnerabilities

SCADA software security must be viewed holistically. It is achieved through the combination of effective people, process and technology with none of these three on their own capable of fully replacing the other two entities. This also means that just like software quality in general, software security requires that we focus on security throughout the application's life cycle.

Input Validation Vulnerabilities

An input validation attack is any malicious action against a computer system that involves manually entering strange information into a normal user input field. Input validation attacks take place when an attacker purposefully enters information into a system or application with the intentions to break the system's functionality. The best form of defense against these attacks is to test for input validation prior to deploying an application. A few common types of input validation attacks include:
- Buffer Overflow- This is a type of attack that sends too much information for a system to process, causing a computer or network to stop responding. A buffer overflow might also cause excess information to take up memory that was not intended for it, sometimes even overwriting memory.
- Canonicalization attacks- A canonicalization attack takes place when someone changes a file directory path that has digital permissions to access parts of a computer in order to allow access to malicious parties that use this unauthorized entry to steal sensitive information or make unapproved changes.
- XSS attacks Also called cross-site scripting, these attacks involve placing a malicious link in an innocuous place, like a forum, which contains most of a valid URL with a dangerous script embedded. An unsuspecting visitor might trust the site they are on and not worry that a comment or entry on the site contains a virus.

- SQL injection attacks- SQL injection attacks involve taking a public URL and adding SQL code to the end to try to gain access to sensitive information. An attacker might enter code into a field commanding a computer to do something like copy all of the contents of a database to the hacker, authenticate malicious information, reveal hidden entries in a database or delete information without consent.

Software Tampering

Modifying the application code before or while running the application is known as software tampering. Software tampering can lead to override or bypass the security or protective controls. Modifying the unauthorised application's runtime behaviour to perform unauthorised actions, exploitation via binary patching, code substitution,, software licence cracking, trojenisation of applications, etc are the common attacks associated with software tampering vulnerabilities.

Authorization & Authentication:

Authorization and Authentication deal with appropriate mechanisms to enforce access control on protected resources in the system. Authentication vulnerabilities include failure to properly check the authentication of the user or bypassing the authentication system altogether. Login bypassing, Fixed parameter manipulation, Brute force and dictionary attacks, Coockie replay and Pass-the-hash attacks are generally identified as authentication vulnerabilities.

Authorization is the concept that follows access to resources only to those who are permitted to use them. It comes after the successful authentication. Authorization flaws could result in either horizontal or vertical privilege escalation. The usage of strong protocols to validate the identity of a user or components. Further, issues such as the possibility or potential for authentication attacks such as brute-force or dictionary based guessing attacks. Elevation of privileges, disclosure of confidential data, data tampering are some of the authorization vulnerabilities.

Configuration Vulnerabilities

This will consider all issues surrounding the security of configuration information and deployment. It is very crucial in the security of an application. Usually the systems and applications will run with a default configuration as it has been described in the manual. This helps attackers to guess the passwords, bypass login pages and finding setup vulnerabilities. Hence configuration especially the configuration of Firewalls must be done as per the requirement of the security policy of the organization.

User and Session Management Vulnerabilities

This concerns how a user's account and session is managed within the application. The quality of session identifiers and the mechanism for maintaining sessions are some of the considerations here. Similarly, user management issues such as user provisioning and de-provisioning, password management and policies are also covered as part of this category. By mismanaging a session handling, an attacker can guess or reuse a session key and take over the session and the identity of a legitimate user. Session management and session replay are the common session management vulnerabilities associated with session management.

Parameter Management Vulnerabilities

The manipulation of parameters exchanged between a client and the server inorder to modify application data, such as user credentials and permissions, price and quantity of products, etc. Cookie manipulation, form field manipulation and query string manipulation are common parameter manipulation vulnerabilities.

Security requirements in the Software Development Cycle

Developing security architecture and engineered approach to the problem are generally recommended because current technology is not enough to prevent cyber attacks. Developing requirements for control systems with security features and use of simulation models based on a framework could improve the definition of requirements and reveal problems early in the software development cycle.

Auditing and Logging:

This concerns with how information is logged for debugging and auditing purposes. The security of the logging mechanism itself, the need and presence of an audit trail and information disclosure through log files are all important aspects.

Compliance to standards for software development

Software development for control systems can be improved by following documents such as NIST published guidelines SCADA Security, Configuration, Guidelines, general assessment methods and tools for SCADA vulnerabilities and Holzman's rules

Data Protection in Storage & Transit:
This includes handling of sensitive information such as social security numbers, user credentials or credit card information. It is also covers the quality of cryptographic primitives being used, required / minimum key lengths, entropy and usage of industry standards and best practices.

9.8 DEVICE SECURITY

Devices such as routers, firewalls and even network hosts including servers and workstations must be assessed as part of the security testing process. certain high level security vulnerabilities usually found on many network devices can also create many problems, hence one must ensure that HTTP and Telnet interfaces to the routers, switches and firewalls are properly configured and not with a blank, default or easy to guess passwords. If a malicious insider or other attacker gains access to the network devices, he can own the network and can lockout administrative access, setup backdoor user accounts, reconfigure ports, and even bring down the entire network.

When HTTP, FTP and TELNET are enabled in network devices with the help of free tools and a few minutes of time one can sniff the network and capture login credentials as they are sending clear texts. In case of wireless devices one must watch out for unauthorised Access Points (APs) and wireless clients that are attached to the network, else chances for social engineering and there are chances of connecting in to the malicious network and systems.

Mobile computing is convenient for personal, business and hacking. if secured mobile devices are not properly connected to the enterprise networks which represents thousands of unprotected islands as the phones, tablets and laptops which running numerous operating system platforms with a number of applications and infinite number of risks are associated with mobile computers.

Physical attack
- Manipulation of the mechanical and electrical part of an ICS device is always feasible if the attacker gets physical access to the device which is not a matter of IT security. However being able to manipulate the programmed functionality of an ICS device, such an attack can be simultaneously applied to a large number of devices by a single attacker. A manipulated device can also be used by an attacker as a platform to compromise other parts of the system. A device that is compromised by a physical attack might have privileged access to other components and its segments, and there for act as a backend to attack farther devices.

- As a counter measure against physical security problems, ensure that APs,. Antennas and other wireless and network infrastructure equipment are located away in secure closets, ceilings or places which are difficult to access physically. Terminate the APs outside any firewall or any network perimeter security devices wherever possible. placing unsecured wireless equipment inside your secured network, it can negate any benefits which can be obtained from the perimeter security devices such as the firewall.

Device hardening

- One of the areas of the device hardening is disabling unnecessary and unused options and features on ICS devices if the ICS devices do not provide the ability to disable unnecessary and unused options place these behind an industrial firewall and blocking the corresponding service port. Industrial firewalls are available with CISCO, TOFFINO, ROCKWELL, etc.
- Another method for ICS device hardening is restricting physical access to the device. It can be done by both administratively disabling the unused communication ports and physically block those ports from being connected to with block out devices. Keep the ICS devices in an enclosure that can be safely locked is another option. As the availability is more important than integrity and confidentiality, the device hardening of ICS presently have to confirm redundant power supplies, redundant communication port/paths, redundant I/O and redundant computing and controls.

Device patching

- Make sure that the ICS devices are installed with latest firmware and software releases on a consistent basis and within a reasonable time and updates with new releases ensure that the software, firmware, patches and manuals are from reputable reliable sources or from OEMs. also check that the ICS vendors offers cryptographically signed firmware versions for their devices this feature prevents installing and using tampered firmware.
- New firmware images, OS and patches to ICS should be tested in a testing and or developing environment to make sure that the new revisions works with the existing setup approximately before deploying to the production network this indeed save lots of headache and down time.

9.9 MODERN APPROACH TO IT/OT INTEGRATION

Modern advice describes and recommends that one side of the IT/OT DMZ be protected with unidirectional gateways to replicate OT systems to the IT network. The defining feature of a unidirectional gateway is that it is hardware-enforced: a combination of hardware and software that physically moves information in one direction only – meaning no messages whatsoever (including attacks) can enter the protected OT network from external sources, thus fulfilling the mission and purpose of implementing an IT/OT DMZ. The software element of unidirectional gateways replicates industrial servers and applications in two common scenarios with a Modern IT/OT DMZ:

- either replicating a historian or OPC server insider the OT network with a unidirectional gateway to the DMZ network whereby the IT network accesses the replicated server inside the DMZ through a firewall,
- or, replicating the historian or OPC server which sits inside the DMZ with a unidirectional gateway to a replica server sitting on the IT network for corporate use. In both scenarios, the IT replica of the DMZ OPC or historian server is still the focus for IT/OT data exchange – it is now thoroughly protected in its original operational state.

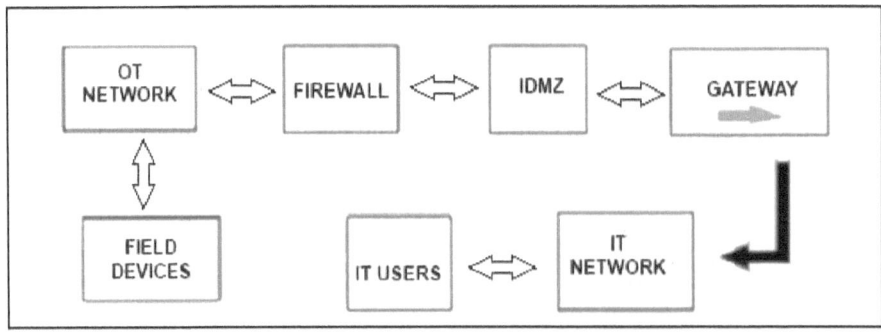

▲ **Figure 9.4:** Modern IT/OT DMZ Scenario No.1

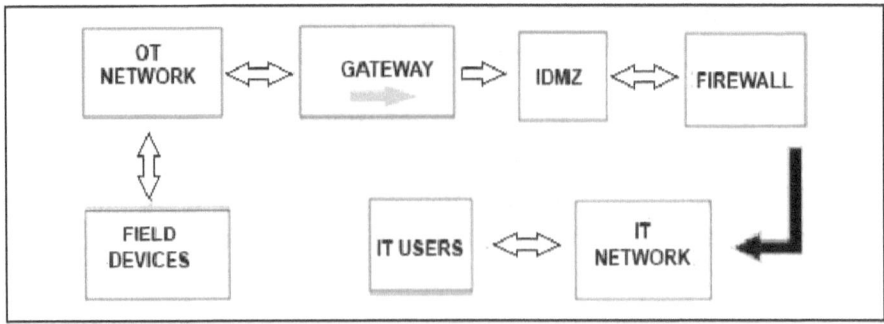

▲ **Figure 9.5:** Modern IT/OT DMZ Scenario No.2

Enterprise Zone

This is in fact the ERP and administrative zone which deals with the complete information regarding the functionality of enterprise, its planning, operations, etc. All IT data protection strategies are expected to be implemented at this zone. A network security engineer in an ICS has to be aware and ensure that

- the devices, such as firewalls and IPS are properly placed on the network and configured.
- should be aware how the attackers perform part scans and exploit vulnerabilities.
- while designing the IDMZ, the security engineers must be aware of the internet connections, remote access capabilities, layered defenses, and placement of ho the network.
- should be aware of ports which are unprotected.

If any attacker, compromises any one of the systems in the network, exploiting vulnerabilities due to the lack of proper security implementation, then

- the attacker can launch a Denial of Service(DoS) whichever can take down the internet connection, or entire network.
- with the use of a network analyser can steal confidential information of emails and files send over the network.
- a hacker can set up a backdoor access to the network,

Hence while integrating IT and OT, a prompt testing must be carried out between IDMZ and internal network of the enterprise. Obtain permissions from the predetermined and remotely connected other network for vulnerabilities on other systems that can affect the enterprise network mainly due to open parts, lack of firewalls and improper configured routers.

9.10 INCREASING RESILIENCY BY SEGMENTING THE OT NETWORK

Industrial Control Systems (ICS) and Supervisory Control and Data Acquisition (SCADA) systems are being increasingly targeted by attackers involved in terrorism and cyber warfare. Besides factories, these systems run critical infrastructure like power plants, water treatment systems, and traffic control systems. At stake is not just financial loss and brand reputation, but these attacks can result in national security threats and even death.

Based on our experience when evaluating ICS and SCADA systems, we almost always find security weaknesses that can be easily taken advantage of.

These weaknesses allow the ICS components to be manipulated and controlled beyond its intended parameter. The root of the problem can be trace to:
- Patching: Unable to install the latest patch because it is not approved by Original Equipment Manufacturer (OEM)
- No or weak authentication: Any device that can get connected to the system can change value and configuration of PLCs easily or bypass its authentication.
- Backdoors: Use of undocumented and unrestricted access to critical functions
- Buffer overflow: Poorly written software that allows restrictions to be bypassed or functions to be manipulated
- No encryption: Data and information can be deciphered easily for manipulation

The previous best-practice for ICS systems was to keep them segregated (air-gapped) from IT networks to prevent outside interference. However, we are seeing ICS networks increasingly being connected to the IT network. This convergence event is exposing ICS systems to threats and vulnerabilities that was previously protected through isolation.

SUMMARY

It can seem a difficult task to keep track of all the ICS security threats that are out there, and the new ones that just keep emerging. Whether the media is creating a culture of fear out of being online and placing trust in leaving our information out for all to see, or whether the threats that wait in the dark corners of the Internet are truly serious and can happen to anyone, the best thing one can all do is to be prepared. There is no way to be completely sure that a system is impenetrable by cyber security threat. The responsibility of the security expert is to ensure that our systems are as secure as much as possible. This chapter give a modern aspects of ICS security as per the *Defense-in-Depth architecture describing the five level security viz.* Physical Security, Network Security, Computer Security, Application Security and Device Security.

CHAPTER 10

SMART GRID CONCEPTS AND APPLICATIONS

10.1 INTRODUCTION

Once the electrical distribution grids became a reality, the demand for devices to measure the electric power consumption accurately, to distribute, price, and monitor their service were also arisen. The path from the first tentative devices used to measure power consumption, to the Smart meters using bidirectional metering technology which can turn appliances on and off according to demand and off-peak electricity prices, has been a long one. Many obstacles had to overcome in order to obtain accurate information about the way the grid behaves, and some of the obstacles to the earliest attempts to devise technologies for monitoring electrical distribution about one hundred years ago are amazingly analogous to the problems facing Smart Grid technologies today.

In 1882, in Pearl Street system at Manhattan, installed by Thomas Alva Edison, the pull of an electromagnet against a carefully-adjusted closed or opened contact spring turn on either a red lamp with voltage rise or a blue lamp with voltage drop. This indication to the operator to turn a hand wheel to control the strength of the electromagnetic field in the generators in order to match the output of the generators to the load. To measure the electricity consumed, Edison devised a meter consisting of two electrodes in an electrolyte. As current passes through the meter, the current causes to transfer the metal of the electrodes. The customer's consumption can be calculated by weighing the two electrodes.

The first known electric meter was patented in 1872 by Samuel Gardiner where an electromagnet starts and stops a clock. This provided information only on the duration of the flow of the current, but not the amount. In 1883, Hermann Aron patented a recording meter which showed the energy used on a series of clock dials. Edward Weston's indicating meter of 1886, which set high standards for precision, was not intended to measure consumption, but rather to measure current.

The recording wattmeter introduced by Elihu Thomson in 1889, immediately became a very popular metering technology which allowed the utilities to measure the amount of electricity supplied to a customer with better accuracy. However the road to accuracy was long. Braking magnets in the meters were sometimes weakened by the power surges and lightning storms which resulted, the meters to run fast. This is analogous to modern consumer complaints about fast-running of smart meters. Wherease older meters tended to run slow under overload conditions.

In the years prior to utilities being able to disconnect customer devices at peak times and reconnect them during periods of low demand, problems of load management sometimes took care of themselves in a somewhat immediate and non-negotiable manner. Transmission lines would simply burn out if demand exceeded the capacity of the line.

As electricity demands on grids increased through the late 20th century, utilities searched for ways of managing peak loads. The capital costs of building generating capacity to handle these peaks led utilities to find ways to study their demand periods, price them accordingly, and prompt the customers to switch consumption from peak to non-peak periods. The goal of matching consumption to generation, required meters which could measure the time of day of the consumption, in addition to the cumulative consumption. Automatic meter reading devices introduced in the 1970s were the beginning of meters which provided information back to the utility, a basic requirement of any Smart Grid system.

All the above mentioned technologies, together with the other developments in the power sector within this century, pave necessary foundations for building a safer, more efficient, and more reliable electricity distribution network. These technological innovations in the power grid will eventually became the Smart Grid.

10.2 SMART GRID DEFINITION AND DEVELOPMENT

A Smart Grid(SG) is an electricity distribution network that can monitor electricity flowing within itself and, based on this self-awareness, it get adjusted to changing conditions. It does this by automatically reconfiguring the network and/or exerting a level of control over connected demand and generation. In fact it is a perfect power system which meets the following goals.
1. smart, self-sensing, secure, self-correcting and self-healing,

2. sustain the failure of individual components without interrupting the service,
3. capable to focus on regional, specific area needs,
4. capable to meet consumer needs at a reasonable cost with minimum resource utilization,
5. have minimal environmental impact, and
6. enhance quality of life and improve economic productivity.

The development of the perfect power system is based on the effective integration of devices such as smart appliances, storage devices, Building Management Systems (BMS) and Distributed Energy Resources (DER)). This is followed by the integration of an efficient Distribution Management System (DMS), set up a fully integrated power system which is referred as Smart Grid. Further the Smart Grid concept is not a static concept rather it is a dynamic one. With the advancement of technology, it will continue to evolve and the type, configuration, and implementation of the new technologies will decide the new shape and characteristics but may introduce new concerns of security. At present the key characteristics of Smart Grid are mentioned below.

1. Enabling cognizant participation by customers,
2. Enabling new products, service, and markets,
3. Accommodating all generation and storage options,
4. Providing the power quality as required economy,
5. Optimizing asset utilization and operating efficiency,
6. Addressing disturbances through automation, prevention, suppression, and
7. Operating resiliently against various hazards.

Though Smart Grid has not been developed into a full-fledged technology rather embryonic, will definitely be the biggest technological revolution since the invention of the Internet and will play an important role in tomorrow's societies.

10.3 CONSTRAINTS OF OLD GRID

Traditionally, power has been generated by a few numbers of large power stations. It is then transported at very high voltages to areas of demand on a transmission system and delivered at lower voltages to end users via a distribution network. Flows on the distribution network are generally one way, only with power taken off the high voltage transmission network and supplied to the end consumer. Transmission systems have always been relatively smart but, at the distribution network side, things dumb down rapidly.

Focusing the distribution, grids are built on a build and connect principle. When new house is built, the network is estimated for the likely maximum anticipated load by applying tried and trusted design principles. The infrastructure is then built, homes connected.

However, the global warming-induced pressure for countries to move towards low-carbon economies is now challenging this traditional build and connects culture. While discussing the business drivers for Smart Grids, it is observed that, electricity distributors are now being forced to move from a build and connect to a connect and manage culture. Distribution networks can no longer be left to their own devices but need to be actively managed, along with the consumers they serve, to cope with rapidly changing demands on the network.

10.4 BENEFITS OF SMART GRID

The Smart Grid has seven key benefits for consumers, business, utilities and the Nation and they are,

1. *Self-Healing:* A Smart Grid automatically detects and responds to routine problems and quickly recovers if they occur, minimizing downtime and financial loss,
2. *Motivates and Includes the Consumer:* A Smart Grid gives all consumers visibility into real-time pricing, and affords them the opportunity to choose the volume of consumption and price that best suits their needs,
3. *Resists Attack:* A Smart Grid has security built-in features so that any attack vectors will be appropriately detected and prevented,
4. *Provides Power Quality:* A Smart Grid is expected to provide quality power free of sags, spikes, disturbances and interruptions. It is suitable for use by the data centers, computers, electronics and robots which may decides the future economy,
5. *Accommodates All Generation and Storage Options:* A Smart Grid enables *plug-and-play* interconnection to multiple and distributed sources of power and storage such as wind, solar, battery storage, etc.
6. *Enables Energy Markets:* By providing consistently dependable operation, a Smart Grid supports energy markets that encourage both investment and innovation,

7. *Optimizes Assets and Operates Efficiently:* A Smart Grid enables to build lesser new infrastructure, transmit more power through existing systems, and thereby reduce the cost of the O & M of the Smart Grid, and
8. *Reducing CO_2 emission:* Smart Grid has the potential to reduce carbon dioxide emissions through the integration of distributed renewable energy sources, energy storage, and electric vehicles.

10.5 SCADA: HEART AND BRAIN OF SMART GRID

The Supervisory Control and Data Acquisition (SCADA) technology is the heart and brain of Smart Grid for monitoring, decision making and control. Line sensors and other connected equipment on a Smart Grid provide a stream of data back to a central control room where the information is analyzed and decisions are automatically made and executed, regulating voltage levels, optimizing efficiency, routing and generation. The SCADA system in the control room is able to make these automated decisions in real-time, by running algorithms based on the data it receives and orchestrate adjustments to optimize voltages and self-heal any disruption issues. Today utilities are reinventing the advantages of SCADA of every component in the power sector from generation, to transmission, to distribution, to the customers. Across the world, deregulation and restructuring have created a healthy competition and modernization of aging infrastructure, and most of the utilities are embracing the Smart Grid concepts with great enthusiasm.

Over time, SCADA processing moved to Personal Computer (PC) and laptop environments where information was shared and processed on LAN networks. Modern system networks use always-on connections and are able to communicate via the Internet. While the next generation of SCADA will rely heavily on cloud computing to have the processing power necessary to analyze continuous streams of data from thousands of sources simultaneously from larger grids. The more sources of information the SCADA system is providing, the more effective it is, at making optimal decisions.

Most modern electric utilities have SCADA systems to manage their grids, but all systems are not created equally. Many current systems operate primarily at the substation level. Hence upgrades are necessary, to calculate data from line sensors that are becoming more common to expand control to a more granular level. Significant upgrades will be required not only to analyze the data provided by these sensors, but also to process the complex algorithms that effectively launch responses to the continuous stream of data they provide. By definition, a true Smart Grid is capable of distribution automation, which is mainly made up of two components viz. volt/VAR optimization and self-healing. The accompanying SCADA systems are called upon to make the necessary decisions that carry out these tasks. Presently huge investments are being made for implementing Advanced Metering Infrastructure (AMI) not only for industrial and commercial customers, but also for domestic customers by many utilities because of the various advantages what AMI offer to the utilities. Transmission utilities are installing Wide Area Measurement Systems (WAMS) with Phasor Measurement Units (PMU), Phasor Data Concentrators (PDC) and associated data integration with the existing SCADA systems to make the transmission systems more secure and avoid breakdown.

10.6 STAKEHOLDERS OF SMART GRID

In fact Smart Grid is one way or other, a reality in every utility, sometimes with partial capabilities. A full-fledged Smart Grid requires the involvement of a number of stakeholders who has to work in harmony with a vision to improve the grid and be prepared for meeting the challenges ahead. The various stakeholders who can transform the Smart Grid concept into a reality is listed below.

1. Appliance and consumer electronics providers,
2. Equipment manufacturers and automation vendors,
3. Residential, commercial, and industrial consumers,
4. Electric transportation industry Stakeholders,
5. Investor Owned Utilities (IOU),
6. Rural Electric Association (REA),
7. Electricity and financial market traders,
8. Independent Power Producers (IPP),
9. Information and communication Technologies (ICT) Infrastructure and Service Providers,
10. Information Technology (IT) application developers and integrators,
11. Power equipment manufacturers,
12. Professional societies and users groups,
13. R and D organizations and academia,
14. Relevant Government Agencies,
15. Renewable Power Producers,
16. Retail Service Providers,
17. Standards and specifications development organizations (SDOs),
18. State and local regulators,

19. Testing and Certification Authorities,
20. Transmission Operators and Independent System Operators, and
21. Venture Capital and Government Roles in Smart Grid.

10.7 TRANSFORMING TO SMART GRID

The basic philosophy behind the Smart Grid concept is to develop a perfect balance among reliability, availability, efficiency, and cost, in fact optimize the traditional grid. Grid optimization help
1. to improve the utilization of current infrastructure and defer investments in new generation, transmission, and distribution facilities,
2. to reduce the overall cost of delivering power to end users,
3. to improve the reliability of power grid, and
4. to reduce resource usage and emissions of greenhouse gases and other pollutants.

Smart Grid also conceives the following concepts,
1. *Demand Response and Demand Side Management:* Incorporate automated mechanisms that enable utility customers to reduce electricity consumption during periods of peak demand and help utilities to manage their power loads,
2. *Advanced utility control:* Monitor essential components, enable rapid diagnosis and precise solutions,
3. *Energy storage:* Add technology to store electrical energy to meet demand when the need is greatest,
4. *Plug-in hybrid electric vehicle (PHEV):* Smart charging and vehicle-to-grid technologies: enable electric and plug-in hybrid vehicles to communicate with the power grid and store or feed electricity back to the grid during periods of high demand,
5. *Advanced metering:* Collect usage data and provide energy providers and customers with this information via two-way communications,
6. *Home Area Networks (HAN):* Allow communication between digital devices and major appliances so customers can respond to price signals sent from the utility, and
7. *Renewable energy and distributed generation sources:* Reduce greenhouses gas emissions; provide energy independence, and lower electricity costs.

10.8 SMART GRID INTEROPERABILITY FRAMEWORK

Inorder to transform into Smart Grid, a set of interoperable technical solutions are essential and they are:
1. Asset Optimization,
2. Consumer optimization,
3. Distribution Optimization,
4. Smart Meter and Communications,
5. Transmission Optimization, and
6. Human Resource Development and Optimization

In fact these set of solutions forms an integrated grid optimization and is a perfect balance between reliability, availability, efficiency and cost. Grid optimization ranges from generation to transmission and from distribution to end-user. All of these rely on secured two-way communication, intelligent sensing and Advanced Metering Infrastructure (AMI).

10.8.1 Asset Optimization

With the deployment of Intelligent Electronic Devices (IEDs), Smart Grid improves the features and functions of the legacy grid equipment. These IEDs will not only provide more detailed information but also bi-directional communications. The new multifunction electronic meters, intelligent relays and control units are able to exchange information with the utilities central control systems. Asset optimization also include the proactive maintenance by frequent monitoring the equipment conditions, focused maintenance, and asset optimization which in turn reduces the outage and risk of failure.

10.8.2 Customer Side Optimization

Customers or Prosumers are the end users of the power grid. Hence consumer side or end use optimization depends on the demand optimization and load forecasting. Load forecasting can be at Substation transformer level or Feeder level or Section level. Demand management works to reduce electricity consumption in homes, offices, and factories by frequently monitoring and actively managing consumption of electrical appliances. It consists of demand-response programs, smart meters and variable electricity pricing, smart buildings with smart appliances, and energy dashboards. Combined, these innovations allow utility companies and consumers to manage and respond to the variances in electricity demand more effectively.

Currently most retail consumers are charged a flat price for electricity regardless of the *Time of Day* or actual demand. So they have little incentive to lower their energy use to reduce their energy bill, while helping utility companies meet the demand. Smart meters give customers the opportunity to choose variable-rate pricing based on the time of day. By seeing the real cost of energy, consumers can respond accordingly by shifting their energy consumption from high-price to low-price periods. This process can have the joint benefit of reducing costs for typical consumers while lowering demand peaks for utility companies.

10.8.3 Distribution Optimization

Automation of distribution assets and integration of renewable energy are the main functions of distribution optimization. The distribution automation brings added advantage to the utility as it opens the pathway to the Smart Grid implementation and all the associated functionalities. The integration of the renewable energy brings the additional advantages like reduction in peak demand, deferred capital investment, satisfied customers, etc. A proper Distribution optimization will ensure less energy waste, with higher profit margins by reducing the losses in the distribution systems.

Plug-in Hybrid Electric Vehicles (PHEVs) are powered by conventional or alternative fuels and by electrical energy stored in a battery and Electric vehicles (EV) use energy stored in the battery exclusively. Integration of these vehicles brings a demand of adequate charging facilities and infrastructure to cope with the growth. This is also a concern of distribution optimization.

The distribution optimization improves the utilization of current infrastructure and defer investments distribution facilities. It also reduces the overall cost of delivering power to end users, improve the reliability of power grid, and reduce resource usage and emissions of greenhouse gases and other pollutants.

10.8.4 Customer Automation And Communications

Whatever automation is implemented, ultimately it must be useful to the end user. Smart meter which is the heart of the customer automation accomplishes this requirement. A smart meter with well-defined functionalities realizes the benefits of automation to both customers and utility. The basic challenge lies on the secured communication infrastructure required for the interaction between customer and utility. Network connectivity, customer enablement, demand and communication optimization, etc. are some of the features of smart meter. At the same time it is very important to note that giving everyone a smart meter won't deliver a Smart Grid rather it is an essential component of a Smart Grid. Retrofitting millions of customer premises would be a financial and logistical challenge.

10.8.5 Transmission Optimization

Transmission of electric power is achieved through substations and transmission lines, which are spread over a wide area. Hence wide area monitoring, remote control and protection activities have to be optimized to achieve reliability and efficiency. Today with the advent of phasor measurement techniques and PMU, state estimation and related applications produces more reliable and faster data processing. Another area of optimization in transmission is the integration of large renewable sources to the grid. The uncertainty in renewable energy generation is a challenging issue in transmission optimization.

10.8.6 Human Resource Development And Optimization

As the present Grid is aged and requires innovation, fifty percent of the present work force with outdated skill-set are expected to retire worldwide and the remaining should be appropriately trained. New technology on the network will require new skills within the power utility workforce. But the changes that a Smart Grid will bring to a distributor are more fundamental than just the technology. Becoming a distribution system operator in the age of the Smart Grid will require a new way of operating. Other aspects of the business such as investment planning and asset maintenance are likely to change dramatically as well. Distribution companies are likely to face a period of significant up-skilling and recruitment of engineers, technicians and other professionals who are acquainted with the modern technologies.

For Smart Grids to work, consumers need to become much more engaged in the electricity industry. A certain faction of the consumers is expected to become prosumers. They need to be aware of their own energy use and its associated cost. The Smart Grid can provide the awareness and information as well as the means to better manage consumption, so that they can save money through more efficient energy use.

Mobile crew management system enables a utility to allot maintenance jobs to the crews in the field on real-time basis. In the traditional model, crew attending any work in the field will always return to their base station and then they will be dispatched to the next work. Thus their productivity is reduced.

With mobile crew management systems, the work will get allotted to the crew with required skills, tools and spare parts and nearest to the work location. In that scenario, from one work location to another work location they can quickly move, increasing their productivity multiple times. Also information on the type of fault is made available on their mobile to support trouble shooting. Good mobile crew management applications should have real time scheduling engine.

10.9 SMART GRID ROAD MAP

Smart Grids comprises of electricity transmission and distribution networks, Distributed Energy Resources, Distributed Generation, storage and end-users. While many regions have already initiated to *smarten* their electricity system, all regions will require significant additional investment and planning to achieve a smarter grid. Smart Grids are an evolving set of technologies that will be deployed at different rates in a variety of settings around the world, depending on local commercial attractiveness, compatibility with existing technologies, regulatory developments and investment frameworks. The Smart Grid Road map aims to,

1. increase understanding among a range of stakeholders of the nature, function, costs and benefits of Smart Grids,
2. identify the most important actions required to develop Smart Grid technologies and policies that help to attain global energy and climate goals, and
3. develop pathways to follow and milestones to target based on regional conditions.

10.10 STANDARDS AND INTEROPERABILITY

If the Smart Grid is to succeed it is vital that all the diverse devices and systems that make up the grid should be capable to inter-operate. The current lack of interoperability is one of the primary factors preventing the deployment of new systems by the utilities. It also prevent the development of controllable system interfaces and devices by interested third parties. Standards not only eliminate incompatibilities between vendor products but also simplify testing, implementation, system security, maintenance, and development of monitoring and operational practices necessary to manage grid resources. Standards also help to,

1. establish technical specifications to achieve the required level of compatibility, interchangeability, or commonality to obtain interoperability between system components and systems,
2. establish a common perception and understanding of system operations,
3. provide data compatibility and eliminate data incompatibilities, and
4. facilitate collaboration between units within an organization and between organizations, which facilitates system interoperability.

10.11 SMART DISTRIBUTION

The needs and requirements of the customer are also changing faster, and the utility has to provide innovative solutions with quality power to the consumers. Today vendors are coming up with various solutions of Distribution Automation (DA). The distribution network facilitates network communication of DA system and connects DA main stations, substations, and terminals with a backbone network having low latency and high reliability, access network flexibility and auto-adaptation. The following are the major building blocks of smart distribution.

1. Demand side energy management,
2. Integration of renewable energy Sources,
3. Distributed energy resource and energy storage,
    - Distributed Generation
    - Energy Storage
4. Advanced Metering Infrastructure (AMI),
5. Energy Efficient Smart Homes(EESHs),
6. EVs and PHEVs in a Smart Grid, and
7. Multiple Distributed Microgrids.

10.11.1 Demand Side Energy Management

Concerns about climatic changes around the world, continuous hike in the energy prices, and the reliable availability of energy supply stresses, for a new energy supply chain with a modified demand supply philosophy. The present energy consumption pattern shows an increase, but with unpredictable fluctuations. This results in the decrease of the efficiency of the conventional plants, which in turn affects the generation capacity and the grid requirements. International agreement between the Nations demands, for reduction of the CO_2 emission, by increasing the renewable energy generation significantly. This raises new concerns and challenges of grid integration of RES to provide a reliable, dependable, and affordable electricity supply. Hence new technologies such as integration of the large scale RESs, bulk storage of energy, etc. are required to achieve an efficient use of generated electricity.

The present demand-supply philosophy based on centralized generation, in which electricity is generated in a few large central power plants and is transported top-down and one-way to the consumers. Consumers switch on the devices and generation side has to cater the demand, as the supply chain of the consumer side is static. But to increase the efficiency of the existing power plants and introducing the renewable sources, the consumer side of the supply chain should be more flexible. This means the consumption should be adjusted to generation. The best way to achieve this is the introduction of the Smart Grid technology and the static consumers should be active participants in the energy supply chain. One of the main goals of the Smart Grid is to accommodate the renewable generation and keep up with the growing electricity demand while maintaining the stability and reliability. The new technologies should be, capable of adjusting the consumption to the generation, should be capable of compensating the decrease in flexibility of the generation with a more flexible electrical grid and more flexible consumers. A Smart Grid is a system that monitors and manages all components of the grid, hence the smartening of the grid and updating the electricity supply chain got an enthusiastic momentum which triggered numerous initiatives worldwide. However to realize the Smart Grid, a number of technical, economical, political, and ethical challenges have to be addressed. As far the technical challenge of monitoring and managing the grid is concerned, the best solution is the integration of ICT, the key enabler.

10.11.2 Renewable Energy Sources (Res)

Wind, solar, and water are the alternate source of energy to the conventional fossil fuel central power generating stations.These renewable energy sources are rapidly becoming the mainstream power sources and obtained wide acceptance. But it also opened a set of new challenges to be addressed such as integration of RES, variability and un-predictive nature of wind, Photovoltaic generation etc.

Distributed Energy Resources (DER) are small-scale power generation sources located close to the load to provide an alternative to or an enhancement of the traditional electric power grid. DER is a faster, less expensive option to the construction of large, central power plants and high-voltage transmission lines. They offer consumers the potential for lower cost, higher service reliability, high power quality, increased energy efficiency, and energy independence. The use of renewable distributed energy generation technologies and green power such as wind, photovoltaic, geothermal, biomass, or hydroelectric power can also provide a significant environmental benefit.

10.11.3 Distributed Generation And Technologies

Distributed Generation (DG) are electric generation units (typically in the range of 3 kW to 50 MW) located within the electric distribution system at or near the end user. They are parallel to the electric utility or stand-alone units. DG have been available for many years, and are known by different names such as generators, back-up generators, or on-site power systems. Within the electric industry the terms that have been used include distributed generation, Distributed Power (DP), and DES.

As described the Distributed Generation(DG) is the use of small-scale power generation technologies located close to the load being served, capable of lowering costs, improving reliability, reducing emissions and expanding energy options. This encompasses a range of technologies including fuel cells, micro turbines, reciprocating engines, load reduction, and other energy management technologies. DG also involves power electronic interfaces, as well as bi-directional communications and control devices for efficient dispatch and operation of single generating units, multiple system packages, and aggregated blocks of power.

The primary fuel for many distributed generation systems was natural gas for many years, but in future it is expected that hydrogen will play an important role. Renewable energy technologies such as solar electricity, biomass power, and wind energy are presently popular. Some of the primary areas where the DG find useful are described below.

1. *Premium power:* Reduced frequency variations, voltage transients, surges, dips, or other disruptions,
2. *Back-up power:* Used in the event of an outage, as a back-up to the electric grid,
3. *Peak shaving:* The use of DER during times when electric use and demand charges are high,
4. *Low-cost energy:* The use of DER as base load or primary power that is less expensive to produce locally than it is to purchase from the electric utility, and
5. *Combined heat and power (cogeneration):* Increases the efficiency of on-site power generation by using the waste heat for existing thermal process.

Generally, DG provides the consumer with greater reliability, adequate power quality, and the possibility to participate in competitive electric power markets. DG also have the potential to mitigate overloaded transmission lines, control price fluctuations, strengthen energy security, and provide greater stability to the electricity grid.

Thousands of grid-connected and off-grid DER systems are being used today in many places, including office complexes, national park facilities, and employee townships. The DGs are an integral part of most of the distribution systems. They are owned and operated by both utilities and customers.

10.11.4 Grid Energy Storage

Grid energy storage or large-scale energy storage is a collection of methods used to store electrical energy on a large scale within an electrical power grid. Electrical energy is stored during times when production exceeds consumption, and returned to the grid when production falls below consumption. The widespread use of renewable energy sources such as PV and wind power brings energy storage an essential need of today.

Energy storage systems provide a wide array of technological approaches to manage the power supply in order to create a more resilient energy infrastructure and bring cost savings to utilities and consumers. To help understand the diverse approaches currently being deployed around the world, and divided them into six main categories and they are,
1. Electrical,
2. Chemical,
3. Electrochemical,
4. Thermal,
5. Mechanical(Kinetic Energy based), and
6. Mechanical (Potential Energy based).

Presently the largest form of grid energy storage is dammed hydroelectricity, with both conventional hydroelectric generations. In fact energy storage gives an option for the electrical system to mitigate many problems created by the dynamic power systems to maintain load generation balance and to ensure power quality.

10.11.5 Advanced Metering Infrastructure (Ami)

AMI are comprised of state-of-the-art electronic/digital hardware and software, which combine interval data measurement with continuously available remote communications. These systems enable measurement of detailed, time-based information and frequent collection and transmittal of such information to various parties. AMI typically refers to the full measurement and collection system that includes meters at the customer site, communication networks between the customer and a service provider, such as an electric, gas, or water utility, and data reception and management systems that make the information available to the service provider.

10.11.6 Energy Efficient Smart Homes (EESHS)

Energy management is a broader term, which applies differently in different scenarios, but Energy Efficient Smart Homes (EESHs) are concerned about the one which is related with energy saving in homes, government organizations or business. In this scenario the process of monitoring, controlling and conserving energy in an organization or building may be termed as energy management. In Smart Grid where the consumers can generate local energy from different distributed generation units and there is a plenty of space for different pricing schemes, and the need for energy management programs.

Demand Side Management (DSM) can aid to reduce emissions, provide reliable power supply and lower the energy cost. Current grid has DSM programs for consumers like, commercial buildings and industrial plants, however it does not have any such scheme for domestic consumers due to the reasons of lack of effective communication, efficient automation tools and sensors. Secondly implementation costs of various Demand Response programs are higher when compared with its impact. However in Smart Grid, smart loads, low cost sensors, smart meters and the information and communication technology open a window for domestic energy management programs. Different techniques for energy management in Smart Grid are:

1. Optimization of Residential Energy Management (OREM),
2. Home Energy Management (HEM),
3. Decision Support Tool (DST), and
4. Automatic Home Energy Consumption Scheduler (AHECS).

10.11.7 E-Mobility

Electric Vehicles (EVs) and Plug-in Hybrid Electric Vehicles (PHEVs) are being developed around the world, and lots of research work are going on to optimize engine and battery operations for efficient operation, during discharge and recharge. If the recharging of EVs and PHEVs are during the off-peak hours and the number of vehicles are limited, the grid will not be greatly affected by the use of these vehicles. But as the number of vehicles increases, the recharging of these vehicles will not be confined to off-peak hours, rather the time of recharging will be as per the convenient of the end users and not the time which utilities prefer. It is expected that the number of EVs and PHEVs are poised to grow significantly in the coming years and their dependencies in the power grid will have a major impact in integration of these vehicles.

10.12 MULTIPLE DISTRIBUTED MICROGRIDS

Microgrid is a localized coordinated cluster of DER units where the loads connected to normally operate in synchronous with the traditional centralized electrical grid which is referred to as a Macrogrid, but can disconnect and function autonomously with the help of an appropriately designed intelligence for management and control. By this way, Microgrid can be connected to the main power grid or be cut off and run in an isolation mode. It paves a way to effectively integrate various sources of Distributed Generation (DG), especially Renewable Energy Sources (RES). It also provides a good solution for supplying power in case of an emergency, by having the ability to change between islanded mode and grid-connected mode. On the other hand, control and protection are big challenges in this type of network configuration, which is generally treated as a hierarchical control.

10.12.1 Characteristics of A Microgrid

While describing the Microgrid, it is better to have an awareness of the main characteristics of a Microgrid which are briefly described below.
1. It has well defined geographical and electrical boundaries,
2. It has connectivity to the utility main grid and capable of exchanging power as and when required,
3. It has the capability of automatically islanding from grid connected mode and continues to function, in case of the failure of main utility grid,
4. It has DERs, including renewables; fossil fuel based generators such as diesel gensets, and/or integrated energy storage
5. It has the capability to perform real-time switching among various generation and load sources to balance supply and demand quickly, manage power exchanges and participate in demand response, and
6. It has information exchanges capability between the Microgrid and the main utility grid (Macrogrid).

10.12.2 Microgrid Components

The major components of a Microgrid are: Distributed generation resources, controller, and communication infrastructure and storage system. These are described in the following sections. The Figure 10.1 illustrates major components in a Microgrid.

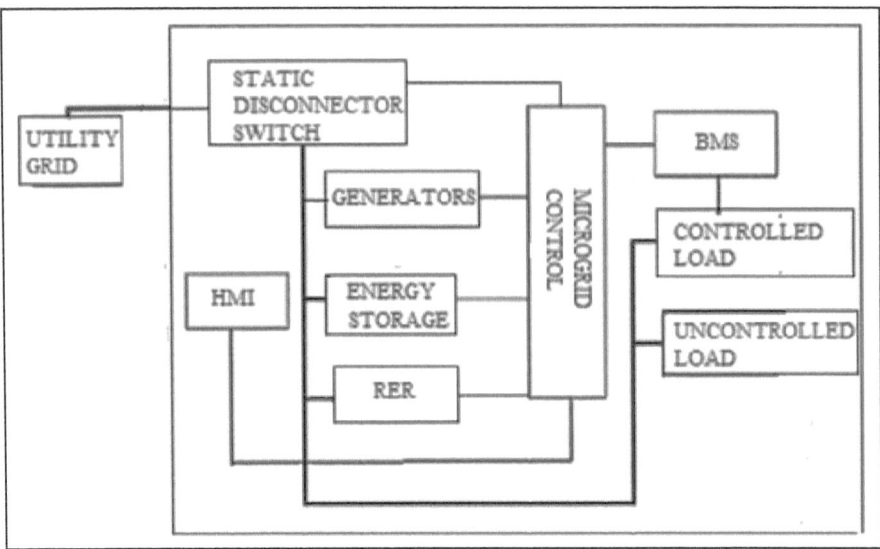

▲ **Figure 10.1:** Components of a Microgrid

Microgrid Master Controller: The basic functions of the master controller are,
1. Match the load with Generation- in both island or grid-connected mode,
2. Optimize integration, dispatching and control of DER and loads,
3. Ensures combination of Distributed Energy Resources (DERs) and hence improves economics,
4. Maintains reliability and manages frequency and voltage,
5. Real-time response and fault protection,
6. Connect and disconnect from the grid, and
7. Predictive and forecasting analysis.

Fast and Secure Communication: The Basic Functions of the Fast and Secure Communication are,
1. optimize operations and control of DERs. and loads,
2. connect to buildings via EMS,
3. continued monitoring and trends of Microgrid component's health,

4. smart metering to obtain load and DERs. profile,
5. electricity pricing and demand response capabilities, and
6. continuous communication to utilities and energy markets.

Advanced Metering Infrastructure: Advanced Metering Infrastructure (AMI) is the collective term to describe the whole infrastructure from Smart Meter to bi-directional communication network to control center equipment and all the applications that enable the gathering and transfer of energy usage information in near real-time. AMI makes bi-directional communications possible with customers and is the backbone of Microgrids. This infrastructure built extensible and persistent information network in Microgrids.

Distributed Energy Resources (DER): Distributed Energy Resources (DERs) provides an on-site generation for electricity consumers. Microgrid not only allow DERs. to operate as dispatchable assets to provide power to the Microgrid when appropriate, but also to aggregate several DERs. and energy storage devices to optimize their use on behalf of the electricity users. When one DER goes down, the Microgrid can reduce energy usage in other locations and utilize other DERs. to adjust for the electricity generation capacity loss.

Energy Storage in Microgrids: To counteract power imbalances between demand and supply, energy storage unit functions as energy buffer or backup. The main Energy storage technologies used in Microgrids are,
1. Different type of batteries,
2. Flywheel energy storage (FES), and
3. Superconducting magnetic energy storage (SMES).

10.12.3 Benefits of Microgrid

Microgrid paves a simple way to integrate Wind, Solar, and Micro-hydroelectricity, etc. to the main grid with lesser complexities of control. It is capable of operating in grid-connected and stand-alone modes and handling the transitions between these two modes. It helps to supply power in case of an emergency and power shortage during power interruption in the main grid. Microgrid has two modes of operation and they are grid-connected mode and island mode.

In the grid-connected mode, ancillary services can be provided by trading activity of Microgrid and the main grid. But in the islanded mode of operation instead, the real and reactive power generated within the Microgrid, including

the help of energy storage system should be in balance with the demand of local loads. Islanding mode can be intentional or unintentional. In intentional islanding, situations can occur such as,
1. for scheduled maintenance,
2. when degraded power quality of the host grid can endanger Microgrid operation, and
3. for economical operations.

On the other hand, unintentional islanding can occur due to faults and other unscheduled events that are unknown to the Microgrid. Both of those situations can be dealt actively by using Microgrid. All the points mentioned above and by means of modifying energy flow through Microgrid components, Microgrid allows and facilitates integration of renewable energy generation such as photovoltaic, wind and fuel cell generations without requiring re-design of the distribution system. Modern optimization methods can also be incorporated into the Microgrid energy management system to improve the efficiency, economics, and resiliency. Microgrids are of mainly three types and they are,
1. Microgrid owned by private, industrial or commercial organizations which requires better control, reliability and economy and the new breed of Microgrids which are coming up in University and college campus as a part of innovation,
2. Microgrid owned by Government organizations such as military base, Airports and Seaports, which are strategically important, etc. and which require resilient energy reliability and city and municipal Microgrids which are coming up as a part of Smart City vision, and
3. Microgrids owned by electric utilities inorder to serve certain elite consumers with special, localized and high power quality electricity supply. Usually utilities operates in synchronous with the main grid of the utility but prudently exploiting the possibility of exploiting the local DER.

10.12.4 Multiple Microgrids And Hierarchical Control

Regarding control system architecture of the Microgrid or any control problem there are two different approaches and they are centralized and decentralized. A fully centralized control relies on big data exchange between involving units and based on these information suitable decisions are made at a single point.

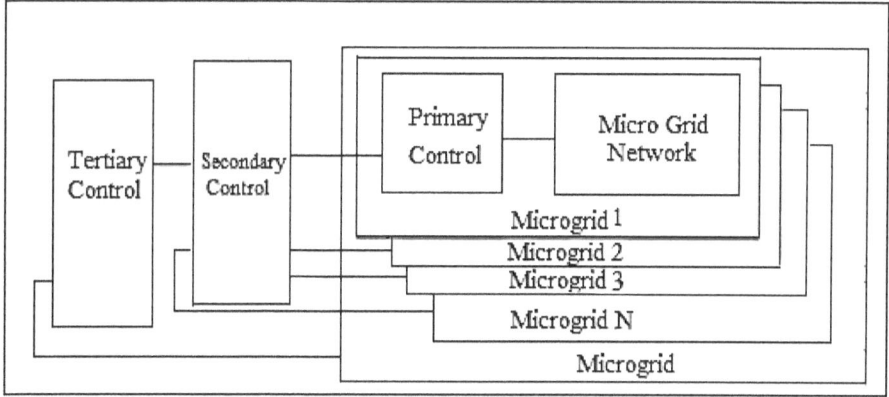

▲ **Figure 10.2:** Hierarchical control scheme for Microgrid

This will create big problems in implementation since interconnected power systems usually cover extended geographic and involves enormous number of units. Hence this fully centralized control is not currently considered as viable solution. On another hand, in a fully decentralized control each unit is controlled by its local controller without viewing the system and situation holistically. It implies that the fully decentralized control is also irrelevant in this context due to strong coupling between the operations of various units in the system. A compromise between those two extreme control schemes can be achieved by means of a hierarchical control scheme consisting of three control levels viz. primary, secondary, and tertiary which is shown in Figure 10.2.

Primary control: The primary control is designed to satisfy the following requirements,
1. To stabilize the voltage and frequency,
2. To offer plug and play capability for DERs. and properly share the active and reactive power among them, preferably, without any communication links, and
3. To mitigate circulating currents that can cause over-current phenomenon in the power electronic devices.

The primary control provides the set points for a lower controller which is the voltage and current control loops of DERs. These inner control loops are commonly referred to as zero-level control.

Secondary control: Secondary control has lesser sampling rate, when compared to sampling rate of primary control, typically seconds to minutes sampling time which justifies the decoupled dynamics of the primary and the secondary control loops and facilitates their individual designs. Set point of primary control is given by secondary control in which as a centralized controller, it restores the Microgrid voltage and frequency and compensates for the deviations caused by the primary control. The secondary control can also be designed to satisfy the power quality requirements, such as voltage balancing at critical buses.

Tertiary control: Tertiary control is the final and the slowest control level which consider economical concerns in the optimal operation of the Microgrid. The sampling rate is from minutes to hours, and manages the power flow between Microgrid and main grid. This level involves the prediction of weather, grid tariff, and load requirements in the coming hours or day to design an economic generation dispatch plan.

10.12.5 DC Microgrids

A DC Microgrid is a localized grouping of generation, storage, and loads in its native DC form preferably at voltage levels less than 1.5 kV. It may have the capability of connecting to the central grid whenever required. Usually DC Microgrids are expected to providing sufficient and continuous energy to a significant portion of internal load demand. It also possesses independent controls, and intentional islanding takes place with minimal service interruption. It is estimated that more than 80% of the electric power consumed in commercial buildings AC-DC conversion and incur losses. Most of the modern equipment such as computers smart phones, LED lighting, data centers, electric vehicles, variable speed drives, etc. use DC power. Usually the AC power is converted at device level and incur loss of power. Further the PV produces the DC which is converted to AC for distribution and again converted back to DC at device level. In general the load profile has shifted from AC to more and more DC loads over the past several decades and seems continuing down that path. The driving factor in efficiency for many of today's devices is the conversion from AC power to DC power. It is not possible to get a 100% efficient conversion from one form to the other. The energy is lost to either heat, magnetism, or shunted to ground depending on the conversion technique. As loads go more and more to DC the amount of these losses is increasing at the same rate. These conversions are happening all over buildings and they are often not considered when designing a system. In such situations DC Microgrids found useful.

However there are drawbacks associated with the use of a DC Microgrid, and most of these problems can be overcome as these systems become more familiar to contractors and designers. As familiarity is achieved, the system cost should continue to decrease leading to even more familiarity and so on. As with all new technologies it takes time to be adopted and the full potential of the system to be utilized. In future, DC Microgrid use may increase with the world becoming energy conscious and it will be interesting to watch where this technology can go, in the next decade.

10.12.6 Microgrid Challenges

Microgrids, involves integration of many DER units, but a number of operational challenges are to be addressed in the design of control and protection systems in order to ensure the required levels of reliability. Further the potential benefits of Distributed Generation units must be fully harnessed. Some of these challenges arise from the inappropriate assumptions typically applied to conventional distribution systems, while others are due to the result of stability issues formerly observed only at a transmission system level. The most relevant challenges in Microgrid protection and control include,

1. *Bi-directional power flows:* The presence of DG units in the network at low voltage levels can cause reverse power flows, that may lead to complications in protection coordination, undesirable power flow patterns, fault current distribution, and voltage control,
2. *Stability issues:* Integration of control system of DG units may create local oscillations, requiring, a thorough small-disturbance stability analysis. Moreover, transition activities between the grid-connected and stand-alone modes of operation in a Microgrid can create, transient stability. Recent studies have shown that Direct-Current (DC) Microgrid interface can result in significantly simpler control structure, more energy efficient distribution and higher current carrying capacity for the same line ratings,
3. *Modeling:* Many characteristic in traditional scheme such as prevalence of three-phase balanced conditions, primarily inductive transmission lines, and constant-power loads are not necessarily valid for Microgrids, and consequently models need to be revised,
4. *Low inertia:* The Microgrid shows low-inertia characteristic that are different to Bulk Electric Systems (BES) where, high number of synchronous generators ensures a relatively large inertia. This is very dominant when operated with power electronic-interfaced DG units. Further the low inertia in the system can lead to severe frequency deviations in stand-alone operation if, a proper control mechanism is not implemented, and

5. *Uncertainty:* The operation of Microgrid contains very much of uncertainty. The economical and reliable operation of Microgrids relies on isolated Microgrid load profile and weather forecast. These are two main components that make this coordination more challenging. The critical demand-supply balance and typically higher component failure rates demands solving this strongly coupled problem over an extended limit. This uncertainty is much higher than those in BES, due to the reduced number of loads and highly correlated variations of available energy resources.

10.13 SMART TRANSMISSION

Smart transmission involves the installation of PMU and developing wide area measurement systems and related applications. Synchronized PMU measurement is referred to as a power system's *health meter*, which samples voltage and current many times a second at a given location on the network, and giving the utility a near real time view of the power system's behavior. In other way, the traditional SCADA systems provide an X-ray view of the network, while PMUs provide an MRI scan of the wide area power transmission grid.

10.14 SG BIG DATA ANALYTICS AND CHALLENGES

Current and future deployment of Smart Grid devices is generating huge data, which require effective data integration and management systems to integrate and analyze. These systems should also provide demand response and Demand Side Management (DSM) features that can be easily integrated into a customer's HAN so they can take advantage of the various DR or DSM rate structures offered by the electric utility.

The utilities can also use this data to understand critical aspects of the grid such as load factors, usage patterns, equipment condition, and voltage levels, among others. This information can then be analyzed and integrated into customer programs and/or operational and maintenance programs to identify and alert operators to any emerging failures.

Big data is a term applied to data sets whose size or type is beyond the ability of traditional relational databases to capture, manage, and process the data with low-latency. Big data analytics is the use of advanced analytic techniques which operates against very large, diverse data sets that include different types such as structured/unstructured and streaming/batch and different sizes from terabytes(10^{12}) to zettabytes(10^{21}). And it has one or more of the characteristics such as high volume, high velocity, or high variety. Big data comes from sensors, devices, video/audio, networks, log files, transactional applications, web, and social media - much of it generated in real time and in a large scale.

Analyzing big data allows analysts, researchers, and engineers to make better and faster decisions using data that was previously inaccessible or unusable. Using advanced analytics techniques such as text analytics, machine learning, predictive analytics, data mining, statistics, and natural language processing, utilities can analyze, previously untapped data sources independent or together with their existing enterprise data to gain new insights resulting in significantly better and faster decisions. Today Big data is one of the most overused buzz-phrases in the Smart Grid industry today.

10.14.1 AMI Data Analytics-Present Status

Smart metering, or automated metering infrastructure (AMI) brings big data and it's natural for utilities to start their big data efforts in business operations. Over the past few years, it is seen that AMI-centered projects emerge from pilot development, whether in the form of embedded IT systems at large investor-owned utilities, or as a cloud-based Software-as-a-Service (SaaS) models. But the move from probably once-in-a-lifetime, proof-of-concept deployments to broader adoption of analytics is still a work in progress for the utility industry. A host of reports out recently prove that many utilities haven't even captured AMI data's to make sure their core meter-to-cash and communications systems are working at optimum value.

10.15 COMMUNICATION - THE KEY ENABLER

A key enabler for the Smart Grid is the availability of bi-directional data communications. The combination of distributed intelligence and telecommunication technologies enable utilities to improve the stability of the grid, facilitate the integration of renewable energy sources, and give consumers the means to monitor and manage their own energy consumption. Thus, the design and implementation of a modern, reliable telecommunications infrastructure is a fundamental requirement for making the grid smarter.

The architecture of monitoring, control, coordination and communications of the grid as it exists today predates the huge advances made in the last few decades in the fields of computing, networking and telecommunications. The last few decades have seen the development of the Internet and networked communications and the large-scale deployment of wide-area broadband wireless networking technology.

The communications infrastructure for monitoring and control of the power grid today is a patchwork of protocols and systems, often proprietary and mutually-incompatible, including leased lines, fixed RF networks, microwave links and fiber. Furthermore, the legacy paradigm employs a purpose-built communications network for each application system. Most utilities to use separate communications networks for SCADA, advanced metering and mobile workforce access.

Automated meter reading systems for collecting meter data are still predominantly based on one-way low-bandwidth communications technologies, whether based on fixed RF networks or drive-by reading. These one-way communications technologies need to be updated to support the low-latency bidirectional traffic flows to enable applications such as Demand Response (DR). While utilities frequently have fiber to the substations from the control center, it is often not consistently leveraged across applications. The poor communications infrastructure underlying the monitoring of the grid leads to inadequate situational awareness for utility operators who are often blind to disturbances in neighboring control areas and often within their own control areas. There is no unified broadband communications infrastructure today that can simultaneously serve the needs of distribution automation, mobile workforce automation, advanced metering, SCADA and other applications.

Broadband communications reinforces the Smart Grid by introducing many of the newer capabilities, such as demand response and remote disconnects, require real-time bi-directional communications capabilities down to the meter end-points.

10.15.1 Smart Grid Communication Requirements

The major communication requirements of Smart Grid are in the following areas. Substation and distribution automation, Advanced Metering Infrastructure, Wide Area Monitoring System, and E-Mobility. As the communication technology requirements are different for different, each area should be treated individually.

Power System Automation: Communication technologies employed in power system automation especially in transmission and distribution requires high sensitivity, high reliability and low latency. Required bandwidth is approximately 10kbps and the latency requirement is less than 200ms as it deals with high voltage lines and large currents which involve many life-critical operations.

Wide Area Monitoring System (WAMS): As WAMS is a mission critical application, it requires high reliability and very low latency are crucial. Required bandwidth is approximately 1500kbps and the latency requirement is preferably less than 100ms.

E-Mobility: In future with the development and wide spread of Electric Vehicles(EVs) bi-directional communication may become a need for the charging of EVs, but not a mission critical rather needed for billing and meeting the energy demand and supply. Latency requirement is relatively low and can be upto 5minutes. The bandwidth requirement is approximately 50kBPS.

*Advanced Metering Infrastructure (AMI):*The main communication requirements are the following functionalities viz. Automatic Meter Reading, Demand Response and Distributed Energy Resources (DER). In addition, Smart Appliance also needs connectivity. Among these DER requires low latency and high reliability. For all other functionalities require medium reliability and latency. Bandwidth requirement is approximately 100 kbps is satisfactory except Backhaul network which require comparatively higher bandwidth

10.15.2 Networking the Smart Grid

Distribution automation applications uses advanced sensors, that generate larger volumes of data which require real-time high-speed communications links to control centers. Utility field workforces employing bandwidth-intensive productivity applications such as mobile Geographical Information System need a communications network that is high in capacity and supports seamless mobility for standards-based wireless devices. Smart Grid today can be envisaged as a computer network or data network with various control equipment which allows computers to exchange data and control commands. The computer networks, and networked computing devices exchange data using a data link. The connections between nodes are established using either cable media or wireless media. The best-known computer network is the Internet and today Smart Grid rely on Internet considerably.

Network devices are components used to connect computers or other electronic devices together so that they can share files or resources like printers or fax machines. Devices used to setup a LAN are the most common type of network devices used by the public. A LAN requires a hub, switch, router, firewalls, etc.

10.15.3 Smart Communication for Smart Grid

The Smart Grid differs from the present electrical grid in such a way that it is intelligent and self-healing to the maximum extent possible to provide a reliable, efficient, and clean energy distribution. It achieves these features and functions mostly with the implementation of modern IEDs. These robust IEDs are capable of providing relevant information to ensure quality and reliable power supply with bi-directional communications facility between utility and consumer. Within the utility, these new multifunction electronic meters, intelligent relays and control units exchange information with utility's central control systems. These enhanced communication features help both grid operators and consumers to make more informed and more be aware of. Further extend the functionality of the monitoring systems not only to collect information but these gathered information appropriately used to automatically control the grid and minimize the need for human intervention. In fact, these developments mainly depend on the new smart communication technologies and new levels of interconnectivity built into the electrical grid, and cooperation among various organizations and big data analysis.

10.15.4 Communication Technologies For Smart Grid

The communications technologies available today are divided into three categories:
1. *Guided or landlines*: including analogue subscriber lines, digital subscriber lines, coaxial cables, and fiber optical cables.
2. *Unguided or Wireless*: including cell phone communications systems, Wi-Fi, etc.
3. *Utility owned radio:* including trunk mobile dispatching channels and meshed meter networks.

Power Line Carrier encompasses traditional power line carriers between substations and the new technology of broadband-over-power line at the distribution segments. These different communication technologies have distinct advantages and disadvantages, and selection is based on the site requirements. A detailed description regarding the various communication technologies relevant to Smart Grid are given in the succeeding chapters.

10.16 SMART GRID SECURITY-THE FOREMOST CHALLENGE

While Smart Grid has the capability to provide improved efficiencies in monitoring the power consumption, managing power distribution to serve peak power demands, and increasing efficiency of power delivery, it has opened the way for information security breaches and other type of security issues. Smart Grid potential threats ranges from meter manipulation to nation sponsored catastrophic attacks on critical subsystems that could sabotage regional or national power grids. Hence it is most essential that security measures are put in place to ensure that the Smart Grid should not succumb to those threats, rather Smart Grid should be smart enough to safeguard this critical infrastructure. A brief elaboration of the Smart Grid security aspects are given below.

10.16.1 Cyber-Physical Security

The power sector is the critical infrastructure of a nation and other sectors depend directly and indirectly on the power sector. Cyber-physical security is protection of the assets, both hardware and software, from natural and manmade disasters due to intended and unintended activities. Since physical assets are associated with the cyber space of a utility, cyber-physical security completely defines the security paradigm of a utility. This dependency of the physical assets on the cyber-assets and vice versa, has prompted the utilities to inject resiliency and robustness into their grids.

10.16.2 System Security

Although a number of countries have recently experienced large-scale blackouts, their electricity systems are regarded as generally secure, according to industry-specific indices that measure the number and duration of outages. Smart Grid technologies can maintain and improve system security in the case of challenges such as ageing infrastructure, rising demand, variable generation and electric vehicle deployment. By using sensor technology across the electricity system, Smart Grids can monitor and anticipate system faults before they happen and take corrective action. If outages occur, Smart Grids can reduce the spread of the outages and respond more quickly through automated equipment.

10.16.3 Cyber-Security

Smart Grids can improve electricity system reliability and efficiency, but the use of new ICTs can also introduce vulnerabilities that jeopardize reliability, including the potential for cyber-attacks. Smart Grids provide electricity demand from the centralized and distributed generation stations to the customers through transmission and distribution systems. The Smart Grid is operated, controlled and monitored using Information and Communications

Technologies. These technologies enable energy companies to seamlessly control the power demand and allow for an efficient and reliable power delivery at reduced cost. Via digital bi-directional communications between consumers and electric power companies, the Smart Grid system provides the most efficient electric network operations based on the received consumer's information. Security remains to be one of the most important issues in Smart Grid systems which can cause the danger and inconvenience to residents and companies alike, if the grid falls under attack. Three main security objectives which must be incorporated in the Smart Grid system are,
1. Availability of uninterrupted power supply according to user requirements,
2. Integrity of communicated information, and
3. Confidentiality of user's data.

Comprehending the importance and necessity of Smart Grid Cyber security, it is currently being addressed by several international collaborative organizations. The following concerns are reported by them.
1. the existing aspects of the electricity system regulatory environment may make it difficult to ensure the cyber-security of Smart Grid systems,
2. utilities are focusing on regulatory compliance instead of comprehensive security,
3. consumers are not adequately informed about the benefits, costs and risks associated with Smart Grid systems,
4. insufficient security features are being built into certainPower System Automation and Smart Grid systems,
5. the electricity industry does not have an effective mechanism for sharing information on cyber-security, and
6. the electricity industry does not have metrics for evaluating cyber-security.

These findings confirm that cyber-security must be considered as part of a larger Smart Grid deployment strategy.

10.16.4 Smart Grid Vulnerabilities

Smart Grid network introduces enhancements and improved capabilities to the conventional power network with the incorporation of ICT but making it more complex and vulnerable to different types of attacks. These vulnerabilities might allow attackers to access the network, break the confidentiality and integrity of the transmitted data, and make the service unavailable. The following vulnerabilities are the most serious in Smart Grids,

1. *Customer security:* Smart meters autonomously collect massive amounts of data and transport it to the utility company, consumer, and service providers. This data includes private information of the consumer that might be used to infer consumer's activities, devices being used, and times when the home is vacant,
2. *Greater number of intelligent devices:* A Smart Grid has several intelligent devices that are involved in managing both the electricity supply and network demand. These intelligent devices may act as attack entry points into the network. Moreover, the massiveness of the Smart Grid network makes network monitoring and management extremely difficult,
3. *Physical security:* Unlike the traditional power system, Smart Grid network includes many components and most of them are out of the utility's premises. This fact increases the number of insecure physical locations and makes them vulnerable to physical access,
4. *The lifetime of power systems:* Since power systems coexist with the relatively short lived IT systems, it is inevitable that outdated equipments are still in service. This equipment might act as weak security points and might very well be incompatible with the current power system devices,
5. *Implicit trust between traditional power devices:* Device-to-device communication in control systems is vulnerable to data spoofing where the state of one device affects the actions of another. For instance, a device sending a false state makes other devices behave in an unwanted way,
6. *Different Team's backgrounds:* Inefficient and unorganized communication between teams might cause a lot of bad decisions leading to much vulnerability,
7. *IP and commercial off-the-shelf hardware and software:* Using Internet Protocol (IP) standards in Smart Grids offer a big advantage as it provides compatibility between the various components. However, devices using IP are inherently vulnerable to many IP-based network attacks such as IP spoofing, Tear Drop, Denial of Service, and others, and
8. *More stakeholders:* Having many stakeholders might give raise to a very dangerous kind of attack known as insider attacks.

10.16.5 Security Protocols

Like Information Technology, Industrial Control System application layer protocols leveraging IP communications have relied on other application layer protocols or Internet layer protocols to provide network security such as authentication, and encryption rather than designing security into the protocol itself. In Modbus, no authentication required for any request, whether it be a monitor, or control request. If one has the access to an IP network in which a Modbus device is linked, packets can be sent to the device, and as long as the format of the packets are exact Modbus packets, the device will react to the packets. However in Distribution Network Protocol 3(DNP3) has an option of authentication wherein a DNP3 device can be programmed to only respond to requests coming from whitelisted IP addresses. But, this should not be considered a strong form of authentication since IP addresses are easily spoofed. Hence generally automated power system network employ a separate security protocol, such as TLS/SSL, IPsec, SSH, or a custom secure protocol.

Today Internet Protocol Security (IPsec) is almost obligatory and not an optional specification of the IPv6 protocol suite and provides security at the Internet layer of the protocol stack. IPsec is a protocol suite for securing IP communications by authenticating and encrypting each packet of a communication session. One advantage of using IPsec instead of TLS/SSL is that applications do not have to support IPsec, making it a good candidate for use in legacy systems.

Another advantage is that IPsec supports UDP and multicast traffic as well as TCP, while TLS/SSL only supports TCP traffic. However, a disadvantage in using IPsec is since the security is applied at the host kernel level, any application running on the host can use the IPsec tunnel to send and receive traffic, including malicious applications. Other disadvantages could include additional traffic latency, and the effects of which are system and application-dependent.Further any vulnerability that exists at the IP layer in the remote network could be passed to the corporate network across the IPSec tunnel. Making sure that this doesn't happen is possible, but results in higher support costs. By contrast, SSL VPNs run at higher network layers so they don't expose network drives to remote workers, shielding the network against vulnerabilities like worms.

Yet another option for securing communications between control system applications is the Secure Shell (SSH) protocol. SSH is a cryptographic application layer protocol used for secure data exchange, remote shell services and command execution, and other secure network services. Although SSH is an application layer protocol, existing applications can take advantage of its secure communication capabilities without modification through the use of SSH extensible port forwarding and secure tunneling features. The compilation, management, latency, and capability aspects of SSH and IPsec need to be compared when deciding which is best suited for use in the power system SCADA system network.

10.17 OTHER SG IMPLEMENTATION CHALLENGES AND CONSIDERATIONS

A major change usually entails considerable challenges, and the Smart Grid is also no exception. The following are identified as the major barriers to achieving Smart Grid.

Financial Resources: The business case for a self-healing grid is good, particularly if it includes societal benefits. But regulators will require extensive proof before authorizing major investments based heavily on societal benefits.

Government Support: The industry may not have the financial capacity to fund new technologies without the aid of government programs to provide incentives to invest. The utility industry is capital-intensive, but it might have undergone hard times in the marketplace and some utilities have impaired financial ratings.

Compatible Equipment: Some older equipment must be replaced as it cannot be retrofitted to be compatible with Smart Grid technologies. This may present a problem for utilities and regulators since keeping equipment beyond its depreciated life minimizes the capital cost to consumers.

Speed of Technology Development: The solar cells, the basement fuel cell, and the chimney wind generator were predicted 50 years ago as an integral part of the home of the future. This modest historical concepts have to be accelerated.

Policy and Regulation: Utility regulatory bodies and commissions usually follow a parochial view of new construction projects. A critical circuit tie crossing state boundaries has historically met significant resistance. The project, financed by the utilities or local government agencies may not always be the one benefiting most from it. Hence unless an attractive return on Smart Grid investments is assured or encouraged, utilities will remain unenthusiastic to invest in new technologies.

Cooperation: The stakeholders are many and the challenge for these diverse utilities will be in perfect harmony and in cooperative to install critical circuit ties and freely and spontaneously exchange information for a successful implementation of Smart Grid concepts.

SUMMARY

This chapter begins with explaining the need, necessities and definition of Smart Grid and then moving to describe the benefits and road map for implementation of Smart Grid. It further elucidates how the ICT enabled Smart

Grids will enhance bi-directional communications, bi-directional power flow, overall efficiency, reliability, and reduce the costs of electricity services. A brief introduction of Microgrids also has been given. The various other challenges in implementing Smart Grid such as dealing with AMI big data, Smart Grid communication networking, etc. As the ICT innovations to the present power grid makes the it a massive efficient system but unfortunately prone to cyber-attacks. Since the Smart Grid is considered a critical infrastructure of a Nation, any vulnerabilities, it is most essential that it should be identified and addressed with utmost important and sufficient solutions must be implemented to reduce the risks to an acceptable secure level. The risk and challenges of the same also has been introduced with proposed solutions in this chapter.

CHAPTER 11

SCADA SECURITY STANDARDS

11.1 INTRODUCTION

Currently, there are quite a few OT cyber-security standards in existence, but each of these cyber security standards have a specific focus or objective and they are described below.
- ISA 99 / IEC 62443: ISA 99 standards are intended for industrial automation devices. IEC 62443 is a superset of ISA99, incorporating all the specification of ISA 99, but in addition there are ongoing works for new specifications and updates under it.
- NERC (North American Electric Reliability Corporation): This is primarily intended to electrical T&D (Transmission & Distribution) equipment. The Standards are called North American Electric Reliability Corporation-Critical Infrastructure Protection (NERC CIP). ISA 99 / IEC 62443 and NERC CIP have many common requirements, but not complementary each other.
- NIST (National Institute of Standard Technology) standards: NIST is US federal government body and though they have their own set of requirements for control system security, it endorses mainly the ISA99 standards.
- ISO 27001: This is intended for security practice to be implemented at organization level, mainly by IT department.

Whichever standard is selected, it is a fact that no foolproof cyber security protection does exist. But by making a device comply with available standards, its immune system gets strengthened and probabilities of withstanding attacks can be elevated considerability. The best way to keep the entire industrial control system secure is keeping it isolated from the external network but has many practical difficulties. This chapter briefly explain these security standards and its compliance requirements.

11.2 SELECTION OF STANDARDS

For embedded device, especially embedded device that goes into manufacturing, industrial machines, controllers used in industrial automation, ISA99/IEC62443 is most relevant standard to comply with. The following areas such as
- embedded devices used in mining trucks to control the movements,
- controller used in process automation to control a refinery plant,
- embedded device to control the air conditioning of building, and
- embedded device that control elevator/escalator movement, etc. follow the IEC 62443 security standards.

Further it is accepted that anywhere where an embedded device is not a standalone device and whose operation or mal-operation could risk or cause injury to human lives, ISA99 is applicable. Other than the power system devices and automation devices used in transmission and distribution which need to comply with NERC CIP or its equivalent, for all other embedded devices that need to be cyber secure may follow the ISA99 /IEC62443 standards. Following proper security standards and strictly complying the selected standards, the immunity of the device gets strengthened. This improves the capability of withstanding the physical-cyber attacks. The best way to secure an ICS is keeping the entire system *AirGapped* and keeps the system totally isolated from external network.

Today complying the NERC CIP is becoming a mandatory security standard for power system automation especially when implementing automation in power Transmission and Distribution (T&D) sector. Though it is a complex process to ensure NERC CIP requirements, it offers reasonably flawless physical-cyber security for automated power system. Further many nations realizing the need and necessity of complying and implementing NERC CIP standards in power system automation from the lessons learned from the BlackEnergy attack on Ukraine power grid, strictly enforcing NERC CIP compliance.

11.3 ISA 99 / IEC 62443 STANDARD

ISA/IEC-62443 is a series of standards, technical reports, and related information that define procedures for implementing electronically secure Industrial Automation and Control Systems (IACS). This guidance applies to end-users (i.e. asset owner), system integrators, security practitioners, and control systems manufacturers responsible for manufacturing, designing, implementing, or managing industrial automation and control systems. All ISA work products are now numbered using the convention ISA-62443-x-y and previous ISA99 nomenclature is maintained for continuity purposes only. Corresponding IEC documents are referenced as IEC 62443-x-y. The approved IEC and ISA versions are generally identical for all functional purposes.

Compliance Procedure

The procedure to get an embedded device compliant is very much similar to getting a device safety certified. The very first step is to have a certified system/product development process. Though it is not a must to have a mature process, it is an advantage in obtaining the IEC 62443 certification. Compliance to CMMI (Capability Maturity Model Integration), level 3 certification and above quality standards is good benchmark of a mature process.

The device needs to be designed considering the requirements as specified in ISA99 /IEC62443. The processes and design of device have to be approved by an Assessor. The Assessor is usually from an external organization that has track record of certifying compliance to ISA99. Testing the device based on set of test procedures usually carry out in an accredited lab such as Achilles Test Platform.

Organization certifying embedded devices

There are two well recognized organizations certifying embedded devices on cyber security viz. Wurldtech and ISASecure. The choice between the two certification bodies would be purely based on specific project teams or cost or comfort level or previous project experiences.

1. Wurldtech, a GE company, offers the Achilles certification for devices. There are 2 levels of certification viz. level 1 and level 2. Level 2 requires more number of tests to be passed. Wurldtech does not publicly state the requirements that need to pass to obtain the certification rather they recommend the use of Achilles Test Platform during the development stage to test the embedded device.
2. ISASecure, an association of Industrial control system users and manufacturers. ISASecure's EDSA (Embedded Device Security Assurance) certification is intended for embedded devices. ISA Secure EDSA certification comprises of,
   - certifying the processes,
   - the design of embedded device and
   - passing the conformance testing.

The ISA Secure have come out with their own specifications on all 3 above mentioned steps and publicly available on their website. It is largely based on IEC 62443 standards and also some specifications from NERC and NIST. There are 3 levels of EDSA certification with level 3 being the stringent.

One thing to be noted is both the organizations offer their own named certification- Achilles certification or ISA Secure EDA certification and do not explicitly certify complying to ISA 99/IEC 62443. The choice between the 2 certification bodies would be purely based on specific project teams or cost or comfort level or previous project experiences.

11.4 NERC CIP STANDARD FOR BES

The digital innovation of power system automation is booming and shows no signs of slowing down. Advances in Information and Communication Technology (ICT) bring new vulnerabilities which threaten the reliable functioning of the power grid that is critical to any Nation's energy future. With critical infrastructure attacks on the rise, compliance mandates seem timelier than ever. North American Electric Reliability Corporation (NERC)- Critical Infrastructure Protection (CIP) standards are made up of nearly 40 rules and almost 100 sub-requirements are today most preferred security standard in power system automation. These provisions are critical for ensuring that electric systems are prepared for mitigating the cyber threats. Understanding certain definitions such as *Critical Assets and Responsible Entities* are better to conceive the concepts well.

Critical Assets: These assets include but are not limited to: Control systems, data acquisition systems and networking equipment, as well as hardware platforms running virtual machines or virtual storage.

Responsible Entities: They are defined as reliability coordinators, balancing authorities, interchange authorities, transmission service providers, transmission owners, transmission operators, generator owners, generator operators; load servicing entities and NERC/regional entities. All responsible entities are required to adhere to standards as defined by NERC.

The set of standards the NERC CIP standards covers mainly the Electronic Security Perimeter (ESP) and the protection of critical cyber-assets, personnel and employee training, security management and disaster recovery planning which are briefly described below.

CIP-002-5.1a – Cyber-Security Cyber Security- BES Cyber System Categorization

This standard requires the security engineer of the utility to identify and categorize BES cyber systems and their associated BES cyber assets for the application of cyber security requirements commensurate with the adverse impact that loss, compromise, or misuse of those BES cyber systems could have on the reliable operation of the BES. Identification and categorization of BES cyber systems support appropriate protection against compromises that could lead to mal-operation or instability in the BES. During this time the security engineer will identify each critical asset, categorize the asset, prioritize how the asset coincides with compromise or loss and, ultimately, highlight the overall relationship or operating dependency the asset has to the facility. This is helpful when submitting to the NERC Compliance Registry (NCR), and it also aids in creating compliance monitoring objectives.

CIP-003-6-Cyber Security-Security Management Controls

This standard requires to specify consistent and sustainable security management controls that establish responsibility and accountability to protect Bulk Electric Supply (BES) cyber systems against compromise that could lead to mal-operation or instability. This necessitates consistent and sustainable security management controls be enacted by an organization to protect all identified critical cyber assets from compromise, mal-operation or instability. Cyber security policy, leadership, exceptions, information protection, access control, change control and configuration management are all included in CIP-003-6, while adherence to sub-requirements may vary by organization, criticality of assets and impact rating.

CIP-004-6 -Cyber Security-Personnel & Training

This standard requires minimizing of risk against compromise that could lead to mal-operation or instability in the BES from individuals accessing BES cyber systems by requiring an appropriate level of personnel risk assessment, training, and security awareness in support of protecting BES Cyber Systems. This necessitates that all personnel with authorized access to critical cyber assets have an adequate degree of personnel screenings and risk assessments, employee training and security awareness programs. Power utility also needs to maintain a list of credentialed access lists, including service providers and contractors. Moreover, CIP-004-6 also requires the organization to document, review and update such training and programs on an annual basis.

CIP-005-5- Cyber Security-Electronic Security Primeter(s)

This standard requires to manage electronic access to BES cyber systems by specifying a controlled Electronic Security Perimeter in support of protecting BES cyber systems against compromise that could lead to mal-operation or instability in the BES. This standard primarily focuses on the perimeter and efforts to address vulnerabilities encountered during remote access. The perimeter that houses all critical cyber assets should be protected and any and all access points be secured. Key components to this include, but are not limited to, the following: remote session encryption, multi-factor authentication, anti-malware updates, patch updates and using extensible authentication protocol (EAP) to limit access based upon roles.

CIP-006-6-Cyber Security-Physical Security of BES Cyber Systems

This standard requires managing physical access to BES cyber systems by specifying a physical security plan in support of protecting BES cyber systems against compromise that could lead to mal-operation or instability in the BES.

This standard emphasizes the physical security perimeter and tasks the responsible entity with implementing a physical security program. The goal is to address the physical security zone and create preventative controls aimed at protecting and controlling access to cyber assets based upon risk-based security zones. A physical security plan, protection of physical access control systems, protection of electronic access control systems, physical access controls, physical access monitoring, physical access logging, log retention access, and maintenance and testing are all requirements of the security program for CIP-006-6.

CIP-007-6 -Cyber Security-System Security Management

This standard requires managing system security by specifying select technical, operational, and procedural requirements in support of protecting BES cyber systems against compromise that could lead to mal-operation or instability in the BES. This requires that create, implement and maintain processes and procedures for securing systems for both critical and non-critical cyber assets. This also means documenting security measures, including records of test procedures, ports and services, security patch management and malicious software prevention.

CIP-008-5 - Cyber Security-Incident Reporting and Response Planning

This standard requires mitigation of the risk to the reliable operation of the BES as the result of a cyber security incident by specifying incident response requirements. Security incidents related to any critical cyber assets must be identified, classified, responded to and reported in a manner deemed appropriate by NERC. Utility has to create an incident response plan that should include the actions, roles and responsibilities of those involved, as well as details of how incidents should be handled and reported to governing bodies. This plan will need to be updated annually and tested for applicability.

CIP-009-6-Cyber Security-Recovery Plans for BES Cyber Systems
This standard requires that recover reliability functions performed by BES cyber systems by specifying recovery plan requirements in support of the continued stability, operability, and reliability of the BES. Utility critical cyber assets must have recovery plans that align with their energy utilizes organization and adhere to disaster recovery best practices. A recovery plan, change control, backup and restoration processes and testing or backup media are all requirements of CIP-009-6.

CIP-010-2 - Cyber Security -Configuration Change Management and Vulnerability Assessment
This standard requires preventing and detecting unauthorized changes to BES cyber systems by specifying configuration change management and vulnerability assessment requirements in support of protecting BES Cyber Systems from compromise that could lead to mal-operation or instability in the BES.

CIP-011-2-Cyber Security-Information Protection
This standard requires to prevent unauthorized access to BES cyber system Information by specifying information protection requirements in support of protecting BES cyber systems against compromise that could lead to mal-operation or instability in the BES.

CIP-014-2-Cyber Security-Physical Security
This standard requires to identify and protect transmission stations and transmission substations, and their associated primary control centers, that if rendered inoperable or damaged as a result of a physical attack, could result in instability, uncontrolled separation, or cascading within an interconnection.

11.4.1 Achieving NERC CIP Compliance

As NERC CIP is becoming mandatory in power system automation as presently no other equivalent standards are available or developed. Achieving NERC CIP compliance is a complex process, needs in-depth knowledge of ICT. To be NERC CIP compliant, bulk power supply operators must ensure that they've enacted the measures contained in all of the enforceable CIP standards. The steps below outline what an operator would need to do to ensure they're compliant with NERC's program.

Step 1: Categorization
The very first step in achieving NERC CIP compliance is categorizing an organization's assets. CIP-002 outlines the system used to determine which assets are critical. This allows bulk power suppliers to determine what risks pose the most immediate threats to their systems and so rationalize their security operations.

Step 2: Management and Training
The next steps outlined by the CIP are the management of security (CIP-003) and the training of personnel (CIP-004). CIP-003 requires the creation and maintenance of a security plan, which should outline the measures and processes for the other security measures contained in the other standards. For example, CIP-003 requires operators to create procedures for training personnel according to CIP-004, and to maintain documents pertaining to those procedures. CIP-003 recommends review of the security processes once every 15 months. CIP-004 includes requirements about keeping staff up to code, such as going through quarterly reviews of security practices.

Step 3: Creating and Managing Perimeters
In order to achieve NERC CIP compliance, operators must also implement the requirements of CIP-005 and 006.CIP-005's recommendations focus on creating electronic security perimeters, including creating access limitations for both inbound and outbound network traffic, and the use of measures such as password protection, encryption, and firewalls. CIP-006 moves security into the real world, setting out requirements for ensuring physical security, such as limiting unescorted access to the assets. Measures might include requiring someone to sign in and sign out when accessing the facility. CIP-007 provides information on managing system security. Items covered include keeping lists

of listening ports, configuring firewalls, and documenting ports, as well as patch installation and management.

Step 4: Reporting and Recovering from Incidents
CIP-008 and 009 deal with what happens after an incident occurs: how to report it and implement recovery plans. CIP-008 emphasizes the need to adhere to a reporting procedure for incidents, and requires operators to have clear response procedures in place. The plans must be tested once every 15 months. CIP-009 outlines the requirements for recovery plans, which must set forth criteria for triggering a response and the roles and responsibilities of responders, among other things.

Step 5: Changing Environments
CIP-010 addresses change management and vulnerabilities. Operators are required to develop baseline configurations for system assets and use those to monitor and implement changes to the system. They must also document software and patches that are installed on the system. Vulnerability assessments are required once every 15 months.

Step 6: Protecting Data and Physical Assets
Keeping operators' assets, both digital and physical, safe is the goal of the CIP. CIP-011 lays out standards for protecting information and the new CIP-014 addresses the need for physical security. CIP-011 includes requirements for ensuring the staffs to know how to recognize and handle sensitive system information, and for protecting such information. CIP-014 discusses the need to perform vulnerability and risk assessments for the physical operating environment.

The current set of NERC's standards which almost cover all cyber issues of a Bulk Electric System (BES) is flexible and acceptable. Presently it is observed that the power utilities move towards an active consideration of NERC CIP versions 5 & 6 system security, needs rather than just compliance but an effective implementation especially after the cyber-attack on Ukraine power sector. It is largely due to the concerns that some owner/operators are not designating their bulk electric power facilities as *cyber-assets*, leaving potential holes in power system cyber-security.

11.5 ISO/IEC 27001 STANDARD

The ISO released the **ISO/IEC 27000** family of standards in 2005 and has been making periodic updates to the various policies. Ownership of ISO 27001 is shared between the ISO and the International Electrotechnical Commission (IEC). It is part of a set of standards developed to handle information security. It is the leading international standard focused on information security, published. Both **ISO and IEC** are leading international organizations that develop international standards. ISO 27001 is developed to help organizations, of any size or any industry, to protect their information in a systematic and cost-effective way, through the adoption of an Information Security Management System (ISMS). It not only provide companies with the necessary know-how for protecting their most valuable information, but a company can also get certified against ISO 27001 and, in this way, prove to its customers and partners that it safeguards their data.

The certification helps the customers, governments, and regulatory bodies to believe that the organization is secure and trustworthy. This will enhance the company reputation and help to avoid financial damages or penalties through data breaches or security incidents. If the organization once qualified to receive a certification, it could be at risk of failing a future audit and losing the compliance designation. Receiving an ISO 27001 certification is typically a multi-year process that requires significant involvement from both internal and external stakeholders. It is not as simple as filling out a checklist and submitting it for approval. Before even considering applying for certification, one must ensure the ISMS is fully mature and covers all potential areas of technology risk.

The ISO 27001 certification process is typically broken up into three phases:
- The organization hires a certification body who then conducts a basic review of the ISMS to look for the main forms of documentation.
- The certification body performs a more in-depth audit where individual components of ISO 27001 are checked against the organization's ISMS. Evidence must be shown that policies and procedures are being followed appropriately. The lead auditor is responsible for determining whether the certification is earned or not.
- Follow-up audits are scheduled between the certification body and the organization to ensure compliance is kept in check.

The two most important **mandatory certification requirements** when implementing ISO 27001 are scoping the ISMS in which define what information needs to be protected and conducting a risk assessment and defining a risk treatment methodology in which identify the threats to the information. ISO 27001 is broken into 12 separate sections which are explained below.

- Introduction which describes what information security is and why an organization should manage risks.
- Scope which covers high-level requirements for an ISMS to apply to all types or organizations.
- Normative References which explains the relationship between ISO 27000 and 27001 standards.
- Terms and Definitions which covers the complex terminology that is used within the standard.
- Context of the Organization which explains what stakeholders should be involved in the creation and maintenance of the ISMS.
- Leadership which describes how leaders within the organization should commit to ISMS policies and procedures.
- Planning which covers an outline of how risk management should be planned across the organization.
- Support which describes how to raise awareness about information security and assign responsibilities.
- Operation which covers how risks should be managed and how documentation should be performed to meet audit standards.
- Performance Evaluation which provides guidelines on how to monitor and measure the performance of the ISMS.
- Improvement which explains how the ISMS should be continually updated and improved, especially following audits.
- Reference Control Objectives and Controls which provides an annex detailing the individual elements of an audit.

10.5.1 Audit Controls For the ISO 27001

- The documentation for ISO 27001 breaks down the best practices into 14 separate controls. Certification audits will cover controls from each one during compliance checks. Here is a brief summary of each part of the standard has been given below.
- **Information Security Policies** which covers how policies should be written in the ISMS and reviewed for compliance.
- **Organisation of Information Security** which describes what parts of an organization should be responsible for what tasks and actions.
- **Human Resource Security which** covers how employees should be informed about cyber security when starting, leaving, or changing positions.
- **Asset Management** which describes the processes involved in managing data assets and how they should be protected and secured.

- **Access Control** which provides guidance on how employee access should be limited to different types of data.
- **Cryptography** which covers best practices in encryption.
- **Physical and Environmental Security** which describes the processes for securing buildings and internal equipment.
- **Operations Security** which provides guidance on how to collect and store data securely.
- **Communications Security** which covers security of all transmissions within an organization's network.
- **System Acquisition, Development and Maintenance** which details the processes for managing systems in a secure environment.
- **Supplier Relationships** *which* covers how an organization should interact with third parties while ensuring security.
- **Information Security Incident Management** which describes the best practices for how to respond to security issues.
- **Information Security Aspects of Business Continuity Management** which covers how business disruptions and major changes should be handled.
- **Compliance** which identifies what government or industry regulations are relevant to the organization.

Organisations are also required to complete the following mandatory clauses.
- Information security policy and objectives
- Information risk treatment process
- Risk Treatment plan
- Risk assessment report
- Records of training, skills, experience and qualifications
- Monitoring and measurement results
- Internal Audit Program
- Results of internal audits
- Results of the management review
- Results of corrective actions

10.5.2 Maintaining ISO 27001 Compliance

Obtaining an initial ISO 27001 certification is only the first step to being fully compliant. Maintaining the high standards and best practices is often a challenge for organizations, as employees tend to lose their diligence after an audit has been completed. It is leadership's responsibility to maintain

ISO 27001 compliance. Given how often new employees join a company, the organization should hold quarterly training sessions so that all members understand the ISMS and how it is used and required to pass a yearly test that reinforces the fundamental goals of ISO 27001. In order to remain compliant, organizations must conduct their own ISO 27001 internal audits once every three years. Cyber security experts recommend doing it annually so as to reinforce risk management practices and look for any gaps or shortcomings.

An ISO 27001 task force should be formed with stakeholders from across the organization. This group should meet on a monthly basis to review any open issues and consider updates to the ISMS documentation. One outcome from this task force should be a compliance checklist such as mentioned below.

- **Obtain management support** for all ISO 27001 activities.
- Treat ISO 27001 compliance as an **ongoing project**.
- **Define the scope** of how ISO 27001 will apply to different parts of the organization.
- **Write and update the ISMS policy**, which outlines the cyber security strategy at a high level.
- **Define the Risk Assessment methodology** to capture how issues will be identified and handled.
- **Perform risk assessment and treatment on a regular basis** once issues have been uncovered.
- **Write a Statement of Applicability** to determine which ISO 27001 controls are applicable.
- **Write a risk treatment plan** so that all stakeholders know how threats are being mitigated. Using threat modelling can help to achieve this task.
- **Define the measurement of controls** to understand how ISO 27001 best practices are performing.
- **Implement all controls and mandatory procedures** as outlined in the ISO 27001 standard.
- **Implement training and awareness programs** for all individuals within the organization who have access to physical or digital assets.
- **Operate the ISMS** as part of the organization's everyday routine.
- **Monitor the ISMS** to understand whether it is being used effectively.
- **Run internal audits** to gauge the ongoing compliance.
- **Review audit outcomes** with management.
- **Set corrective or preventive actions** when needed.

11.6 NIST SP 800-53

NIST sets the security standards for agencies and implementers. It's structured as a set of security guidelines, designed to prevent major security issues that are making the headlines nearly every day. It is a non-regulatory agency of the U.S. Commerce Department, tasked with researching and establishing standards across all federal agencies. NIST SP 800-53 defines the standards and guidelines for federal agencies to architect and manage their information security systems. It was established to provide guidance for the protection of agency's and citizen's private data. In many cases, complying with NIST guidelines and recommendations will help federal agencies ensure compliance with other regulations, such as the Health Insurance Portability and Accountability Act (HIPAA), Federal Information Security Management Act (FISMA), or Sarbanes-Oxley Act (SOX). NIST guidelines are often developed to help agencies meet specific regulatory compliance requirements. NIST has outlined the following nine steps toward FISMA compliance.

- Categorize the data and information you need to protect
- Develop a baseline for the minimum controls required to protect that information
- Conduct risk assessments to refine your baseline controls
- Document your baseline controls in a written security plan
- Roll out security controls to your information systems
- Once implemented, monitor performance to measure the efficacy of security controls
- Determine agency-level risk based on your assessment of security controls
- Authorize the information system for processing
- Continuously monitor your security controls

SUMMARY

This chapter briefly explains the various security standards presently existing in the industrial sector and their selection requirements. IEC 62443, NIST 800, NERC CIP and ISO 27001 standards are briefly explained. The compliance procedure for IEC 62443 and NERC CIP are also described as they are most pertinent today.

CHAPTER 12

DOCUMENTED SCADA CYBER ATTACKS

12.1 INTRODUCTION

Cyber security is becoming an increasingly challenging in process automation. Continuous innovation with modern technologies has taken manufacturing processes from initial industrial age into the information age. Hence there is a greater scope of cyber threats from various sources. Internet helped in boosting productivity and efficiency to unimaginable levels. Industrial automation thundered into today's data-driven, Internet-associated world, it sped past computerized security without taking its foot off the accelerator. For many years, malicious cyber-actors have been targeting the industrial control systems (ICSs) that manage our critical infrastructures. Most of these events are not reported to the public, and the threats and incidents to ICSs are not as well-known as enterprise cyber-threats and incidents. This chapter briefly explain certain publically reported cyber threats to critical infrastructure, which sheds light on the growing cyber threats to ICS devices.

12.2 DOCUMENTED POWER SYSTEM CYBER INCIDENTS

Crashed Ohio Nuke Plant Network by Slammer Worm (2003)

In January 2003, a Slammer worm penetrated a private computer network at Ohio's Davis-Besse nuclear power plant and disabled a safety monitoring system for more than 5 hours, despite a belief by plant personnel that the network was protected by a firewall. The Slammer worm spread from the enterprise network to the ICS network by exploiting the vulnerabilities of the MS-SQL. It was reported that process computers had crashed for hours, aggravating the system operators.

Taum Sauk Hydroelectric Power Station Failure (2005)

The Taum Sauk incident on December 14, 2005, was not an attack but instead a failure of a hydroelectric power station. Various explanations, including design or construction flaws, instrumentation malfunction, and human error, have been attributed to the catastrophic failure of an upper reservoir. Investigation revealed that the sensors failed to indicate that the reservoir was full and the pumps were not shut down until the water overflew for about 5 to 6 min. This overflow damaged the parapet wall, resulting in the collapse of the reservoir. Apparently this incident was (apparently) not an attack, but the idea behind it could be exploited to perform undetectable attacks in safety-critical infrastructures. This can be a means for a stealthy attack on a SCADA by sending compromised sensor measurements to the control center.

Cyber Incident on Georgia Nuclear Power Plant (2008)
In July 2008, a nuclear power plant in Georgia was forced into an emergency shutdown for 48 hours because a computer used to monitor chemical and diagnostic data from the corporate network rebooted after a software update. When the updated computer restarted, it reset the data on the control system. The safety systems interpreted the lack of data as lowering of the levels in the water reservoirs that cool the plant's radioactive nuclear fuel rods, and triggering a system shutdown.

PSS Giant Telvent Compromised in Canada (2013)
A breach on the internal firewall and security systems of Telvent, Canada, one of the most reliable company which supplies remote monitoring and control tools to the energy sector, was discovered on September 10, 2012. After penetrating the network, the intruders stole project files related to the OASyS SCADA product, a highly sophisticated remote administration tool allowing companies to combine older IT equipment with modern Smart Grid technology. It is very likely that the adversaries gathered information about this new product to find its vulnerabilities and to prepare for future stealthy attacks against PSS.

Ukraine Power Grid Cyber-attack December (2015)
The Ukraine power grid Cyber-attack took place on 23 December 2015 and is considered to be the first known successful cyber-attack on a power grid. Hackers were able to successfully compromise information systems of three energy distribution companies in Ukraine and temporarily disrupt electricity supply to the end consumers. The cyber-attack was complex and comprised of the following steps,
- prior compromise of corporate networks using spear fishing emails with BlackEnergy malware, seized the ICS under control, and switched off substations remotely,
- disabled IT infrastructure components such as uninterruptable power supplies, modems, RTUs, commutators,

- destroyed files stored on servers and workstations with the KillDisk malware, and
- denial-of-service attack on call-center to deny consumers up-to-date information on the blackout.

A brief account on the Ukraine attack has been already described in chapter one.

Stuxnet attack on Iran's Natanz nuclear facility (2012 & 2017)

In 2012 Iranian nuclear facilities were attacked and infiltrated by the first *cyber-weapon* or the *digital missile* of the world known as the Stuxnet. It is believed that this Natanz nuclear facility attack was initiated by a random worker's USB drive. By current estimations this attack results in 30% decrease in enrichment efficiency. It is reported that in 2017, once again Iranian critical infrastructures and networks have been vehemently attacked by a new variant of the lethal Stuxnet which is many fold sophisticated than its former variant leaving extreme concerns to the security of the industrial automation world. A brief account on the Iranian nuclear facility attack has been described in chapter one.

12.1.1 ICS Cyber Attacks On US

It has been reported by IBM X-Force that the the number of events targeting the ICS assets in 2019 increased over 2000 percent since 2018. In fact, the number of events targeting OT assets in 2019 was greater than the activity volume observed in the past three years combined. As the US is the most cyber targeted nation they have stepped up security measures by strictly implementing the mandatory standards in ICSs. In 2019 June US grid regulator NERC issued a warning that major hacking group with suspected Russian ties was conducting reconnaissance into the networks of electrical utilities. Some of the significant successful aatacks are described below.

Penetration of Electricity Grid of US by Spies (2009)

On August 14, 2003, the Northeast and Midwest regions of the United States and some provinces in Canada suffered a serious blackout because of a software bug. These incidents have raised concerns about the security of electric power grids, because disrupting national power systems might cause catastrophic damage. *The Wall Street Journal* reported that on April 8, 2009, Cyber-attackers have penetrated the US electric power grid and left behind a software program that can be used to disrupt the system.

Water Tower Decoy in US (2012)

In December 2012, a malicious computer virus concealed in an MS Word document sent from Chinese hacking group, APT1, successfully took over a water tower control system in the US. Fortunately for anyone nearby, the tower was actually a trap set up to attract such would-be industrial attacks. So, while nothing was hurt or destroyed in this incident, it did demonstrate the frightening reality of these attacks.

US Wind farm attack (2019)

The largest renewable energy developer in US located at Utah was hit by a cyber attack that briefly break contact of a number of wind and solar farms. Later it has been reported that it was a DoS attack that left grid operators temporary blinded generation sites totalling a loss of 500MW. The cyber attack took advantage of a known weakness in Cisco firewalls to trigger a series of a five minute communication outage over a span of about 12 hours.

12.1.2 Cyber Threats to Indian ICS

The US was the most cyber targeted nation in the world in 2019 for power sector cyber attacks. However India surpassed US and topped in the list for three months and remained within the top 5 cyber targeted countries in 2019 leaving extreme concerns to India's power sector automation and Smart Grid projects which are struggling to comply the relevant physical-cyber security standards. If any cyber security flaws are introduced by chance to the Indian power grid by the implementation agencies, likelihood of exploiting these vulnerabilities by our enemy nations are very high. Government of India (GoI) and Central Electricity Authority (CEA) are giving directions to implement the proper cyber security measures without any lapse to the various power Transmission and Distribution utilities while moving to digital energy. Realising the possibility supply chain attacks, the GoI recently cautioned the power utilities to test all the Chinese products especially the industrial networking products used in power sector automation.

Indian North-Eastern Grid Blackout (2012)

In 2012 one of the world's worst blackout has occurred in India. The exact cause for this blackout is yet to be confirmed as many speculations/reports of a cyber attack from enemy nations are live though official explanation denies it. But the Nation has realised the consequences of a blackout if it happens due to a cyber attack. The security experts point out it may occur at any times as India is facing 30 cyber attacks daily on its power sector mostly from China, Slovania, Mexico, Ukraine and Pakistan unless India identify and fix the flaws of the imported industrial networking and automation products. The 2012 blackout left almost 710 million people without power. The noted other impacts due to this blackout are briefly described below.

More than 700 million people in India have been left without power in the world's worst blackout of recent times, leading to fears that protests and even riots could follow if the country's electricity supply continues to fail to meet growing demand. First to fail was India's northern grid, leaving an estimated 350 million people in the dark for up to 14 hours. It was quickly followed by the eastern grid, which includes Kolkata, then the north-eastern grid. Twenty of India's 28 states were hit by power cuts, along with the capital, New Delhi, when three of the country's five electricity grids failed at lunchtime.

As engineers struggled for hours to fix the problem, hundreds of trains failed, leaving passengers stranded along thousands of miles of track from Kashmir in the north to Nagaland on the eastern border with Burma. Traffic lights went out, causing jams in New Delhi, Kolkata and other cities. Surgical operations were cancelled across the country, with nurses at one hospital just outside Delhi having to operate life-saving equipment manually when back-up generators failed. Electric crematoriums stopped operating; some with bodies left half burnt before wood was brought in to stoke the furnaces. As Delhiites sweated in 89% humidity and drivers honked their horns even more impatiently than usual, in West Bengal the power cut left hundreds of miners trapped underground for hours when their lifts broke down. All the state's government workers were sent home after the chief minister announced it would take 10 to 12 hours for the power to return. There were some agitations, riots and protests in urban areas which were very reliant on electricity. TVs and computers were not worked for a week and one-third of India's households do not even have electricity to power a light bulb. By early evening, 50 of the trapped miners in West Bengal had been rescued and power had been restored to the north-east of the country, as well the most affluent areas of Delhi.

India has five electricity grids viz. northern, eastern, north-eastern, southern and western. All are interconnected. The blackout and its impacts were an eye opener to the Indian energy engineers that the present power grid of India needs to be cyber secured in every respect with appropriate architecture with approved standards as the all the five regional grids are interconnected, operated and controlled by NLDC, New Delhi. It also given a lesson to the grid security to engineers on the need of PMU based power system monitoring and control when national grids integrates and operates as a single grid.

Breach at the Kudankulam Nuclear Power Plant (2019)

Towards the end of October 2019, reports were spreading in social media regarding a cyber attack at Kudankulam Nuclear Power Plant. On October 29, the Plant authorities, denied such an attack and informed that both the reactors (1000MW and 600MW) were running without any operational or safety concerns. But unexpectedly, Nuclear Power Corporation of India Limited (NPCIL) within 24 hours admitted that there indeed was an incident and Dtrack RAT has been located within the plant. Source code included hard coded credentials to KNPP indicating that it was a targeted attack. It also contains methods to collect browser history, passwords, host IPs, running processes and all files on disk volumes.

The matter was conveyed to Computer Emergency Response Team (CERT-In) as soon as it was noticed. CERT-In said that they had identified a malware attack that breached India's largest nuclear power facility's administrative network on September 4. They further emphasised that the nuclear plant's operational systems were safe they are air gapped and the administrative network was not connected to it. Hence there was nothing to be panic.

This clearly exposed the lack of the technical awareness of the plant authorities regarding the air gap jumpers. Further the malware, DTRACK, was developed by a North Korean hacker group called LAZARUS who specialised in stealing information from a system. It is suspected that a large amount of data was stolen during the breach. This data could be used to plan the next attack more efficiently. As per the cyber security experts and ethical hackers, Domain Controller level access has been gained by the hackers at KNPP.

In the modern environment, most of the ICS networks are air gapped/stand-alone having no connectivity to cyber-space to make it secure. Hence the ICS threat actors usually carried out the attacks to ICS in two stages as explained in previous chapters. Hence it can be observed that the stage I attack has been carried out successfully by the threat actors. The chances of the stage II attack which will be the real objective behind the ICS attack cannot be ruled out. The CERT-In and the NPCIL authorities might have taken all necessary actions without any compromise to prevent such attack else it will be extremely disastrous.

12.3 COMMON ICS SECURITY ISSUES

Security solutions developed for traditional IT networks are not adequate and effective in DCS because of the major differences between them. Their security objectives are different in the sense that security in IT networks aims to enforce the three security principles viz. confidentiality, integrity and availability, while the security in DCS networks aims to provide human safety, equipment and the mission critical operations, etc. Furthermore, the security architecture of IT networks is different from that of the DCS network since security in IT networks is achieved by providing more protection at the center of the network where the data is kept, while the protection in automation networks is done both at the network center and end nodes or edge. Their underlying topology is also different as IT networks use a well-defined set of operating systems and protocols, while DCS networks use multiple propriety operating systems and protocol specific to vendors. Hence Quality of Service (QoS) metrics are different in the sense that it is acceptable in IT networks to reboot devices in case of failure or upgrade, while this is not at all acceptable in DCS networks since services must be available 24x7. These major differences between the IT and DCS network security objectives necessitate the need for new security solutions specific for the ICS network. As explained in the previous chapters, the development of DCS security solutions faces many challenges and mainly they are,

- many ICS components use propriety operating systems to control functionality without security features,
- most of the legacy ICS network was designed without regard to security,
- security should be integrated with existing systems without relegation the latency and efficiency and performance,
- remote access to grid devices must be monitored and controlled, and
- the new protocols selected should have the capability of incorporating future security solution.

12.4 MITIGATION STRATEGIES

Employing zone based architecture is one of the means of managing different parts of the DCS to provide protection. The defense-in-depth zone based architecture has been explained in previous chapter, can be used to categorize as customer operations, business, communication and control systems. Each zone is independently comprised of field devices, systems, communication media, and data centers which serve the specific operational functions of that zone. Implementing a set of security features common to all zones and security features required explicit to that zone can form an effective, modular and segmented security system. The following modular security zones are minimum suggested for DCS.

- Customer operations security zone,

- Corporate security zone, and
- Communication and control systems security zone.

Customer operations security zone: The customer operations security zone contains devices and processes that extend product management to customers. Smart Grid is a typical example of a complex DCS, where security has to be provided, access to home systems and the data gathered by Advanced Metering Infrastructure (AMI) must be limited to authorized people and devices. Customer energy management systems must ensure integrity of command and meter data, authenticate devices, and protect the grid from compromised devices.

Corporate security zone: The corporate security zone includes all the features and functions of the customer operations security zone plus security for IT functions vital to a business, such as email, internet, telephony, messaging, and a wide variety of corporate applications. To meet these requirements, an IDMZ with consistent security policies must be applied across the entire product line of the DCS.

Communication and control systems security zone: The communication and control systems security zone defines the processes used to manage the routing of product from the manufacture area to the end user. It contains data centers involved in the production, testing, marketing and reliable distribution of products. Information collected and processed here supports equipment maintenance, troubleshooting, load capacity, and supply chain re-routing in the event of break down.

To protect the integrity of transmitted data and control, utilities must ensure individual and device authentication, computer health verification, correlation of alarm data with other sensors to prevent false positives, regulatory compliance and enable forensic analysis. Data encryption, intrusion detection and prevention and secure sharing of information between various data centers are critical and must be addressed with utmost care.

As utilities innovate with ICT based modern automation technologies, they need a foundation of converged IP networks, proven security principles, industry-leading networking equipment and software with integrated security capabilities. The maturity, reliability, and success of these products and services can shorten the learning curve for utilities to evolve operations to meet new standards and regulations.

Follow Standards and Guidelines

An OT security expert must strictly follow security the standards and guidelines to ensure the ICS security requirements without any compromise.

Following a security standard with proper updating will indeed help to make ICS smarter against cyber-attacks. The chief cyber-security officer of the utility must be very keen in enforcing the following steps without compromise while designing and implementing the ICS.
- Develop a security policy,
- Establish physical security,
- Lockdown perimeter security,
- Enable existing security features,
- Secure operational traffic,
- Secure management traffic,
- Properly manage the system configuration,
- Eliminate security shortfalls.
- Continuous security training, and
- Perform security audits.

Moreover to build a secure ICS strictly follow the key reminders without any compromise as shown in Figure 12.1.
- Be always on the defensive,
- Be always vigilant for cyber-attacks,
- Have safety, security and disaster recovery plans,
- Implement physical-cyber-security without any compromise, and
- Proper documentation and reporting of cyber incidents, attempts and attacks.

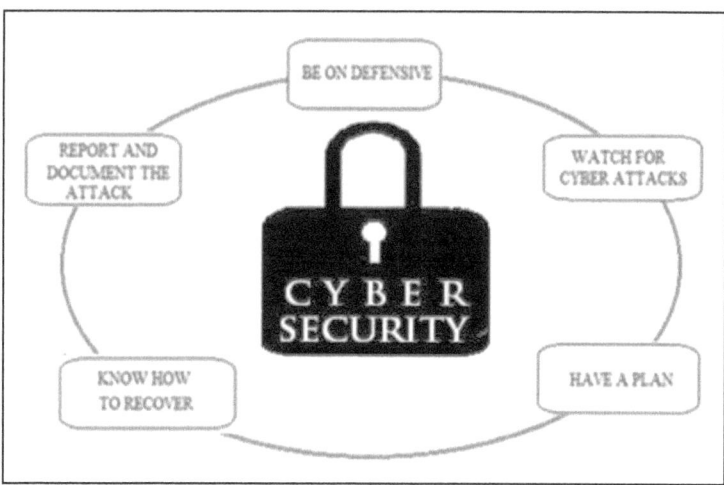

▲ **Figure 12.1:** ICS Security Key Reminders

SUMMARY

In recent years ICS is one of the main target of cyber-attackers and a number of cyber incidents have occurred to the power systems. Chapter twelve describes some of the cyber-incidents as well as different sources of attacks are described to make the ICS engineers aware of the severity of the attacks. It also briefly describes the different types of threats that ICS may encounter from the perspective of the utility and implementation agencies. It then explains the lethal malware threats like Stuxnet which is a nightmare for Power System SCADA and ICS as it mainly exploits the Zero Day Vulnerabilities of the Windows OS. Various proposals for making the automated power system smarter than cyber-attacks by cyber-security expert groups also have been discussed.

INDEX

A
Adjustable Speed Drives, xxvii
Advanced Assets Management, xxvii
Advanced Distribution Operation, xxvii
Advanced Metering Infrastructure, xx, xxvii, 109, 136, 318, 320, 324, 331, 338-339, 368
Advanced Persistent Threat, xxvii, 283, 286
Advanced Transmission Operation, xxvii
Alarm handling, 61-62
American National Standard Institute, xxvii
Analog input (AI) 31, 33, 39
Analog output (AO) 35, 40
Anti-aliasing, 27-28
Application Programming Interface, xxvii, 36, 157
Application server, 55, 254, 295, 297
Aramid, 123-124
Asynchronous Transfer Mode, xxvii, 256
Asynchronous, xv, xxvii, 89-90, 94, 113, 115, 184, 219-220, 226, 256
Attenuation, xvi, 27, 114, 120, 122, 125, 135, 186
Authentication Header, xxvii, 283
Auto Re-closure, xxvii
Automatic Generation Control, 56, 87
Auxiliary power supply, 47

B
Bandwidth, 9, 48, 113-116, 119, 121, 129, 131-132, 134-137, 139-142, 147, 176-177, 179-180, 196, 210-211, 221, 248, 261, 338-339

Baseband, xv, xxv, 113, 118-119
Binary coding, 25
Binary search, 30
Border Gateway Protocol, xxvii, 164-165, 258
Bridge amplifier, 27
British Naval Connector, 125
Broadband over Power Line (BPL), xvi, 134
Broadband Power Line Communication, xxvii
Broadband, xv-xvi, xxv, xxvii, 113, 118-119, 127, 129, 132-137, 145, 147, 338, 340
Broadcast, xv, 113, 115-116, 127, 141, 144, 179-181, 183, 186-189, 202
Building Management System, xxvii, 315
Bulk Electricity System, xxvii

C
Capacitor, 14-15, 27-29, 91-92, 102, 104
Central Processing Unit (CPU), 39
Channel redundancy, xiv, xxiii, 53, 76, 80-81
Co-axial, xv, xxiv, 119, 121
Code Division Multiple Access, 131
Common Cause Failure, 73
Common Information Model, xxvii, 236
Common Mode Failure, 73
Common Mode Rejection Ratio, 32
Communication architecture, xiv, xvii, xxxiv, 51, 65, 81, 224
Communication clouds, xxiii, 81
Communication Front End, xxvii, 18, 55, 57, 255, 266

Companion Specification for Energy Metering, xxvii, 238
Conversion algorithm, 28, 30
Copper UTP, xvi, 134
Counting type ADC, 7, 29, 31
Critical cyber-assets, 262, 350
Critical Infrastructure Protection, xviii, xxvii, xxxv, 10, 44, 260-261, 273, 350
Crosstalk xvi, 32, 120, 125–126
Current excitation, 27-28
Customer Information System, xv, xxvii, xxxiv, 55, 101, 110, 112
Cyber-attacks xxxvi, 11, 247, 260, 280, 283, 285, 287, 341, 346, 369–370

D

Data Acquisition System, vi, xiii, xxiii, xxxiii, xxxv, 4, 23-25, 36-37, 64, 222, 284, 350
Data Concentrator (DC) 18, 50–51, 64, 97, 217, 263
Data Control Unit, xxvii
Data conversion, 24
Data forwarding and communication, xiv, 34
Data handling, 24
Data processing, 5, 24, 61, 322
Data translation, 37
Decision Support Tools, xxviii
Decrypting, 45
Demand Response Management System, xxviii
Demand Side Management, xxviii, 136, 319, 328, 336
Denial Of Service, xxviii, 169, 201, 231, 266, 272, 311, 343
Development server, 55-56, 303
Device Language Message Specification, xxvii, 238
Dial up telephone line, 41
Digital Format, 23, 33, 39, 51
Digital Input (DI) 45
Digital Output (DO) 32, 40, 48
Digital Panel Meter, 25
Direct Load Control, xxvii

Discrete input, 47, 219
Dispatcher Training Simulator, 55-56, 79, 88
Distributed Energy Resources, xxvii, 315, 323, 325, 330-331, 339
Distributed Generation, xx, xxvii, 94, 101, 319, 323-324, 326, 328-330, 335, 341
Distributed Management System, xxvii
Distributed Network Protocol, xvii, 216-217
Distributed SCADA, xxxv, 4, 17, 34, 65, 127, 147
Distribution Automation, v, xv, xxvii, xxxiv, 83, 85, 87, 89, 91, 93, 95, 97, 99, 101-103, 105, 107, 109, 111, 236, 321, 324, 338-339
Distribution Load Forecast, xv, xxvii, 107, 112
Distribution load forecast, xv, xxvii, 107, 112
Distribution management system, xv, xxxiv, 11, 13, 34, 57, 79, 99, 102-103, 112, 235, 315
Distribution Network Model or Dynamic Mimic Diagram, xxvii, 102
Distribution Network Protocol, xxviii, 344
Distribution network, xv, xxvii-xxviii, 11-13, 102-103, 106-107, 136, 314-316, 323-324, 344
DNP3, vii, xvii, xxv, 10, 37, 41, 51, 84, 215-217, 221-222, 230, 241, 344
Door Alarm, 45

E

Electric Vehicle, xxviii, xxx, 317, 319, 321, 328, 334, 339, 341
Electromagnetic interference, 42, 119, 121, 135
Electronic Security Perimeter, xxviii, 350, 352, 354
Encrypting, xxxv, 45, 344
End Node Security, xix, xxviii, 11, 59, 65, 228, 244, 250, 281, 292

Energy Control System, xxviii
Energy Management System, xv, xxviii, xxx, 11, 13, 19, 34, 55, 57, 83, 85, 136, 235, 332, 368
Enhanced Power Architecture, xxviii
Equipment Condition Monitor, xxviii
Equipment Under Control, xxviii, 236, 298
European Committee for Electrotechnical Standardization, xxvii
Exception reporting, xiv, 68, 71-72, 205
Excitation, 27-28
External Access Point, xxviii, 11, 249, 285

F
Fail Safe System, xiv, 65, 73
Fault Management and System Restoration, xv, xxviii
Fault Management and System, xv, xxviii
Fault Passage Indication, xxviii
Federal Communication Commission, xxviii
Feeder Reconfiguration, xv, xxix, 106-107, 112
Fibre Optical Current Sensor, xxviii
Fibre To The Home (FTTH) xxviii
Fibre To The Home, xxviii
Fibre Transmission Transfer Protocol, xxviii
Fibre-To –he- Home (FTTH), 122
Fidelity, 88, 114
Filtering, 27-28, 62-63, 112, 117, 146, 162, 188, 190-196, 270
Firewall, xviii, xxiv-xxv, 20-21, 51, 54-55, 76-77, 81, 146, 162, 171, 188, 190-196, 212, 221, 231, 239, 244, 247, 249, 254, 256-257, 266-271, 275, 292, 295-296, 301-304, 307-311, 339, 354-355, 361-362, 364
Flexible AC Transmission System, xxviii, xxxiv, 90, 98
Flywheel Energy Storage, xxviii, 331
Fourth generation SCADA, xxiii, 6
Frequency Division Multiple Access, 131-132, 205

Front End Processor, xxviii, 266
Full duplex, xv, 65, 113, 117-118, 172, 175

G
General Object Oriented Substation Event, xxviii
General Routing Encapsulation, xxviii, 252
Generalizes Multi-Protocol Label Switching, xxviii
Geographical Information System, xv, xxviii, xxxiv, 101-102, 109, 112, 339
Geostationary satellite, 58
Global Positioning System (GPS), 96
Global Positioning System, xiv, xxviii, 37, 58, 96
Global System For Mobile Communication, xxviii, 131
Guided media technology, xvi, 134-135
Guided media, xv-xvi, 113, 119, 134-135

H
Half duplex, xv, 113, 117, 175
Hand-Held Unit, xxviii, 239
High Availability Seamless Redundancy, xxviii
High Availability, xiv, xxviii, xxxiv, 39, 53-54, 75, 81-82, 86, 257
High Efficiency Distribution Transformer, xxviii
High Level Data Link Control, xxviii, 210, 239
High Voltage Direct Current, xxix, xxxiv, 89
HMI software, 8-9, 54, 60-61, 286
Home Area Network, xxviii, 319
HomePlug, 134
Human Machine Interface, xiv, xxxiii, 8, 37, 58, 64, 218, 263-264
Human Machine Interfaces, xxviii
Hybrid Electronic Volt, xxviii

I
ICCP server, 57, 233-234
IEC 60870, vii, xvii-xviii, xxiv, 9, 37, 41, 84, 215-217, 225-228, 231, 237-238, 241

IEC 60870-5-101, xvii, 9, 217, 225
IEC 60870-5-103, xvii, 226
IEC 60870-5-104, xvii, 217, 225, 227
IEC 60870-6-ICCP, 41
IEC 61508, xviii, 236-237, 241
IEC 61850, vii, xvii, xxiv, 10, 41, 48, 51-52, 85, 215-216, 228-231, 237-238, 240-241
IEC 61968, xviii, 235, 241
IEC 61970, xviii, 235, 241
IEC 62056, xviii, 238-239, 241
IEC 62056-21, xviii, 239, 241
IEC 62325, xviii, 236, 241
IEC 62351, xviii, 225, 227, 237-238, 241
IEC 62443, xx, xxxvi, 45, 347-350, 360
Independent System Operation, xxix
Industrial automation, xi, 74, 347-349, 361, 363
Industrial Control System, 4, 19-20, 44, 243, 245, 282, 290, 311, 343, 347, 349, 361
Information and Communication, xxix, 17, 240, 243, 260-261, 318, 328, 341, 350
Information Storage and Retrieval, 55-56
Information Technologies, xxix
In-Home Energy Management, xxix
Input impedance, 29
Input Range, 30, 32, 34-35, 39
Institute Of Electrical and Electronics Engineering, xxix
Institute of Electrical and Electronics, xxix, 43, 138
Intelligent alarm filtering, 62
Intelligent Electronic Device (IED), 40, 87
Intelligent Electronic Device, vii, xiv, xxix, xxxiii, 1, 37, 40-41, 46, 55, 64, 87, 217, 320
Intercontrol Centre Communication Protocol, xxix
Interline Power Flow Controller, xxix, 92
Internal Combustion Engine, xxix
International Electro technical, 43
International Electrotechnical Commission, xxix, 238, 356
International Society for Automation, xxix

Internet of Things, 5
Internet Protocol Security, xxix, 165, 171, 251, 283, 344
Internet Protocol, xxix, 153, 165, 168, 171-172, 251, 256, 283, 343-344
Intrusion Detection System, 21, 194, 231, 299, 301
IP Fast Reroute, xxix, 259

L
Label Distribution Protocol, xxix, 259
Label Switched Path, xxix, 256
Ladder logic, 7, 49-50
Latency, 5, 12, 32, 45, 211, 230, 245, 250, 257, 301, 324, 336, 338-339, 344, 367
Line of Sight, 58
Linearity error, 33
Linearization, 27-28
Lithium Ferro Phosphate, xxix
Lithium Iron Battery, xxix
Lithium Nickel Manganese Cobalt, xxix
Load Balancing via Feeder Reconfiguration, xv, xxix, 107, 112
Load balancing, xv, xxix, 107, 112
Load Flow Application, xv, xxix, 105, 112
Load Flow Applications, 105
Load Shed Application, xv, xxix, 106, 112
Load Shed Applications, 87
Local Area Network, xvii, xxix, 127, 179, 186, 189, 208-209, 227, 289
Local Data Monitoring System, xxix, 51, 263
Long Term Evaluation, xxix
Loss Minimization via Feeder Reconfiguration, xv, xxix, 106, 112
Loss minimization Via Feeder, xv, xxix, 106, 112

M
Man In The Middle, xxix, 222, 228
Manchester Bus Powered, xxix, 223-224
Manufacturing Messaging Specification, xxix, 232

Master Control Centre, xiv, xxiii, xxix, 53, 60-61, 63
Master Control Station, 23, 54, 57-58, 77, 79, 239
Master RTU, xxiii, 66
Master slave, xiv, 68-69, 225
Media Access Control, xxix, 166, 203, 230
Merging Unit (MU), 51
Merging Unit, xiv, xxix, xxxiii, 23, 37, 50-52, 54, 64, 72, 85
Meter Data Management, xxix, 140
Microprocessor, 37, 39, 51-52
Microwave communication, 127
Mimic diagram, xxvii, 8, 18, 59-60, 102
Mobile communication, xvi, xxviii, 129, 131, 133-134, 140
Modbus, xvii, 9, 20, 37, 41, 215-216, 218-222, 240-241, 269, 281, 344
Modem, 37-38, 43, 72, 79, 115, 118, 120, 140, 147, 152, 169, 183-185, 263-264, 268, 275, 281, 292, 362
Monolithic SCADA, 4
Monotonicity, 33
Multicast, xv, 85, 113, 115-117, 153, 230, 344
Multi-channel data acquisition, xiii
Multiplexer, 25, 28, 35, 39
Multiplexing, xxx, 24, 119, 131, 203, 211, 256
Multi-Protocol Label Switching, xxviii-xxix

N
National Institute For Standard and Technology, xxix
National Load Dispatch Centre, 57
Network Access Server, xxix, 250
Network Connectivity Analysis, xv, xxix, 102
Network Management Server, 55-56
Non-volatile memory, 38, 43
Normally Closed, 47, 74
Normally Open, 47, 74
North American Electric Reliability Corporation, xxix, 10, 44, 261, 347, 350

O
Oil and gas industry, 44
Op-amp 31
Open System Interconnection, xxx
Operating System (OS), 75
Operation and Maintenance, vii, xxx, 13-14, 24, 83-84, 262
Operator console, 18, 59, 63, 291
Operator training simulator, 56
Optical coupling, 27
Outage Management System, xv, xxx, xxxiv, 99, 101, 108-109, 112

P
Parallel Comparator (flash type) ADC xxiii, 7, 29, 31
Peer to peer, 220
Peripheral device, 57, 59-60, 73, 292
Personal Computer Memory Card International Association, xxx, 184
Phase Shifting Transformer, xv, xxx, xxxiv, 111-112
Phasor Data Concentrator, xxx, 97, 318
Phasor Measurement Unit, xv, xxx, xxxiv, 93, 95, 97, 217, 235, 318
Photo Voltaic, xxx
Plug In Hybrid Electric Vehicle, xxx
Point To Point Tunnelling Protocol, xxx
Point-To-Point Protocol, xxx, 167-168, 227, 253
Polling plus CSMA, xiv
Power Line Carrier Communication, xvi, xxx, 15, 134
Power System SCADA, i, vi, xiii-xiv, xix, xxiii, xxx, xxxiii, xxxv-xxxvii, 10, 17-19, 23, 34, 37, 44, 54, 56-57, 59, 61, 64, 73, 76, 79-81, 134, 184, 211, 215-216, 225, 229, 234, 236, 239, 243-244, 246-247, 264, 266, 273, 280-282, 344, 370
Printed Circuit Board, 42
Protocol Data Unit, xxx, 219, 230

Q
Quantizing uncertainty, 33

R

Radio Frequency Interference, 42, 119
Radio navigation, 58
Radio via trunked/VHF/UHF, 41
Real Time Operating System, xxx
Regional Load Dispatch Centre, 57
Relative accuracy, 33
Reliable cloud, xxiii, 80
Remote Telemetry Unit, 9, 43
Remote Terminal unit (RTU), xiv, 23, 37, 54, 248
Remote Terminal Unit, xiv, xxx, xxxiii, 7, 9, 23, 37, 44, 54, 64, 87, 94, 108, 217, 248
Renewable Energy Sources, xx, xxx, 11, 85, 89, 94, 97, 285, 317, 324-325, 327, 329, 337
Repeater, xiv, xxiii-xxiv, 67-68, 124-125, 127, 147, 168, 176, 179-180, 183, 185-187, 206
Resolution, 27, 31, 33, 35, 39-40, 116, 168
Resource Reservation Protocol, xxx, 259
Response, xvi, xxviii, 28, 43, 47, 57, 70, 79, 136, 146, 213, 218, 222, 245, 267, 271, 275, 299-300, 317, 319, 328-331, 336, 338-339, 352-353, 355, 366
RJ-11, 125
RJ-45, 125
Router, 54, 76, 116, 137, 142, 165, 178-179, 181-183, 186-187, 191, 195, 201, 206, 227, 250, 256, 259, 261, 281, 303, 308, 311, 339
RS 232, 41
RTD Module, 29
RTU Architecture, xxiii, 38

S

Safety Integrity Level, xxx, 237
Satellite communication, xvi, xxiv, 128-129, 145, 147-148
SCADA architecture, 23, 65
SCADA communication protocol, xvii, 215-216
SCADA server, 23, 55-56, 218
Scaling, 27, 116, 192

Scanner, 37, 146
Secured Shell, xxx
Secured Socket Layer, xxx, 250
Security in wireless communication, xvi, 145
Security through obscurity, 5, 21, 151, 247, 282
Sensors, xiii, xxv, xxxiii, 6-7, 9, 16, 23-25, 27-28, 37, 64, 101, 142, 284-285, 294, 298, 317, 328, 337, 339, 362, 368
Sequence of Events, 13, 24, 45
Servers, xiv, 5, 9-10, 23, 37, 53-55, 61, 75-77, 87, 179, 196, 206, 208, 212, 232-233, 246, 250, 254, 256, 261, 269, 277, 280-282, 286, 289, 295-297, 303, 308, 310, 363
Signal conditioner, xxxiii, 7, 25, 27, 39, 64
Signal conditioning, xiv, xxv, 7, 24, 27-28, 35
Simplex, xv, 65, 113, 117-118, 161, 172
Single channel Data acquisition, xiii, xxiii, 24-25, 65
Situational awareness, xxxi, xxxiv, 61, 64, 72, 98, 338
Smart Grid Communication Network, xxx
Smart Grid, i, v-vi, xi, xix-xx, xxx, xxxiii-xxxvi, 11-13, 16, 48-49, 58, 77, 82, 85, 93-95, 101, 110, 136, 138, 140, 147-148, 184, 215, 234, 243-244, 246-248, 262-264, 269, 272, 279-285, 293, 313-325, 327-329, 331, 333, 335-343, 345-346, 362, 364, 368
Smart Home Energy Management System, xxx
Smart Meter, xxx, 11, 16, 19, 110, 140-142, 266, 280-281, 285, 313-314, 320-322, 328, 331, 343
Social State Transformer, xxx
Special Protection System, xxx, 98
Standards and certification, 36
State Estimation, xv, xxx, 55, 88, 94, 97-98, 103-104, 112, 284, 322
State Load Dispatch Centre, 57
State Of Charge, xxx
Storage and display, xiv, 24

Substation data access 47
Successive approximation ADC, xxiii, 7, 29-30
Super Conducting Magnetic Energy Sources, xxx
Supervisory Control and Data Acquisition, xiii, xxx, 4, 11, 13, 23, 83, 87, 243, 311, 317
Supervisory Control and Data, xiii, xxx, 4, 11, 13, 23, 83, 87, 243, 311, 317
Synchronous, xv, 91-92, 113, 115, 169, 209-211, 329, 332, 335
System redundancy, xiv, 53, 76, 79-81
System reliability and availability, xiv, 73, 93
System SCADA software, xiv, 54

T

Technical Committee, xxx, 225, 227, 235-236
Terrestrial communication, xvi, xxiv, 127-128
Thermocouple, 28-29
Third party software, 41
Time Division Multiple Access, 131, 206
Time Division Multiplexing, xxx, 203, 256
Time Of Use, xxx, 110
Topology Analyser, xxx
Total Harmonic Distortion, xxx, 35
Transducer, 6, 16, 18, 27-28, 34-35, 47, 94
Transmission Control Protocol, xxx, 153, 160, 164, 174
Transmission System Operation, xxx
Transport Layer Security, xxx, 164, 255
Twisted pair, xv, xxiv, 119-122, 134, 167-169, 176-177, 189, 220, 223

U

Unguided media, xvi, xxxiv, 79, 113, 126-127, 137, 148, 207
Unicast, xv, 113, 115-116
Unified Power Flow Controllers, xxxi, 92
United Nation, xxxi
Universal Serial Bus, xxxi
User Data Protocol, xxx

V

Variable Frequency Drives, xxxi, 298
Vehicle To Grid, xxxi
Video Projection System, 55, 57-58, 63
Virtual Private Dial-Up Network, xxxi, 250
Virtual Private Network, xxxi, 248, 257
Virtual Ring, xxxi
Visual Display Unit (VDU), 59
Volatile memory, 38, 43
Volt VAR Control, xxxi
Volt VAR Optimization, xxxi, 105
Voltage Ampere Reactive, xxxi
Volt-VAR Control, 14
VSAT, xvi, 8, 10, 138, 145, 147

W

WAN connectivity, 50
Water distribution, 11
Wide Area Adaptive Protection Control and Automation, xxxi
Wide Area Monitoring Protection and Control, xxxi
Wide Area Network, xxxi, 166, 209, 227, 231, 259
Wide Area Situational Awareness, xxxi, 98
Wide-Area Monitoring System, xv, xxxi
Wi-Fi, 154, 186, 277
Wimax, xvi, 132-133, 138, 141-143
Wireless standards, xvi, 138
Wireless technology, 127, 132, 137-138, 143, 145
Working Group, xxxi, 229, 235-236

Z

Zero Dynamic Attack, xxxi
ZigBee, xvi, 138, 141-143
Zone based architecture, 367
ZWave alliance, 144-145
ZWave security, 145
ZWave, xvi, 138, 143-145

OTHER USEFUL TITLE FROM NOTION PRESS

Salient Features of the Book
- Covers power grid networks, including how they are developed and deployed for power delivery and other Smart Grid services.
- Discusses power systems, advanced communications, and required machine learning that define the Smart Grid.
- Clearly differentiates the Smart Grid from the traditional power grid as it has been for the last century.
- Provides the reader with a fundamental understanding of both physical-cyber-security and computer networking.
- Presents the complexity and operational requirements of the evolving Smart Grid to the ICT professional and presents the same for ICT to the power system engineers.
- Provides a detailed description of the cyber vulnerabilities and mitigation techniques of the Smart Grid.
- Provides essential information for technocrats to make progress in the field and to allow power system engineers to optimize communication systems for the Smart Grid.

SIZE: 9" x 6", Pages:424
ISBN: 978-1-64650-999-7
MRP: 850/-

www.ingramcontent.com/pod-product-compliance
Lightning Source LLC
Chambersburg PA
CBHW020721180526
45163CB00001B/64